ORGANIC CHEMISTRY AS A SECOND LANGUAGE, 3e

First Semester Topics

DAVID KLEIN
Johns Hopkins University

JOHN WILEY & SONS, INC.

Associate Publisher	Petra Recter
Editorial Assistant	Lauren Stauber
Senior Production Editor	Elizabeth Swain
Marketing Manager	Kristine Ruff
Senior Cover Designer	Wendy Lai

Cover Credits: Background: ©William Hopkins/iStockphoto; test tube: Untitled X-Ray/Nick Veasey/Getty Images, Inc.; bicycle: Igor Shikov/Shutterstock.

This book was typeset by Prepare and printed and bound by Courier Westford. The cover was printed by Courier Westford.

This book is printed on acid free paper. ∞

Founded in 1807, John Wiley & Sons, Inc. has been a valued source of knowledge and understanding for more than 200 years, helping people around the world meet their needs and fulfill their aspirations. Our company is built on a foundation of principles that include responsibility to the communities we serve and where we live and work. In 2008, we launched a Corporate Citizenship Initiative, a global effort to address the environmental, social, economic, and ethical challenges we face in our business. Among the issues we are addressing are carbon impact, paper specifications and procurement, ethical conduct within our business and among our vendors, and community and charitable support. For more information, please visit our website: www.wiley.com/go/citizenship.

Evaluation copies are provided to qualified academics and professionals for review purposes only, for use in their courses during the next academic year. These copies are licensed and may not be sold or transferred to a third party. Upon completion of the review period, please return the evaluation copy to Wiley. Return instructions and a free of charge return mailing label are available at www.wiley.com/go/returnlabel. If you have chosen to adopt this textbook for use in your course, please accept this book as your complimentary desk copy. Outside of the United States, please contact your local sales representative.

ISBN   978-1-118-01040-2

Printed in the United States of America

10 9

INTRODUCTION

IS ORGANIC CHEMISTRY REALLY ALL ABOUT MEMORIZATION?

Is organic chemistry really as tough as everyone says it is? The answer is yes and no. Yes, because you will spend more time on organic chemistry than you would spend in a course on underwater basket weaving. And no, because those who say it's so tough have studied inefficiently. Ask around, and you will find that most students think of organic chemistry as a memorization game. *This is not true!* Former organic chemistry students perpetuate the false rumor that organic chemistry is the toughest class on campus, because it makes them feel better about the poor grades that they received.

If it's not about memorizing, then what is it? To answer this question, let's compare organic chemistry to a movie. Picture in your mind a movie where the plot changes every second. If you're in a movie theatre watching a movie like that, you can't leave even for a second because you would miss something important to the plot. So you try your hardest to wait until the movie is over before going to the bathroom. Sound familiar?

Organic chemistry is very much the same. It is one long story, and the story actually makes sense if you pay attention. The plot constantly develops, and everything ties into the plot. If your attention wanders for too long, you could easily get lost.

OK, so it's a long movie. But don't I need to memorize it? Of course, there are some things you need to memorize. You need to know some important terminology and some other concepts that require a bit of memorization, but the amount of pure memorization is not that large. If I were to give you a list of 100 numbers, and I asked you to memorize them all for an exam, you would probably be very upset by this. But at the same time, you can probably tell me at least 10 telephone numbers off the top of your head. Each one of those has 10 digits (including the area codes). You never sat down to memorize all 10 telephone numbers. Rather, over time you slowly became accustomed to dialing those numbers until the point that you knew them. Let's see how this works in our movie analogy.

You probably know at least one person who has seen one movie more than five times and can quote every line by heart. How can this person do that? It's *not* because he or she tried to memorize the movie. The first time you watch a movie, you learn the plot. After the second time, you understand why individual scenes are necessary to develop the plot. After the third time, you understand why the dialogue was necessary to develop each scene. After the fourth time, you are quoting many of the lines by heart. *Never at any time did you make an effort to memorize the lines.* You know them *because they make sense* in the grand scheme of the plot. If I were to give you a screenplay for a movie and ask you to memorize as much as you can in 10 hours, you would probably not get very far into it. If, instead, I put you in a room for 10 hours and played the same movie over again five times, you would know most of the movie by heart, without even trying. You would know everyone's names, the order of the scenes, much of the dialogue, and so on.

Organic chemistry is exactly the same. It's not about memorization. It's all about making sense of the plot, the scenes, and the individual concepts that make up our story. Of course you will need to remember all of the terminology, but with enough practice, the terminology will become second nature to you. So here's a brief preview of the plot.

THE PLOT

The first half of our story builds up to reactions, and we learn about the characteristics of molecules that help us understand reactions. We begin by looking at atoms, the building blocks of molecules, and what happens when they combine to form bonds. We focus on special bonds between certain atoms, and we see how the nature of bonds can affect the shape and stability of molecules. At this point, we need a vocabulary to start talking about molecules, so we learn how to draw and name molecules. We see how molecules move around in space, and we explore the relationships between similar types of molecules. At this point, we know the important characteristics of molecules, and we are ready to use our knowledge to explore reactions.

Reactions take up the rest of the course, and they are typically broken down into chapters based on categories. Within each of these chapters, there is actually a subplot that fits into the grand story.

HOW TO USE THIS BOOK

This book will help you study more efficiently so that you can avoid wasting countless hours. It will point out the major scenes in the plot of organic chemistry. The book will review the critical principles and explain why they are relevant to the rest of the course. In each section, you will be given the tools to better understand your textbook and lectures, as well as plenty of opportunities to practice the key skills that you will need to solve problems on exams. In other words, you will learn the language of organic chemistry. *This book cannot replace your textbook, your lectures, or other forms of studying.* This book is not the Cliff Notes of Organic Chemistry. It focuses on the basic concepts that will empower you to do well if you go to lectures and study in addition to using this book. To best use this book, you need to know how to study in this course.

HOW TO STUDY

There are two separate aspects to this course:

1. Understanding principles

2. Solving problems

Although these two aspects are completely different, instructors will typically gauge your understanding of the principles by testing your ability to solve problems. So you must master both aspects of the course. The principles are in your lecture notes, but *you* must discover how to solve problems. Most students have a difficult time with this task. In this book, we explore some step-by-step processes for analyzing problems. There is a very simple habit that you must form immediately: *learn to ask the right questions.*

If you go to a doctor with a pain in your stomach, you will get a series of questions: How long have you had the pain? Where is the pain? Does it come and go, or is it constant? What was the last thing you ate? and so on. The doctor is doing two very important and very different things. First, he has learned the right questions to ask. Next, he applies the knowledge he has together with the information he has gleaned to arrive at the proper diagnosis. Notice that the first step is asking the right questions.

Let's imagine that you want to sue McDonald's because you spilled hot coffee in your lap. You go to an attorney and she asks you a series of questions that enable her to apply her knowledge to your case. Once again, the first step is asking questions.

In fact, in any profession or trade, the first step of diagnosing a problem is always to ask questions. Let's say you are trying to decide if you really want to be a doctor. There are some tough, penetrating questions that you should be asking yourself. It all boils down to learning how to ask the right questions.

The same is true with solving problems in this course. Unfortunately, you are expected to learn how to do this on your own. In this book, we will look at some common types of problems and we will see what questions you should be asking in those circumstances. More importantly, we will also be developing skills that will allow you to figure out what questions you should be asking for a problem that you have never seen before.

Many students freak out on exams when they see a problem that they can't do. If you could hear what was going on in their minds, it would sound something like this: "I can't do it . . . I'm gonna flunk." These thoughts are counterproductive and a waste of precious time. Remember that when all else fails, there is always one question that you can ask yourself: "What questions should I be asking right now?"

The only way to truly master problem-solving is to practice problems every day, consistently. You will never learn how to solve problems by just reading a book. You must try, and fail, and try again. You must learn from your mistakes. You must get frustrated when you can't solve a problem. That's the learning process. Whenever you encounter an exercise in this book, pick up a pencil and work on it. Don't skip over the problems! They are designed to foster skills necessary for problem-solving.

The worst thing you can do is to read the solutions and think that you now know how to solve problems. It doesn't work that way. If you want an A, you will need to sweat a little (no pain, no gain). And that doesn't mean that you should spend day and night memorizing. Students who focus on memorizing will experience the pain, but few of them will get an A.

The simple formula: Review the principles until you understand how each of them fits into the plot; then *focus all of your remaining time on solving problems*. Don't worry. The course is not that bad if you approach it with the right attitude. This book will act as a road map for your studying efforts.

CONTENTS

CHAPTER 1

BOND-LINE DRAWINGS

To do well in organic chemistry, you must first learn to interpret the drawings that organic chemists use. When you see a drawing of a molecule, it is absolutely critical that you can read all of the information contained in that drawing. Without this skill, it will be impossible to master even the most basic reactions and concepts.

Molecules can be drawn in many ways. For example, below are three different ways of drawing the same molecule:

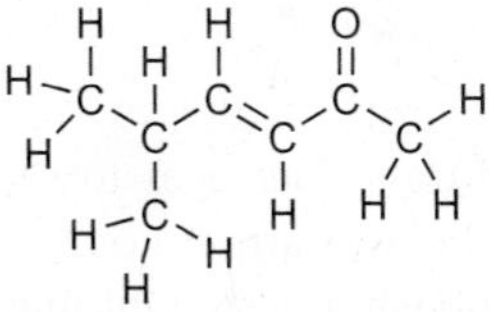

$(CH_3)_2CHCH=CHCOCH_3$

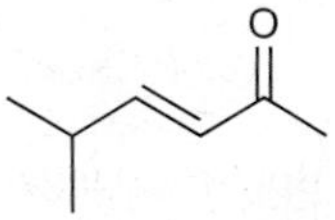

Without a doubt, the last structure (bond-line drawing) is the quickest to draw, the quickest to read, and the best way to communicate. Open to any page in the second half of your textbook and you will find that every page is plastered with bond-line drawings. Most students will gain a familiarity with these drawings over time, not realizing how absolutely critical it is to be able to read these drawings fluently. This chapter will help you develop your skills in reading these drawings quickly and fluently.

1.1 HOW TO READ BOND-LINE DRAWINGS

Bond-line drawings show the carbon skeleton (the connections of all the carbon atoms that build up the backbone, or skeleton, of the molecule) with any functional groups that are attached, such as $-OH$ or $-Br$. Lines are drawn in a zigzag format, where each corner or endpoint represents a carbon atom. For example, the following compound has 7 carbon atoms:

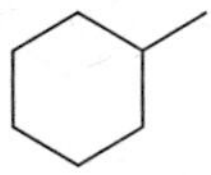

It is a common mistake to forget that the ends of lines represent carbon atoms as well. For example, the following molecule has six carbon atoms (make sure you can count them):

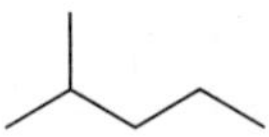

Double bonds are shown with two lines, and triple bonds are shown with three lines:

When drawing triple bonds, be sure to draw them in a straight line rather than zigzag, because triple bonds are linear (there will be more about this in the chapter on geometry). This can be quite confusing at first, because it can get hard to see just how many carbon atoms are in a triple bond, so let's make it clear:

is the same as   so this compound
                        has 6 carbon atoms

It is common to see a small gap on either side of a triple bond, like this:

is the same as

Both drawings above are commonly used, and you should train your eyes to see triple bonds either way. Don't let triple bonds confuse you. The two carbon atoms of the triple bond and the two carbons connected to them are drawn in a straight line. All other bonds are drawn as a zigzag:

$$H-\overset{\overset{H}{|}}{C}-\overset{\overset{H}{|}}{C}-\overset{\overset{H}{|}}{C}-\overset{\overset{H}{|}}{C}-H$$

is drawn like this:

BUT

$$H-\overset{\overset{H}{|}}{C}-C\equiv C-\overset{\overset{H}{|}}{C}-H$$

is drawn like this:

EXERCISE 1.1   Count the number of carbon atoms in each of the following drawings:

Answer   The first compound has six carbon atoms, and the second compound has five carbon atoms:

PROBLEMS Count the number of carbon atoms in each of the following drawings.

1.2 Answer: _6_   **1.3** Answer: _6_   **1.4** Answer: _5_   **1.5** Answer: _7_

1.6 Answer: _6_   **1.7** Answer: _8_   **1.8** Answer: _4_   **1.9** Answer: _10_

1.10 Answer: _9_   **1.11** Answer: _10_

Now that we know how to count carbon atoms, we must learn how to count the hydrogen atoms in a bond-line drawing of a molecule. Most hydrogen atoms are not shown, so bond-line drawings can be drawn very quickly. Hydrogen atoms connected to atoms other carbon (such as nitrogen or oxygen) must be drawn:

But hydrogen atoms connected to carbon are not drawn. Here is the rule for determining how many hydrogen atoms there are on each carbon atom: *neutral carbon atoms have a total of four bonds*. In the following drawing, the highlighted carbon atom is showing only two bonds:

We only see two bonds
connected to this carbon atom

Therefore, it is assumed that there are two more bonds to hydrogen atoms (to give a total of four bonds). This is what allows us to avoid drawing the hydrogen atoms and to save so much time when drawing molecules. It is assumed that the average person knows how to count to four, and therefore is capable of determining the number of hydrogen atoms even though they are not shown.

So you only need to count the number of bonds that you can see on a carbon atom, and then you know that there should be enough hydrogen atoms to give a total of four bonds to the carbon atom. After doing this many times, you will get to a point where you do not need to count anymore. You will simply get accustomed to seeing these types of drawings, and you will be able to instantly "see" all of the hydrogen atoms without counting them. Now we will do some exercises that will help you get to that point.

EXERCISE 1.12   The following molecule has nine carbon atoms. Count the number of hydrogen atoms connected to each carbon atom.

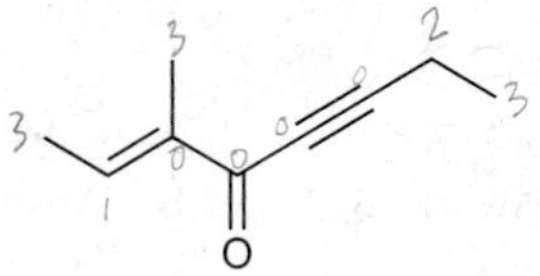

Answer:

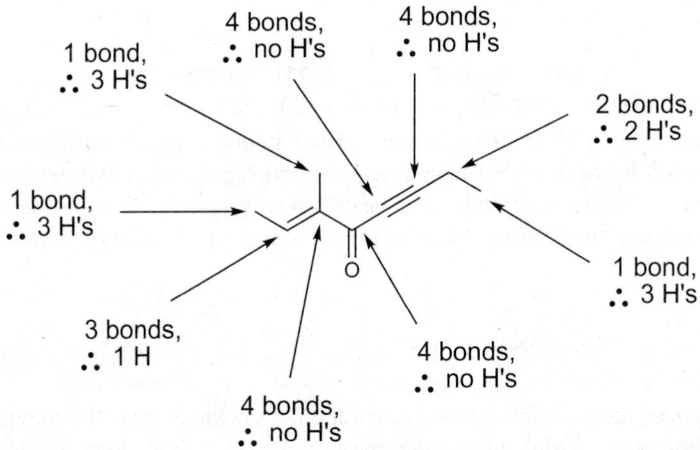

1 bond,
∴ 3 H's

4 bonds,
∴ no H's

4 bonds,
∴ no H's

2 bonds,
∴ 2 H's

1 bond,
∴ 3 H's

1 bond,
∴ 3 H's

3 bonds,
∴ 1 H

4 bonds,
∴ no H's

4 bonds,
∴ no H's

PROBLEMS   For each of the following molecules, count the number of hydrogen atoms connected to each carbon atom. The first problem has been solved for you (the numbers indicate how many hydrogen atoms are attached to each carbon).

1.13

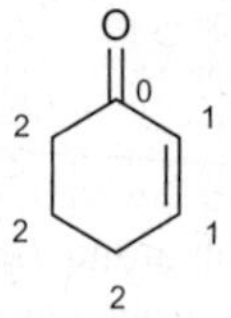

1.14

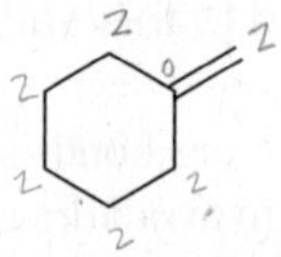

1.15

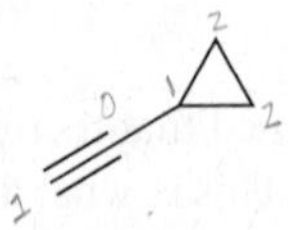

1.16

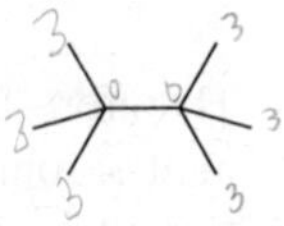

1.17

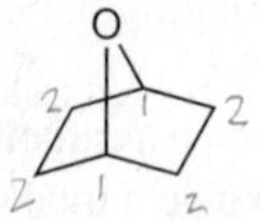

1.18

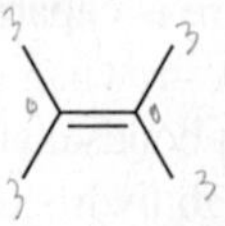

1.19

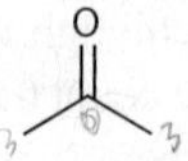

1.20

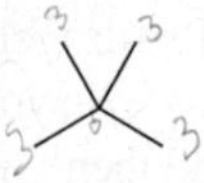

Now we can understand why we save so much time by using bond-line drawings. Of course, we save time by not drawing every C and H. But, there is an even larger benefit to using these drawings. Not only are they easier to draw, but they are easier to read as well. Take the following reaction for example:

$$(CH_3)_2C=CHCOCH_3 \xrightarrow[\text{Pt}]{H_2} (CH_3)_2CHCH_2COCH_3$$

It is somewhat difficult to see what is happening in the reaction. You need to stare at it for a while to see the change that took place. However, when we redraw the reaction using bond-line drawings, the reaction becomes very easy to read immediately:

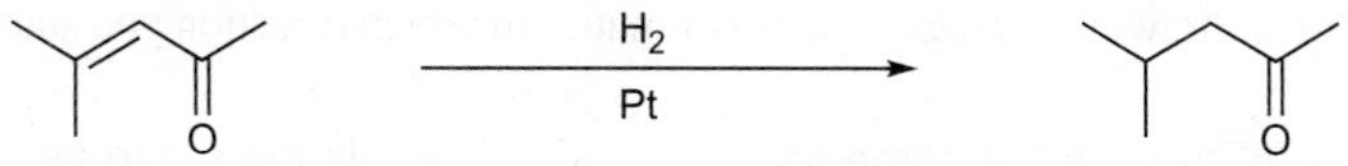

As soon as you see the reaction, you immediately know what is happening. In this reaction we are converting a double bond into a single bond by adding two hydrogen atoms across the double bond. Once you get comfortable reading these drawings, you will be better equipped to see the changes taking place in reactions.

1.2   HOW TO DRAW BOND-LINE DRAWINGS

Now that we know how to read these drawings, we need to learn how to draw them. Take the following molecule as an example:

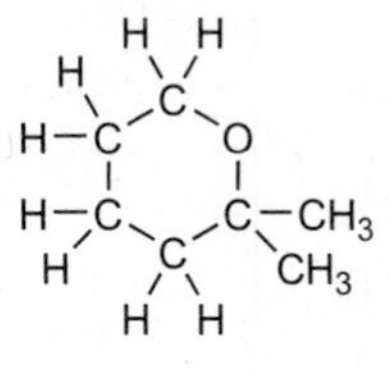

To draw this as a bond-line drawing, we focus on the carbon skeleton, making sure to draw any atoms other than C and H. All atoms other than carbon and hydrogen *must* be drawn. So the example above would look like this:

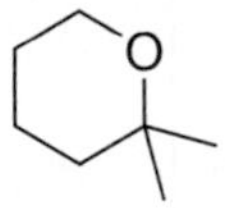

A few pointers may be helpful before you do some problems.

  1. Don't forget that carbon atoms in a straight chain are drawn in a zigzag format:

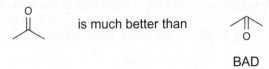

  2. When drawing double bonds, try to draw the other bonds as far away from the double bond as possible:

is much better than

BAD

  3. When drawing zigzags, it does not matter in which direction you start drawing:

is the same as   is the same as

PROBLEMS   For each structure below, draw the bond-line drawing in the box provided.

1.21

1.22

1.23

$$\begin{array}{c} \overset{Br}{|} \quad \overset{H}{|} \quad \overset{Br}{|} \\ H-C-C-C-H \\ \overset{|}{H} \quad \overset{|}{H} \quad \overset{|}{H} \end{array}$$

1.24

1.3 MISTAKES TO AVOID

1. *Never* draw a carbon atom with more than four bonds. This is a big no-no. Carbon atoms only have four orbitals; therefore, carbon atoms can form only four bonds (bonds are formed when orbitals of one atom overlap with orbitals of another atom). This is true of all second-row elements, and we discuss this in more detail in the chapter on drawing resonance structures.

2. When drawing a molecule, you should either show all of the H's and all of the C's, or draw a bond-line drawing where the C's and H's are not drawn. You *cannot* draw the C's without also drawing the H's:

$$\begin{array}{c} \overset{C}{|} \\ C-C-C-C-C \\ \overset{|}{C} \end{array} \qquad \text{NEVER DO THIS}$$

This drawing is no good. Either leave out the C's (which is preferable) or put in the H's:

3. When drawing each carbon atom in a zigzag, try to draw all of the bonds as far apart as possible:

is better than

1.4 MORE EXERCISES

First, open your textbook and flip through the pages in the second half. Choose any bond-line drawing and make sure that you can say with confidence how many carbon atoms you see and how many hydrogen atoms are attached to each of those carbon atoms.

Now try to look at the following reaction and determine what changes took place:

added 2 H

Do not worry about *how* the changes took place. You will understand that later when you learn the mechanism of the reaction. For now, just focus on explaining what change took place. For the example above, we can say that we *added* two hydrogen atoms to the molecule (one on either end of the double bond).

Consider another example:

Br

In this example, we have *eliminated* an H and a Br to form a double bond. (We will see later that it is actually H^+ and Br^- that are eliminated, when we get into the chapters on mechanisms). If you cannot see that an H was eliminated, then you will need to count the number of hydrogen atoms in the starting material and compare it with the product:

H
Br
H
H

H
H

Now consider one more example:

Br

Cl

In this example, we have *substituted* a bromine with a chlorine.

PROBLEMS   For each of the following reactions, clearly state what change has taken place. In each case your sentence should start with one of the following opening clauses: we have added . . . , we have eliminated . . . , or we have substituted . . . .

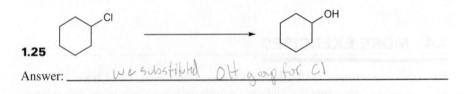

1.25

Answer: _____ we substituted OH group for Cl _____

1.26

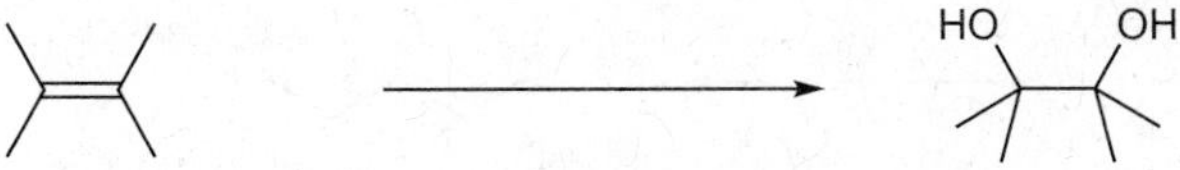

Answer: We added OH groups on each of the double bonded Carbons, breaking the double bond

1.27

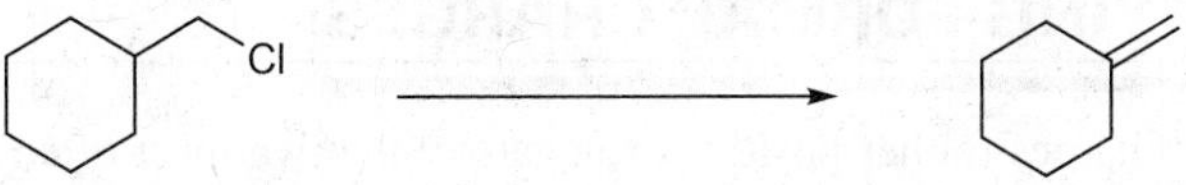

Answer: We eliminated ~~a double bond and added~~ a cl hydrogen and Cl

1.28

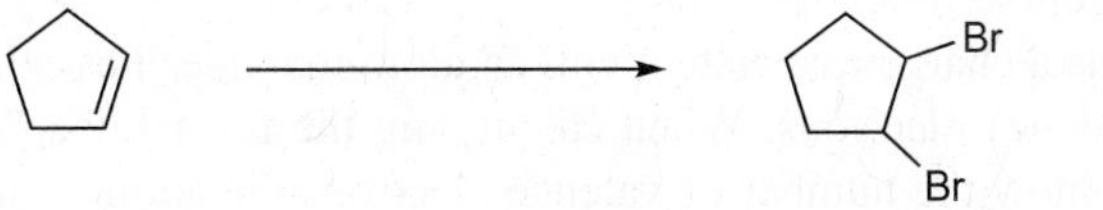

Answer: We eliminated a double bond and added two Br

1.29

Answer: We added a double bond, removed 2 hydrogens

1.30

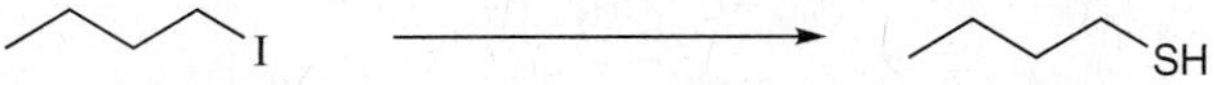

Answer: We substituted SH for I

1.31

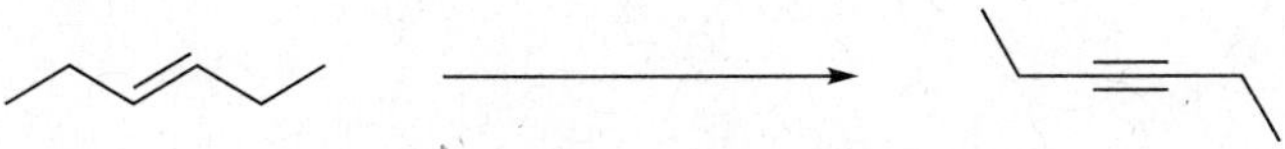

Answer: We removed 2 hydrogens and created a triple bond

1.32

Answer: _____ bC we have eliminated a bond _____

1.5 IDENTIFYING FORMAL CHARGES

Formal charges are charges (either positive or negative) that we must often include in our drawings. They are extremely important. If you don't draw a formal charge when it is supposed to be drawn, then your drawing will be incomplete (and wrong). So you must learn how to identify when you need formal charges and how to draw them. If you cannot do this, then you will not be able to draw resonance structures (which we see in the next chapter), and if you can't do that, then you will have a very hard time passing this course.

A formal charge is a charge associated with an atom that does not exhibit the expected number of valence electrons. When calculating the formal charge on an atom, we first need to know the number of valence electrons the atom is *supposed* to have. We can get this number by inspecting the periodic table, since each column of the periodic table indicates the number of expected valence electrons (valence electrons are the electrons in the valence shell, or the outermost shell of electrons— you probably remember this from high school chemistry). For example, carbon is in Column 4A, and therefore has four valence electrons. Now you know how to determine how many electrons the atom is supposed to have.

Next we ask how many electrons the atom *actually has* in the drawing. But how do we count this?

Let's see an example. Consider the central carbon atom in the compound below:

$$H_3C - \overset{\displaystyle :\overset{..}{O}^{-H}}{\underset{\displaystyle H}{C}} - CH_3$$

Remember that every bond represents two electrons being shared between two atoms. Begin by splitting each bond apart, placing one electron on this atom and one electron on that atom:

$$H_3C \cdot \ \cdot \overset{\displaystyle :\overset{..}{O}^{-H}}{\underset{\displaystyle \cdot H}{C}} \cdot \ \cdot CH_3$$

Now count the number of electrons immediately surrounding the central carbon atom:

There are four electrons. This is the number of electrons that the atom actually has.

Now we are in a position to compare how many valence electrons the atom is *supposed* to have (in this case, four) with how many valence electrons it *actually* has (in this case, four). Since these numbers are the same, the carbon atom has no formal charge. This will be the case for most of the atoms in the structures you will draw in this course. But in some cases, there will be a difference between the number of electrons the atom is supposed to have and the number of electrons the atom actually has. In those cases, there will be a formal charge. So let's see an example of an atom that has a formal charge.

Consider the oxygen atom in the structure below:

Let's begin by determining the number of valence electrons that an oxygen atom is *supposed* to have. Oxygen is in Column 6A of the periodic table, so oxygen should have six valence electrons. Next, we need to look at the oxygen atom in this compound and ask how many valence electrons it *actually* has. So, we redraw the stucture by splitting up the C–O bond:

In addition to the electron on the oxygen from the C–O bond, the oxygen also has three lone pairs. A lone pair is when you have two electrons that are not being used to form a bond. Lone pairs are drawn as two dots on an atom, and the oxygen above has three of these lone pairs. You must remember to count each lone pair as two electrons. So we see that the oxygen atom actually has seven electrons, which is one more electron than it is supposed to have. Therefore, it will have a negative charge:

EXERCISE 1.33 Consider the nitrogen atom in the structure below and determine if it has a formal charge:

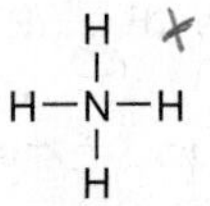

$$\begin{array}{c} H \\ | \\ H-N-H \\ | \\ H \end{array}$$

Answer Nitrogen is in Column 5A of the periodic table so it should have five electrons. Now we count how many it actually has:

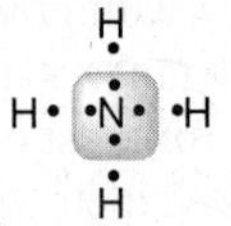

It only has four. So, it has one less electron than it is supposed to have. Therefore, this nitrogen atom has a positive charge:

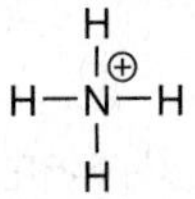

$$\begin{array}{c} H \\ | \oplus \\ H-N-H \\ | \\ H \end{array}$$

PROBLEMS For each of the structures below determine if the oxygen or nitrogen atom has a formal charge. If there is a charge, draw the charge.

1.34

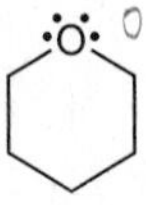

1.35

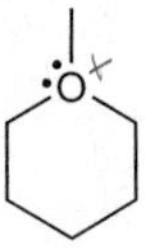

1.36

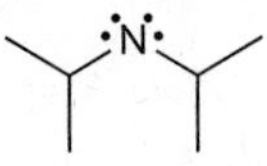

1.37

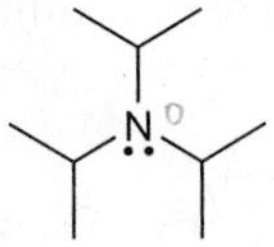

1.38

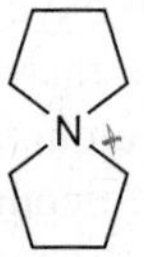

1.39

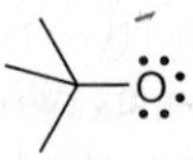

1.40

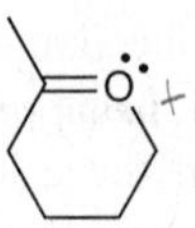

1.41

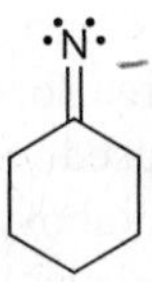

1.42

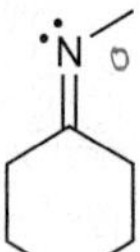

1.43

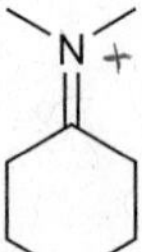

1.44

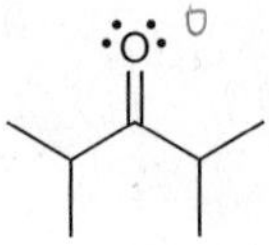

1.45 $H_3C-C\equiv N$:

This brings us to the most important atom of all: carbon. We saw before that carbon always has four bonds. This allows us to ignore the hydrogen atoms when drawing bond-line structures, because it is assumed that we know how to count to four and can figure out how many hydrogen atoms are there. When we said that, we were only talking about carbon atoms without formal charges (most carbon atoms in most structures will not have formal charges). But now that we have learned what a formal charge is, let's consider what happens when carbon has a formal charge.

If carbon bears a formal charge, then we cannot just assume the carbon has four bonds. In fact, it will have only three. Let's see why. Let's first consider C^+, and then we will move on to C^-.

If carbon has a *positive* formal charge, then it has only three electrons (it is supposed to have four electrons, because carbon is in Column 4A of the periodic table). Since it has only three electrons, it can form only three bonds. That's it. So, a carbon with a positive formal charge will have only three bonds, and you should keep this in mind when counting hydrogen atoms:

| No hydrogen atoms on this C^+ | 1 hydrogen atom on this C^+ | 2 hydrogen atoms on this C^+ |

Now let's consider what happens when we have a carbon atom with a *negative* formal charge. The reason it has a negative formal charge is because it has one more electron than it is supposed to have. Therefore, it has five electrons. Two of these electrons form a lone pair, and the other three electrons are used to form bonds:

We have the lone pair, because we can't use each of the five electrons to form a bond. Carbon can *never* have five bonds. Why not? Electrons exist in regions of space called orbitals. These orbitals can overlap with orbitals from other atoms to form bonds, or the orbitals can contain two electrons (which is called a lone pair). Carbon has only four orbitals, so there is no way it could possibly form five bonds—it does not have five orbitals to use to form those bonds. This is why a carbon atom with a negative charge will have a lone pair (if you look at the drawing above, you will count four orbitals—one for the lone pair and then three more for the bonds).

Therefore, a carbon atom with a negative charge can also form only three bonds (just like a carbon with a positive charge). When you count hydrogen atoms, you should keep this in mind:

No hydrogen atoms
on this C⁻

1.6   FINDING LONE PAIRS THAT ARE NOT DRAWN

From all of the cases above (oxygen, nitrogen, carbon), you can see why you have to know how many lone pairs there are on an atom in order to figure out the formal charge on that atom. Similarly, you have to know the formal charge to figure out how many lone pairs there are on an atom. Take the case below with the nitrogen atom shown:

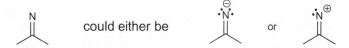

If the lone pairs were drawn, then we would be able to figure out the charge (two lone pairs would mean a negative charge and one lone pair would mean a positive charge). Similarly, if the formal charge was drawn, we would be able to figure out how many lone pairs there are (a negative charge would mean two lone pairs and a positive charge would mean one lone pair). So you can see that drawings must include either lone pairs or formal charges. The convention is to always show formal charges and to leave out the lone pairs. This is much easier to draw, because you usually won't have more than one charge on a drawing (if even that), so you get to save time by not drawing every lone pair on every atom.

Now that we have established that formal charges must *always* be drawn and that lone pairs are usually *not* drawn, we need to get practice in how to see the lone pairs when they are not drawn. This is not much different from training yourself to see all the hydrogen atoms in a bond-line drawing even though they are not drawn. If you know how to count, then you should be able to figure out how many lone pairs are on an atom where the lone pairs are not drawn.

Let's see an example to demonstrate how you do this:

In this case, we are looking at an oxygen atom. Oxygen is in Column 6A of the periodic table, so it is supposed to have six electrons. Then, we need to take the formal charge into account. This oxygen atom has a negative charge, which means one extra electron. Therefore, this oxygen atom must have $6 + 1 = 7$ electrons. Now we can figure out how many lone pairs there are.

The oxygen atom has one bond, which means that it is using one of its seven electrons to form a bond. The other six must be in lone pairs. Since each lone pair is two electrons, this must mean that there are three lone pairs:

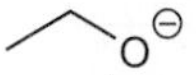

is the same as

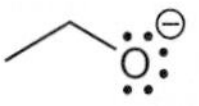

Let's review the process:

1. Count the number of electrons the atom should have according to the periodic table.
2. Take the formal charge into account. A negative charge means one more electron, and a positive charge means one less electron.
3. Now you know the number of electrons the atom actually has. Use this number to figure out how many lone pairs there are.

Now we need to get used to the common examples. Although it is important that you know how to count and determine numbers of lone pairs, it is actually much more important to get to a point where you don't have to waste time counting. You need to get familiar with the common situations you will encounter. Let's go through them methodically.

When oxygen has no formal charge, it will have two bonds and two lone pairs:

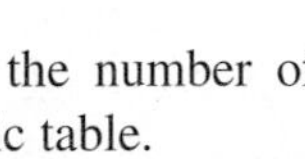

—OH   is the same as   —ÖH

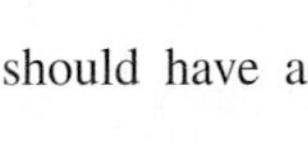

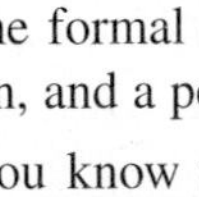

is the same as

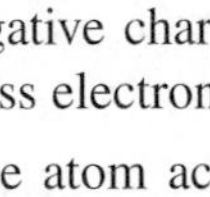

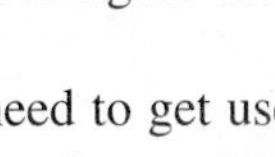

is the same as

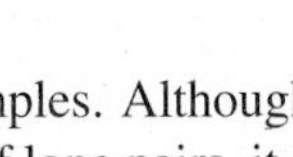

If oxygen has a negative formal charge, then it must have one bond and three lone pairs:

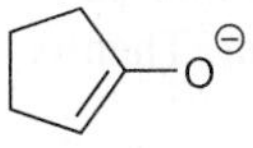

is the same as

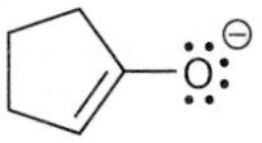

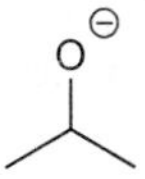

is the same as

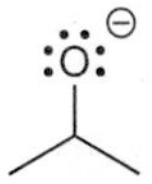

If oxygen has a positive charge, then it must have three bonds and one lone pair:

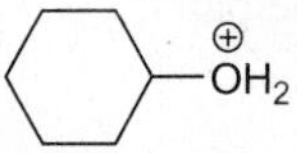 is the same as 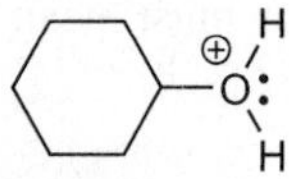

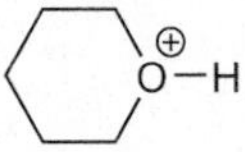 is the same as 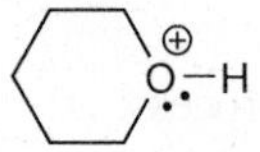

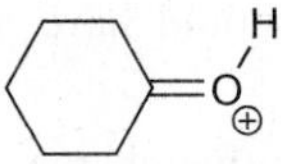 is the same as 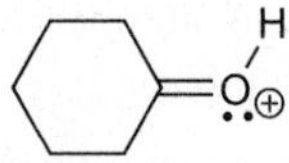

EXERCISE 1.46 Draw all lone pairs in the following structure:

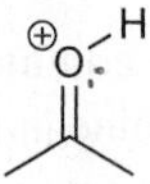

Answer The oxygen atom has a positive formal charge and three bonds. You should try to get to a point where you recognize that this must mean that the oxygen atom has one lone pair:

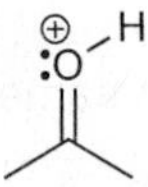

Until you get to the point where you can recognize this, you should be able to figure out the answer by counting.

Oxygen is supposed to have six electrons. This oxygen atom has a positive charge, which means it is missing an electron. Therefore, this oxygen atom must have 6 − 1 = 5 electrons. Now, we can figure out how many lone pairs there are.

The oxygen atom has three bonds, which means that it is using three of its five electrons to form bonds. The other two must be in a lone pair. So there is only one lone pair.

PROBLEMS Review the common situations above, and then come back to these problems. For each of the following structures, draw all lone pairs. Try to recognize how many lone pairs there are *without* having to count. Then count to see if you were right.

1.47 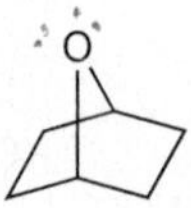 **1.48** 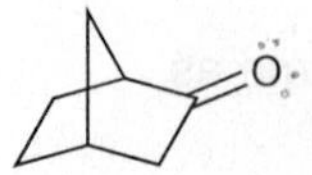 **1.49** 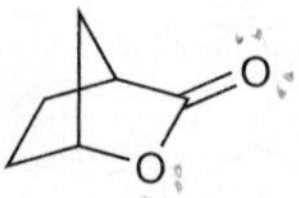

1.50   **1.51**   **1.52**

Now let's look at the common situations for nitrogen atoms. When nitrogen has no formal charge, it will have three bonds and one lone pair:

If nitrogen has a negative formal charge, then it must have two bonds and two lone pairs:

If nitrogen has a positive charge, then it must have four bonds and no lone pairs:

EXERCISE 1.53 Draw all lone pairs in the following structure:

$$\overset{\ominus}{N}=\overset{\oplus}{N}=\overset{\ominus}{N}$$

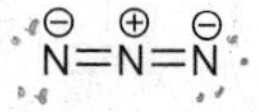

Answer The central nitrogen atom has a positive formal charge and four bonds. You should try to get to a point where you recognize that this nitrogen atom does not have any lone pairs. Each of the other nitrogen atoms has a negative formal charge and two bonds. You should try to get to a point where you recognize that each of these nitrogen atoms has two lone pairs:

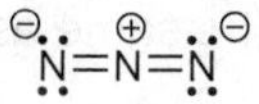

$$\overset{\ominus}{\underset{..}{N}}=\overset{\oplus}{N}=\overset{\ominus}{\underset{..}{N}}$$

Until you get to the point where you can recognize this, you should be able to figure out the answer by counting. Nitrogen is supposed to have five electrons. The central nitrogen atom has a positive charge, which means it is missing an electron. In other words, this nitrogen atom must have $5 - 1 = 4$ electrons. Now, we can figure out how many lone pairs there are. Since it has four bonds, it is using all of its electrons to form bonds. So there is no lone pair on this nitrogen atom.

For each of the remaining nitrogen atoms, there is a negative formal charge. That means that each of those nitrogen atoms has one extra electron, $5 + 1 = 6$ electrons. Each nitrogen atom has two bonds, which means that each nitrogen atom has four electrons left over, giving two lone pairs.

PROBLEMS Review the common situations for nitrogen, and then come back to these problems. For each of the following structures, draw all lone pairs. Try to recognize how many lone pairs there are *without* having to count. Then count to see if you were right.

1.54   1.55   1.56

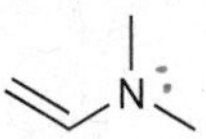

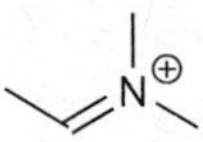

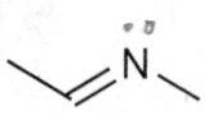

1.57 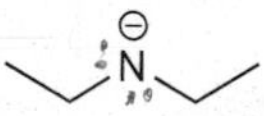   1.58   1.59

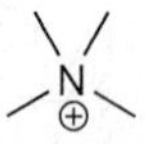

MORE PROBLEMS   For each of the following structures, draw all lone pairs.

1.60

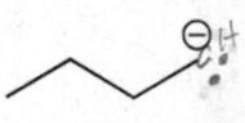

1.61

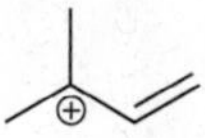

1.62 H—C≡C

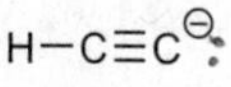

1.63

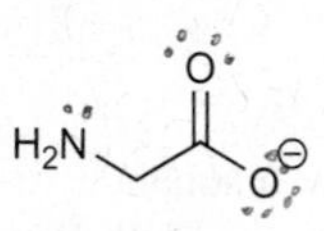

1.64

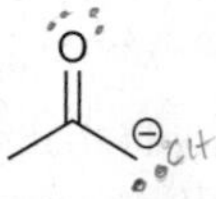

1.65

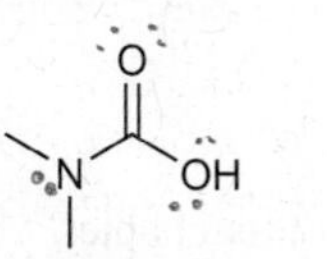

1.66

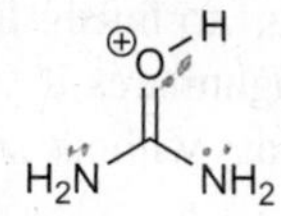

1.67

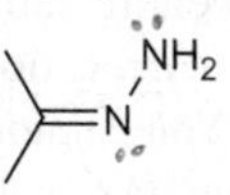

1.68

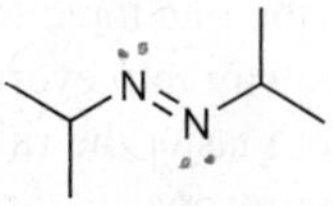

CHAPTER 2

RESONANCE

In this chapter, you will learn the tools that you need to draw resonance structures with proficiency. I cannot adequately stress the importance of this skill. Resonance is the one topic that permeates the entire subject matter from start to finish. It finds its way into every chapter, into every reaction, and into your nightmares if you do not master the rules of resonance. You cannot get an A in this class without mastering resonance. So what is resonance? And why do we need it?

2.1 WHAT IS RESONANCE?

In Chapter 1, we introduced one of the best ways of drawing molecules, bond-line structures. They are fast to draw and easy to read, but they have one major deficiency: they do not describe molecules perfectly. In fact, no drawing method can completely describe a molecule using only a single drawing. Here is the problem.

Although our drawings are very good at showing which atoms are connected to each other, our drawings are not good at showing where all of the electrons are, because electrons aren't really solid particles that can be in one place at one time. All of our drawing methods treat electrons as particles that can be placed in specific locations. Instead, it is best to think of electrons as *clouds of electron density*. We don't mean that electrons fly around in clouds; we mean that electrons *are* clouds. These clouds often spread themselves across large regions of a molecule.

So how do we represent molecules if we can't draw where the electrons are? The answer is resonance. We use the term *resonance* to describe our solution to the problem: we use more than one drawing to represent a single molecule. We draw several drawings, and we call these drawings *resonance structures*. We meld these drawings into one image in our minds. To better understand how this works, consider the following analogy.

Your friend asks you to describe what a nectarine looks like, because he has never seen one. You aren't a very good artist so you say the following:

> Picture a peach in your mind, and now picture a plum in your mind. Well, a nectarine has features of both: the inside tastes like a peach, but the outside is smooth like a plum. So take your image of a peach together with your image of a plum and meld them together in your mind into one image. That's a nectarine.

It is important to realize that a nectarine does not switch back and forth every second from being a peach to being a plum. A nectarine is a nectarine all of the time.

The image of a peach is not adequate to describe a nectarine. Neither is the image of a plum. But by imagining both together at the same time, you can get a sense of what a nectarine looks like.

The problem with drawing molecules is similar to the problem above with the nectarine. No single drawing adequately describes the nature of the electron density spread out over the molecule. To solve this problem, we draw several drawings and then meld them together in our mind into one image. Just like the nectarine.

Let's see an example:

The compound above has two important resonance structures. Notice that we separate resonance structures with a straight, two-headed arrow, and we place brackets around the structures. The arrow and brackets indicate that they are resonance structures *of one molecule.* The molecule is not flipping back and forth between the different resonance structures.

Now that we know why we need resonance, we can begin to understand why resonance structures are so important. Ninety-five percent of the reactions that you will see in this course occur because one molecule has a region of low electron density and the other molecule has a region of high electron density. They attract each other in space, which causes a reaction. So, to predict how and when two molecules will react with each other, we must first predict where there is low electron density and where there is high electron density. We need to have a firm grasp of resonance to do this. In this chapter, we will see many examples of how to predict the regions of low or high electron density by applying the rules of drawing resonance structures.

2.2   CURVED ARROWS: THE TOOLS FOR DRAWING RESONANCE STRUCTURES

In the beginning of the course, you might encounter problems like this: here is a drawing; now draw the other resonance structures. But later on in the course, it will be assumed and expected that you can draw all of the resonance structures of a compound. If you cannot actually do this, you will be in big trouble later on in the course. So how do you draw all of the resonance structures of a compound? To do this, you need to learn the tools that help you: curved arrows.

Here is where it can be confusing as to what is exactly going on. These arrows do not represent an actual process (such as electrons moving). This is an important point, because you will learn later about curved arrows used in drawing reaction mechanisms. Those arrows look exactly the same, but they actually do refer to the flow of electron density. In contrast, curved arrows here are used only as tools to help us draw all resonance structures of a molecule. The electrons are not actually moving. It can be tricky because we will say things like: "this arrow shows the electrons coming from here and going to there." But we don't actually mean that the electrons are moving; they are *not* moving. Since each drawing treats the electrons as particles stuck in one place, we will need to "move" the electrons to get from one drawing to another. Arrows are the tools that we use to make sure that we know how to draw all resonance structures for a compound. So, let's look at the features of these important curved arrows.

Every curved arrow has a *head* and a *tail*. It is essential that the head and tail of every arrow be drawn in precisely the proper place. *The tail shows where the electrons are coming from, and the head shows where the electrons are going* (remember that the electrons aren't really going anywhere, but we treat them as if they were so we can make sure to draw all resonance structures):

Tail _____ Head

Therefore, there are only two things that you have to get right when drawing an arrow: the tail needs to be in the right place and the head needs to be in the right place. So we need to see rules about where you can and where you cannot draw arrows. But first we need to talk a little bit about electrons, since the arrows are describing the electrons.

Atomic orbitals can hold a maximum of two electrons. So, there only three options for any atomic orbital:

# of electrons in atomic orbital	Comments	Outcome
0	Nothing to talk about (no electrons)	-
1	Can overlap with another atomic orbital (also housing one electron) to form a bond with another atom.	*bond*
2	The atomic orbital is filled and is called a lone pair	*lone pair*

So we see that electrons can be found in two places: in bonds or in lone pairs. Therefore, electrons can only come from either a bond or a lone pair. Similarly, electrons can only go to form either a bond or a lone pair.

Let's focus on tails of arrows first. Remember that the tail of an arrow indicates where the electrons are coming from. So the tail has to come from a place that has electrons: either from a bond or from a lone pair. Consider the following resonance structures as an example:

How do we get from the first structure to the second one? Notice that the electrons that make up the double bond have been "moved." This is an example of electrons coming from a bond. Let's see the arrow showing the electrons coming from the bond and going to form another bond:

Now let's see an example where electrons come from a lone pair:

Never draw an arrow that comes from a positive charge. The tail of an arrow must come from a spot that has electrons.

Heads of arrows are just as simple as tails. The head of an arrow shows where the electrons are going. So the head of an arrow must either point directly in between two atoms to form a bond, like this:

or it must point to an atom to form a lone pair, like this:

Never draw the head of an arrow going off into space, like this:

Bad arrow

Remember that the head of an arrow shows where the electrons are going. So the head of an arrow must point to a place where the electrons can go—either to form a bond or to form a lone pair.

2.3 THE TWO COMMANDMENTS

Now we know what curved arrows are, but how do we know when to push them and where to push them? First, we need to learn where we *cannot* push arrows. There are two important rules that you should *never* violate when pushing arrows. They are the "two commandments" of drawing resonance structures:

1. Thou shall not break a single bond.
2. Thou shall not exceed an octet for second-row elements.

Let's focus on one at a time.

1. *Never break a single bond* when drawing resonance structures. By definition, resonance structures must have all the same atoms connected in the same order.

Never break a single bond

There are very few exceptions to this rule, and only a trained organic chemist can be expected to know when it is permissible to violate this rule. Some instructors might violate this rule one or two times (about half-way through the course). If this happens, you should recognize that you are seeing a very rare exception. In virtually every situation that you will encounter, you *cannot* violate this rule. Therefore, you must get into the habit of never breaking a single bond.

There is a simple way to ensure that you never violate this rule. Just make sure that you never draw the tail of an arrow on a single bond.

2. *Never exceed an octet for second-row elements.* Elements in the second row (C, N, O, F) have only four orbitals in their valence shell. Each of these four orbitals can be used either to form a bond or to hold a lone pair. Each bond requires the use of one orbital, and each lone pair requires the use of one orbital. So the second-row elements can never have five or six bonds; the most is four. Similarly, they can *never* have four bonds and a lone pair, because this would also require five orbitals. For the same reason, they can never have three bonds and two lone pairs. The sum of (bonds) + (lone pairs) for a second-row element can never exceed the number four. Let's see some examples of arrow pushing that violate this second commandment:

BAD ARROW   **BAD** ARROW   **BAD** ARROW

In each of these drawings, the central atom cannot form another bond because it does not have a fifth orbital that can be used. *This is impossible.* Don't ever do this.

The examples above are clear, but with bond-line drawings, it can be more difficult to see the violation because we cannot see the hydrogen atoms (and, very often, we cannot see the lone pairs either; for now, we will continue to draw lone pairs to ease you into it). You have to train yourself to see the hydrogen atoms and to recognize when you are exceeding an octet:

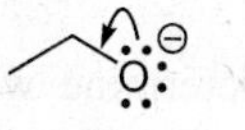

is the same as

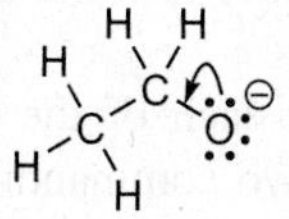

At first it is difficult to see that the arrow on the left structure violates the second commandment. But when we count the hydrogen atoms, we can see that the arrow above would give a carbon atom with five bonds.

From now on, we will refer to the second commandment as "the octet rule." But be careful—for purposes of drawing resonance structures, it is only a violation if we *exceed* an octet for a second-row element. However, there is no problem at all with a second-row element having *fewer* than an octet of electrons. For example:

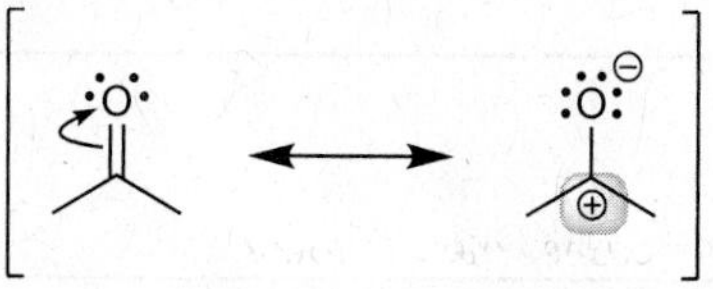

This carbon atom
does not have an octet.

This drawing is perfectly acceptable, even though the central carbon atom has only six electrons surrounding it. For our purposes, we will only consider the "octet rule" to be violated if we exceed an octet.

Our two commandments (never break a single bond, and never violate "the octet rule") reflect the two parts of a curved arrow (the head and the tail). A bad tail violates the first commandment, and a bad head violates the second commandment.

EXERCISE 2.1 For the compound below, look at the arrow drawn on the structure and determine whether it violates either of the two commandments for drawing resonance structures:

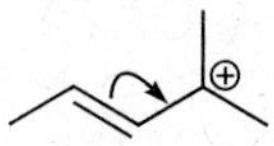

Answer First we need to ask if the first commandment has been violated: did we break a single bond? To determine this, we look at the *tail* of the arrow. If the tail of the arrow is coming from a single bond, then that means we are breaking that single bond. If the tail is coming from a double bond, then we have not violated the first commandment. In this example, the tail is on a double bond, so we did not violate the first commandment.

Now we need to ask if the second commandment has been violated: did we violate the octet rule? To determine this, we look at the *head* of the arrow. Are we forming a fifth bond? Remember that C$^+$ only has three bonds, not four. When we push the arrow shown above, the carbon atom will now get four bonds, and the second commandment has not been violated.

The arrow above is valid, because the two commandments were not violated.

PROBLEMS For each of the problems below, determine which arrows violate either one of the two commandments, and explain why. (Don't forget to count all hydrogen atoms and all lone pairs. You must do this to solve these problems.)

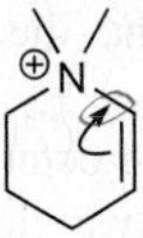

2.2 yes, N would have 5 bonds

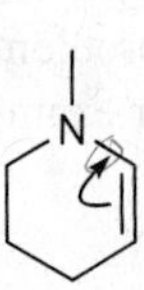

2.3 ~~All~~ yes. N would have 5 bonds

2.4 yes, O would have 5 bonds

2.5 no

2.6 yes, C would have 5

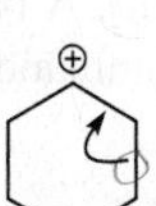

2.7 yes, break single bond

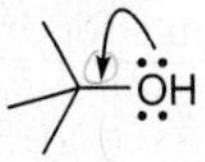

2.8 yes, C would have 5 bonds

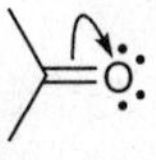

2.9 no

2.10 yes, break single bond

2.11   *[structure]*   yes, C would have 5 bonds

2.12   *[structure]*   no

2.4 DRAWING GOOD ARROWS

Now that we know how to identify good arrows and bad arrows, we need to get some practice drawing arrows. We know that the tail of an arrow must come either from a bond or a lone pair, and that the head of an arrow must go to form a bond or a lone pair. If we are given two resonance structures and are asked to show the arrow(s) that get us from one resonance structure to the other, it makes sense that we need to look for any bonds or lone pairs that are appearing or disappearing when going from one structure to another. For example, consider the following resonance structures:

How would we figure out what curved arrow to draw to get us from the drawing on the left to the drawing on the right? We must look at the difference between the two structures and ask, "How should we push the electrons to get from the first structure to the second structure?" Begin by looking for any double bonds or lone pairs that are disappearing. That will tell us where to put the tail of our arrow. In this example, there are no lone pairs disappearing, but there is a double bond disappearing. So we know that we need to put the tail of our arrow on the double bond.

Now, we need to know where to put the head of the arrow. We look for any lone pairs or double bonds that are appearing. We see that there is a new lone pair appearing on the oxygen atom. This tell us where to put the head of the arrow:

Notice that when we move a double bond up onto an atom to form a lone pair, it creates two formal charges: a positive charge on the carbon atom that lost its bond and a negative charge on the oxygen atom that got a lone pair. This is a very important issue. Formal charges were introduced in the previous chapter, and now they will become instrumental in drawing resonance structures. For the moment, let's just focus on pushing arrows, and in the next section of this chapter, we will come back to focus on these formal charges.

It is pretty straightforward to see how to push only one arrow that gets us from one resonance structure to another. But what about when we need to push more than one arrow to get from one resonance structure to another? Let's do an example like that.

EXERCISE 2.13   For the two structures below, try to draw the curved arrows that get you from the drawing on the left to the drawing on the right:

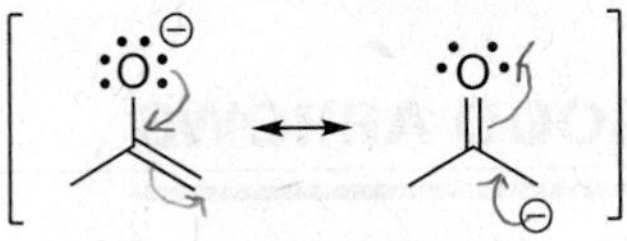

Answer   Let's analyze the difference between these two drawings. We begin by looking for any double bonds or lone pairs that are disappearing. We see that oxygen is losing a lone pair, and the C=C on the bottom is also disappearing. This should automatically tell us that we need two arrows. To lose a lone pair and a double bond, we will need two tails.

Now let's look for any double bonds or lone pairs that are appearing. We see that a C=O is appearing and a C with a negative charge is appearing (remember that a C⁻ means a C with a lone pair). This tells us that we need two heads, which confirms that we need two arrows.

So we know we need two arrows. Let's start at the top. We lose a lone pair from the oxygen atom and form a C=O. Let's draw that arrow:

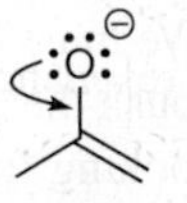

Notice that if we stopped here, we would be violating the second commandment. The central carbon atom is getting five bonds. To avoid this problem, we must immediately draw the second arrow. The C=C disappears (which solves our octet problem) and becomes a lone pair on carbon.

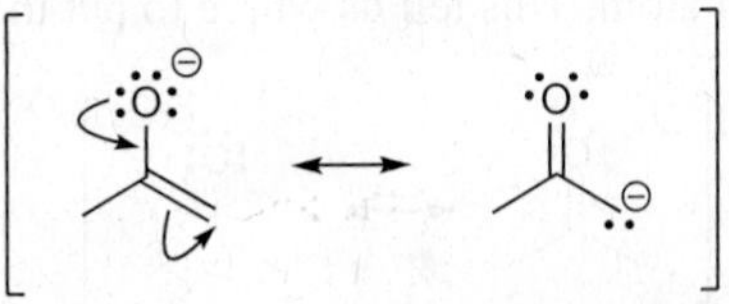

Arrow pushing is much like riding a bike. If you have never done it before, watching someone else will not make you an expert. You have to learn how to balance yourself. Watching someone else is a good start, but you have to get on the bike if you want to learn. You will probably fall a few times, but that's part of the learning process. The same is true with arrow pushing. The only way to learn is with practice.

Now it's time for you to get on the arrow-pushing bike. You would never be stupid enough to try riding a bike for the first time next to a steep cliff. Do not have your first arrow-pushing experience be during your exam. Practice right now!

PROBLEMS   Try to draw the curved arrows that get you from one drawing to the next. In many cases you will need to draw more than one arrow.

2.14

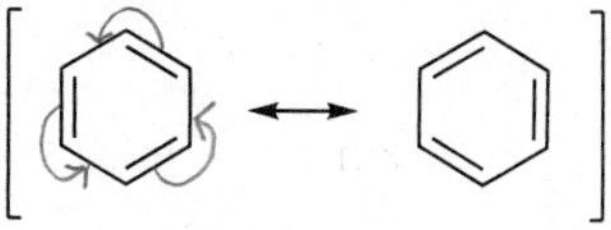

2.15

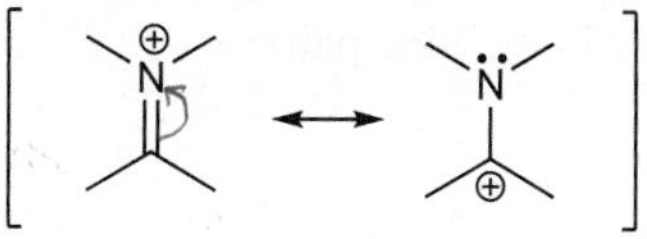

2.16

2.17

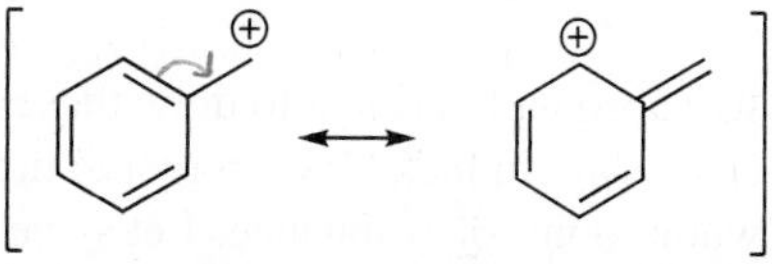

2.18

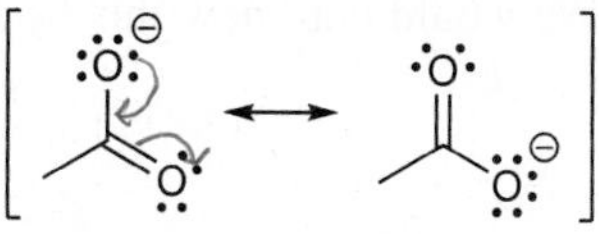

2.19

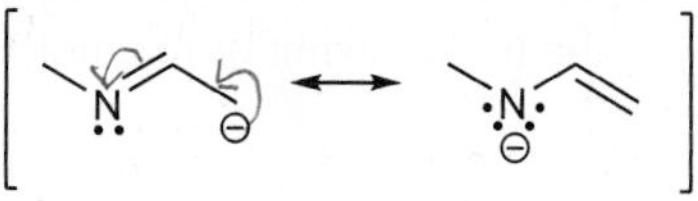

2.5   FORMAL CHARGES IN RESONANCE STRUCTURES

Now we know how to draw good arrows (and how to avoid drawing bad arrows). In the last section, we were given the resonance structures and just had to draw the arrows. Now we need to take this to the next level. We need to get practice drawing the resonance structures when they are not given. To ease into it, we will still show the arrows, and we will focus on drawing the resonance structures with proper formal charges. Consider the following example:

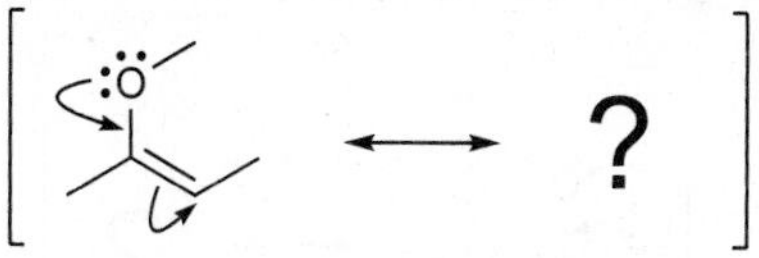

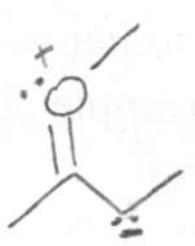

In this example, we can see that one of the lone pairs on oxygen is coming down to form a bond, and the C=C double bond is being pushed to form a lone pair on a carbon atom. When both arrows are pushed at the same time, we are not violating either of the two commandments. So, let's focus on how to draw the resonance structure. Since we know what arrows mean, it is easy to follow the arrows. We just get rid of one lone pair on oxygen, place a double bond between carbon and oxygen, get rid of the carbon–carbon double bond, and place a lone pair on carbon:

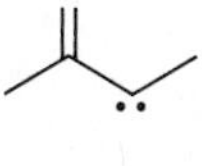

The arrows are really a language, and they tell us what to do. But here comes the tricky part: we cannot forget to put formal charges on the new drawing. If we apply the rules of assigning formal charges, we see that oxygen gets a positive charge and carbon gets a negative charge. As long as we draw these charges, it is not necessary to draw in the lone pairs:

It is absolutely critical to draw these formal charges. Structures drawn without them are *wrong*. In fact, if you forget to draw the formal charges, then you are missing the whole point of resonance. Let's see why. Look at the resonance structure we just drew. Notice that there is a negative charge on a carbon atom. This tells us that this carbon atom is a site of high electron density. We would not know this by looking only at the first drawing of the molecule:

This is why we need resonance—it shows us where there are regions of high and low electron density. If we draw resonance structures without formal charges, then what is the point in drawing the resonance structures at all?

Now that we see that proper formal charges are essential, we should make sure that we know how to draw them when drawing resonance structures. If you are a little bit shaky when it comes to formal charges, you can go back and review formal charges in the previous chapter. More importantly, you should be able to draw formal charges without having to count each time. We saw the common situations for oxygen, nitrogen, and carbon. It is important to remember those (go back and review those if you need to).

Another way to assign formal charges is to read the arrows properly. Let's look at our example again:

Notice what the arrows are telling us: oxygen is giving up a lone pair (two electrons entirely on oxygen) to form a bond (two electrons being shared: one for oxygen and one for carbon). So oxygen is losing an electron. This tells us that it must get a positive charge in the resonance structure. A similar analysis for the carbon atom on the bottom right shows that it will get a negative charge. Remember that the electrons are not really moving anywhere. Arrows are just tools that help us draw resonance structures. To use these tools properly, we imagine that the electrons are moving, but they are not.

Now let's practice.

EXERCISE 2.20   Draw the resonance structure that you get when you push the arrows shown below. Be sure to include formal charges.

Answer   We read the arrows to see what is happening. One of the lone pairs on oxygen is coming down to form a bond, and the C=C double bond is being pushed to form a lone pair on a carbon atom. This is very similar to the example we just saw. We just get rid of one lone pair on oxygen, place a double bond between carbon and oxygen, get rid of the carbon–carbon double bond, and place a lone pair on carbon. Finally, we must draw any formal charges:

There is one subtle point that must be mentioned. We said that you do not need to draw lone pairs—you only need to draw formal charges. There will be times when you will see arrows being pushed on structures that do not have the lone pairs drawn. When this happens, you might see an arrow coming from a negative charge:

The drawing on the left is the common way this is drawn. Just don't forget that the electrons are really coming from a lone pair (as seen in the drawing on the right).

One way to double check your drawing when you are done is to count the total charge on the resonance structure that you draw. This total charge should be the same as the structure you started with. So if the first structure has a negative charge, then the resonance structure you draw should also have a negative charge. If it doesn't, then you know you did something wrong (this is known as *conservation of charge*). You cannot change the total charge when drawing resonance structures.

PROBLEMS   For each of the structures below, draw the resonance structure that you get when you push the arrows shown. Be sure to include formal charges. (*Hint:* In some cases the lone pairs are drawn and in other cases they are not drawn. Be sure to take them into account even if they are not drawn—you need to train yourself to see lone pairs when they are not drawn.)

2.21

2.22

2.23

2.24

2.25

2.26

2.27

2.28

2.6 DRAWING RESONANCE STRUCTURES—STEP BY STEP

Now we have all the tools we need. We know why we need resonance structures and what they represent. We know what curved arrows represent. We know how to recognize bad arrows that violate the two commandments. We know how to draw arrows that get you from one structure to another, and we know how to draw formal charges. We are now ready for the final challenge: using curved arrows to draw resonance structures.

First we need to locate the part of the molecule where resonance is an issue. Remember that we can push electrons only from lone pairs or bonds. We don't need to worry about all bonds, because we can't push an arrow from a single bond (that would violate the first commandment). So we only care about double or triple bonds. Double and triple bonds are called *pi bonds*. So we need to look for lone pairs and pi bonds. Usually, only a small region of the molecule will possess either of these features.

Once we have located the regions where resonance is an issue, now we need to ask if there is any way to push the electrons without violating the two commandments. Let's be methodical, and break this up into three questions:

1. Can we convert any *lone pairs into pi bonds* without violating the two commandments?

2. Can we convert any *pi bonds into lone pairs* without violating the two commandments?

3. Can we convert any *pi bonds into pi bonds* without violating the two commandments?

We do not need to worry about the fourth possibility (converting a lone pair into a lone pair) because electrons cannot jump from one atom to another. Only the three possibilities above are acceptable.

Let's go through these three steps, one at a time, starting with step 1, converting lone pairs into bonds. Consider the following example:

We ask if there are any lone pairs that we can move to form a pi bond. So we draw an arrow that brings the lone pair down to form a pi bond:

This does not violate either of the two commandments. We did not break any single bonds and we did not violate the octet rule. So this is a valid structure. Notice that we cannot move the lone pair in another direction, because then we would be violating the octet rule:

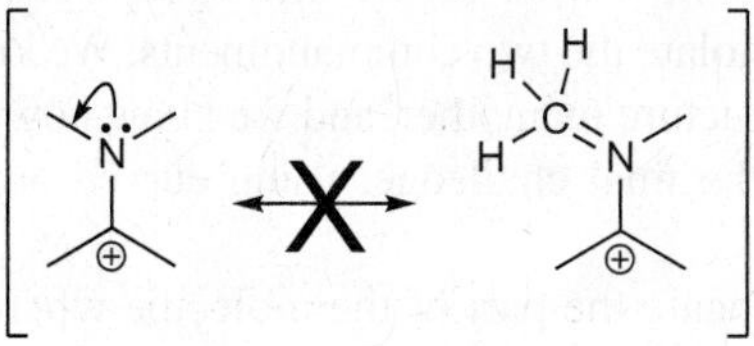

Let's try again with the following example:

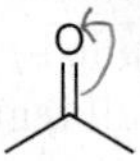

We ask if we can move one of the lone pairs down to form a pi bond, so we try to draw it:

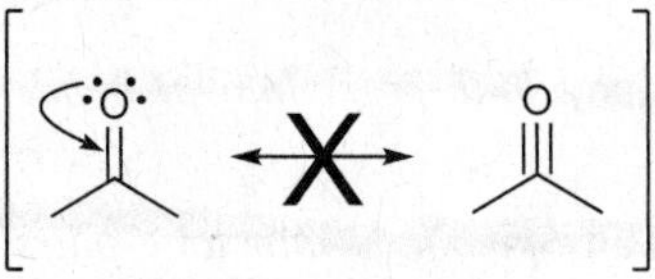

This violates the octet rule—the carbon atom would end up with five bonds. So we cannot push the arrows that way. There is no way to turn the lone pair into a pi bond in this example.

Now let's move on to step 2, converting pi bonds into lone pairs. We try to move the double bond to form a lone pair and we see that we can move the bond in either direction:

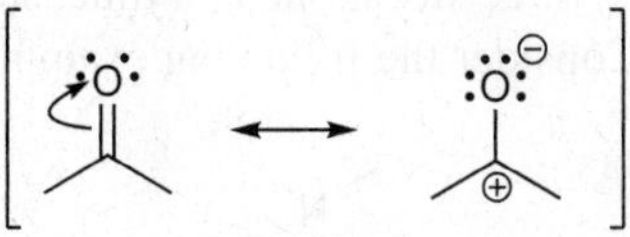

or

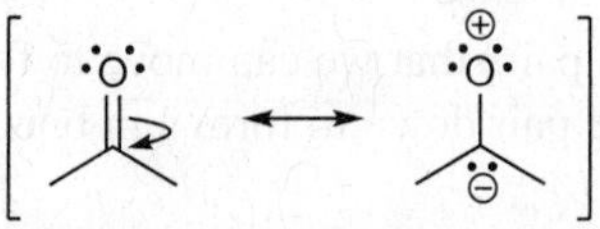

Neither of these structures violates the two commandments, so both structures above are valid resonance structures. (However, the bottom structure, although valid, is not a significant resonance structure. In the next section, we will see how to determine which resonance structures are significant and which are not.)

For step 3, converting pi bonds into pi bonds, let's consider the following examples:

If we try to push the pi bonds to form other pi bonds, we find

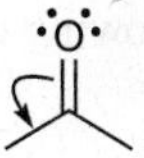

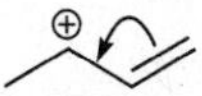

NO. This violates the octet rule

YES. Does not violate the octet rule

The arrow on the top structure violates the octet rule (giving carbon five bonds), and the arrow on the bottom structure does not violate the octet rule. The arrow on the bottom structure will therefore provide a valid resonance structure:

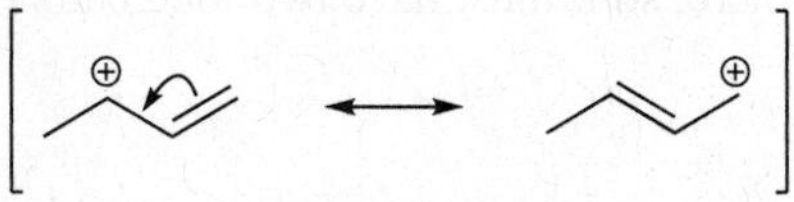

Now that we have learned all three steps, we need to consider that these steps can be combined. Sometimes we cannot do a step without violating the octet rule, but by doing two steps at the same time, we can avoid violating the octet rule. For example, if we try to turn a lone pair into a bond in the following structure, we see that this would violate the octet rule:

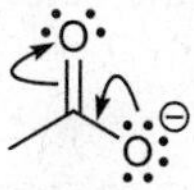

If, at the same time, we also do step 2 (push a pi bond to become a lone pair), then it works:

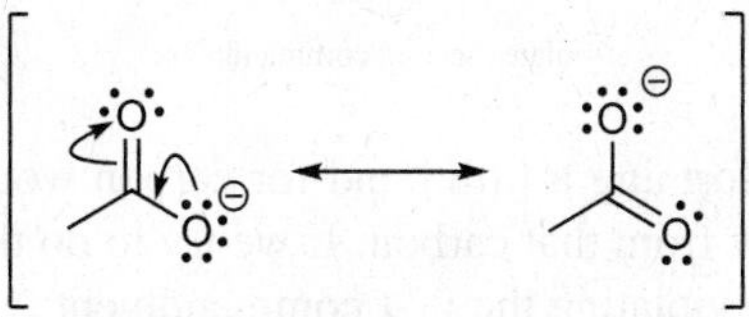

In other words, you should not always jump to the conclusion that pushing an arrow will violate the octet rule. You should first look to see if you can push another arrow that will eliminate the problem.

As another example, consider the following structure. We cannot move the C=C bond to become another bond unless we also move the C=O bond to become a lone pair:

NO   YES

In this way, we truly are "pushing" the electrons around.

Now we are ready to get some practice drawing resonance structures.

EXERCISE 2.29 Draw all resonance structures for the following compound:

Answer Let's start by finding all of the lone pairs and redrawing the molecule. Oxygen has two bonds here, so it must have two lone pairs (so that it will be using all four orbitals):

Now let's do step 1: can we convert any lone pairs into pi bonds? If we try to bring down the lone pairs, we will violate the octet rule by forming a carbon atom with five bonds:

Violates second commandment

The only way to avoid forming a fifth bond for carbon would be to push an arrow that takes electrons away from that carbon. If we try to do this, we will break a single bond and we will be violating the first commandment:

or

Violates first commandment

We cannot move a lone pair to form a pi bond, so we move on to step 2: can we convert any pi bonds into lone pairs? Yes:

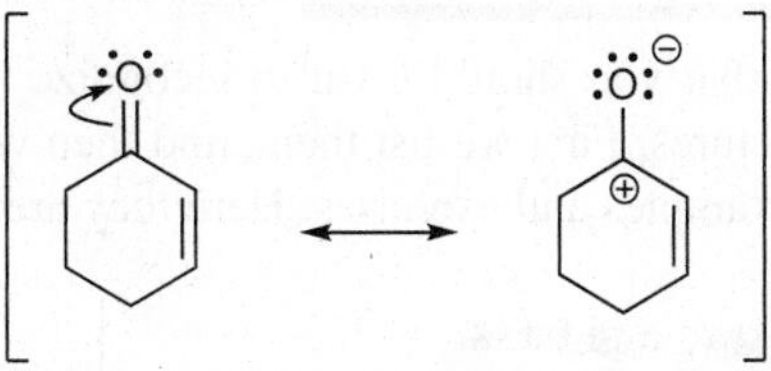

Now we move to step 3: can we convert pi bonds into pi bonds? There is only one move that will not violate the two commandments:

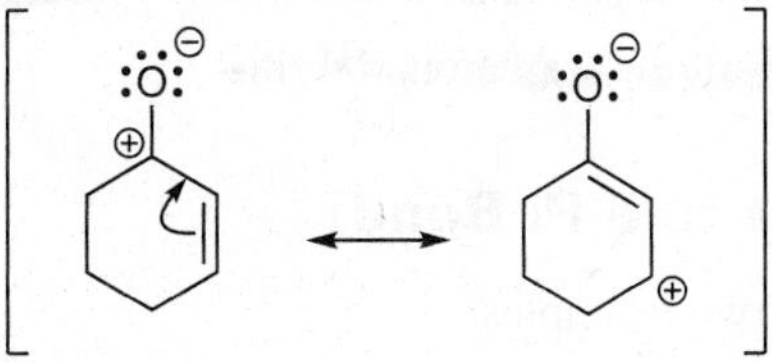

So the resonance structures are

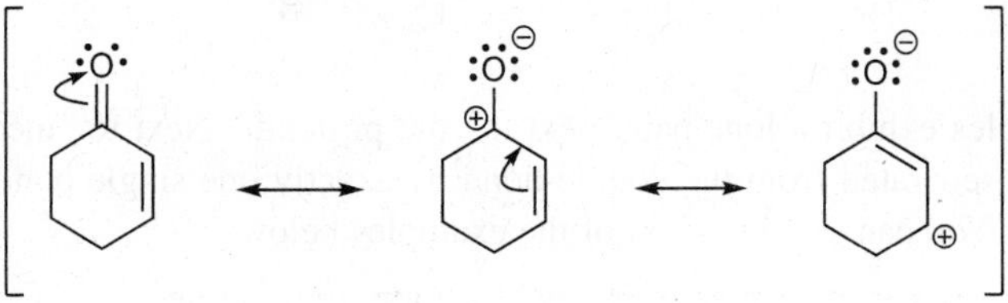

PROBLEM 2.30 For the following compound, go through all three steps (making sure not to violate the two commandments) and draw the resonance structures.

While working through this problem, you probably found that it took a very long time to think through every possibility, to count lone pairs, to worry about violating the octet rule for each atom, to assign formal charges, and so on. Fortunately, there is a way to avoid all of this tedious work. You can learn how to become very quick and efficient at drawing resonance structures if you learn certain patterns and train yourself to recognize those patterns. We will now develop this skill.

2.7   DRAWING RESONANCE STRUCTURES— BY RECOGNIZING PATTERNS

There are five patterns that you should learn to recognize to become proficient at drawing resonance structures. First we list them, and then we will go through each pattern in detail, with examples and exercises. Here they are:

1. A lone pair next to a pi bond.
2. A lone pair next to a positive charge.
3. A pi bond next to a positive charge.
4. A pi bond between two atoms, where one of those atoms is electronegative.
5. Pi bonds going all the way around a ring.

A Lone Pair Next to a Pi Bond

Consider the following two examples:

Both examples exhibit a lone pair "next to" the pi bond. "Next to" means that the lone pair is separated from the double bond by exactly one single bond—no more and no less. You can see this in all of the examples below:

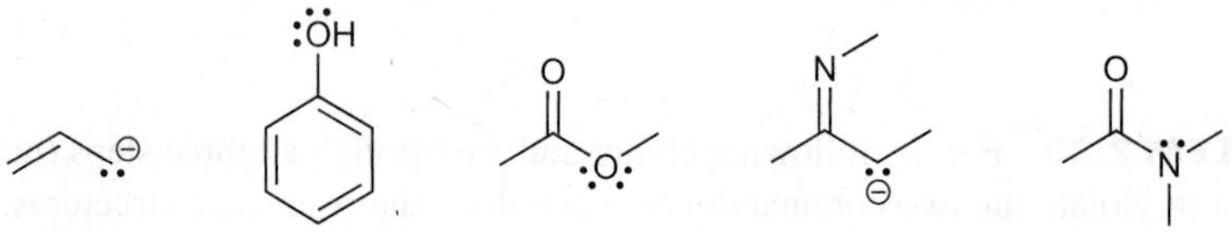

In each of these cases, you can bring down the lone pair to form a pi bond, and kick up the pi bond to form a lone pair:

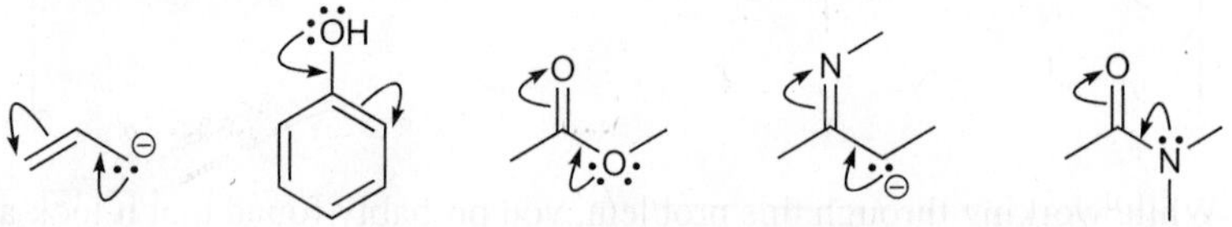

Notice what happens with the formal charges. When the atom with the lone pair has a negative charge, then it transfers its negative charge to the atom that will get a lone pair in the end:

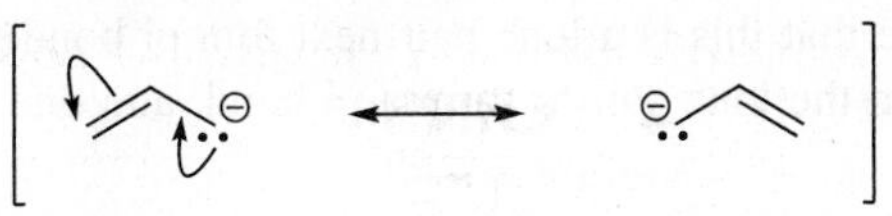

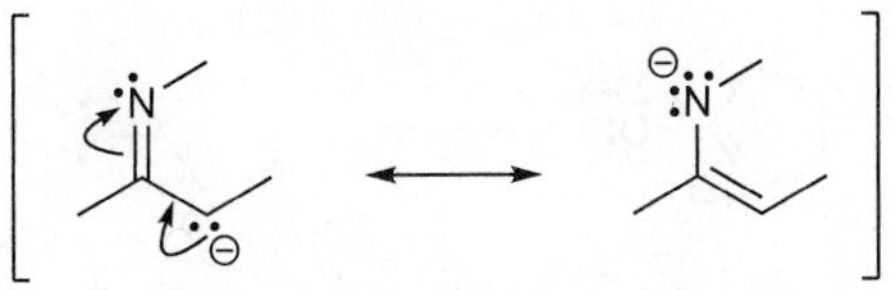

When the atom with the lone pair does not have a negative charge to begin with, then it will end up with a positive charge in the end, while a negative charge will go on the atom getting the lone pair in the end (remember conservation of charge):

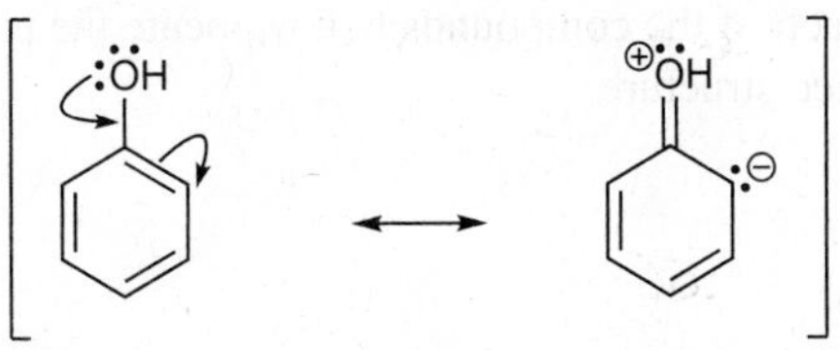

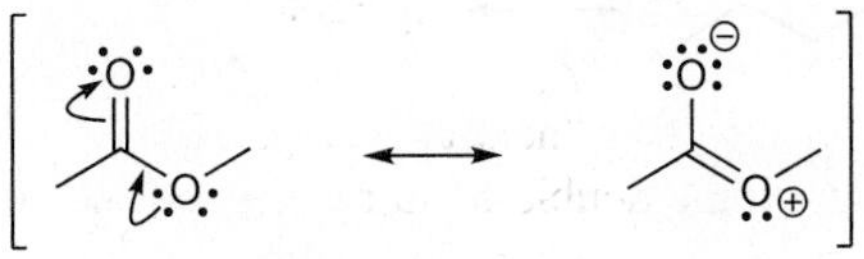

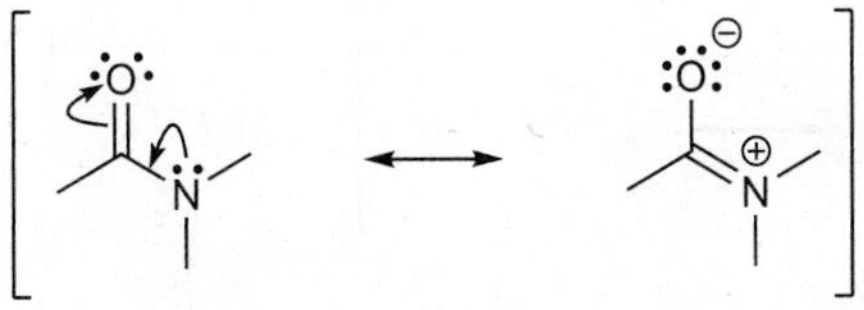

Once you learn to recognize this pattern (a lone pair next to a pi bond), you will be able to save time in calculating formal charges and determining if the octet rule is being violated. You will be able to push the arrows and draw the new resonance structure without thinking about it.

EXERCISE 2.31   Draw a resonance structure of the compound below:

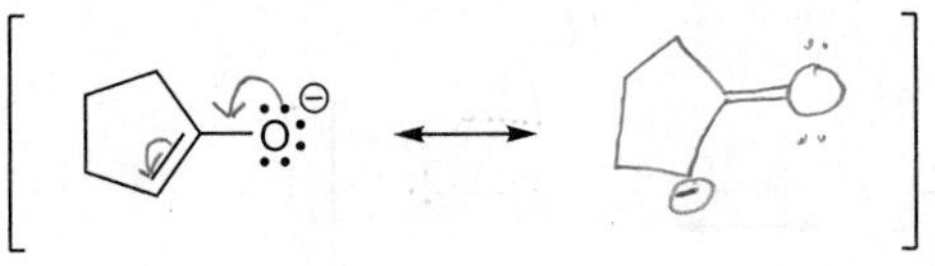

Answer We notice that this is a lone pair next to a pi bond. Therefore, we push two arrows: one from the lone pair to form a pi bond, and one from the pi bond to form a lone pair.

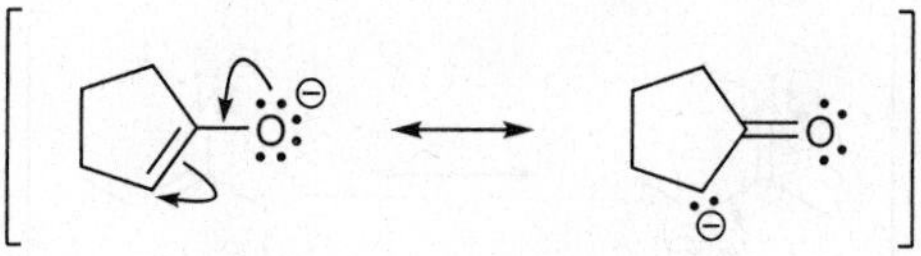

Look carefully at the formal charges. The negative charge used to be on oxygen, but now it moved to carbon.

PROBLEMS For each of the compounds below, locate the pattern we just learned and draw the resonance structure.

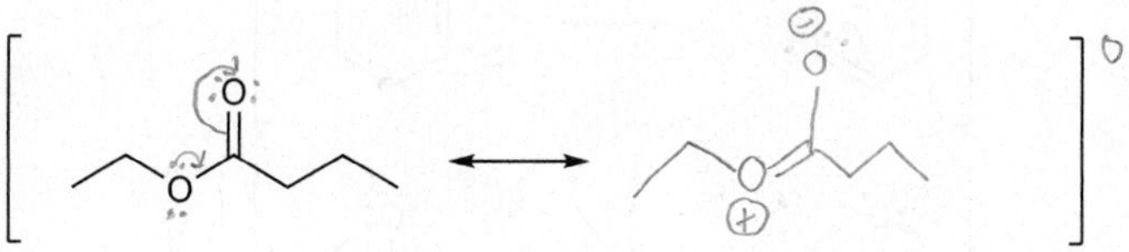

2.32

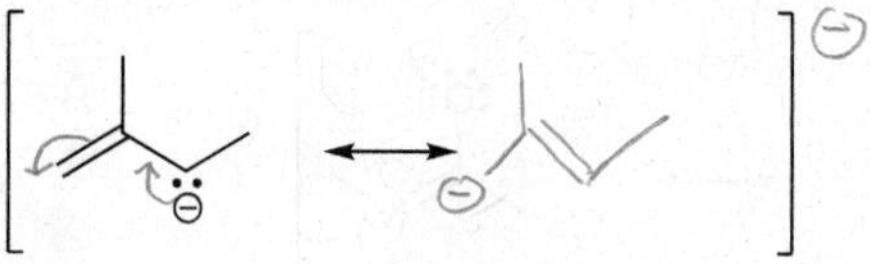

2.33

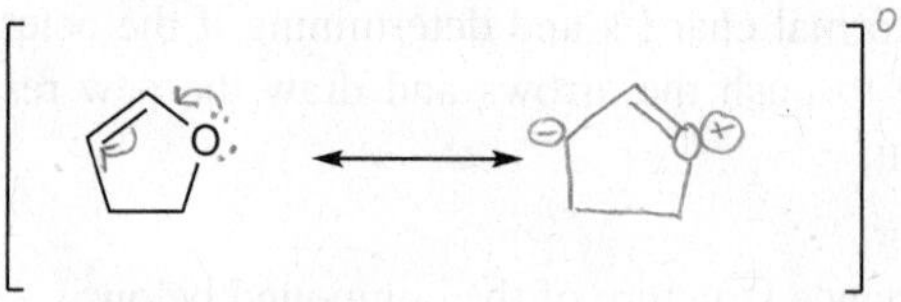

2.34

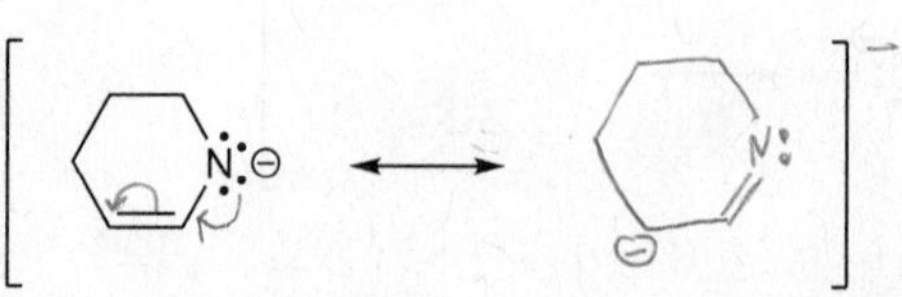

2.35

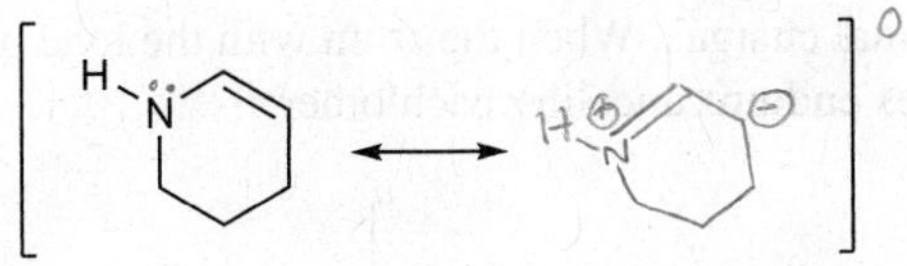

2.36

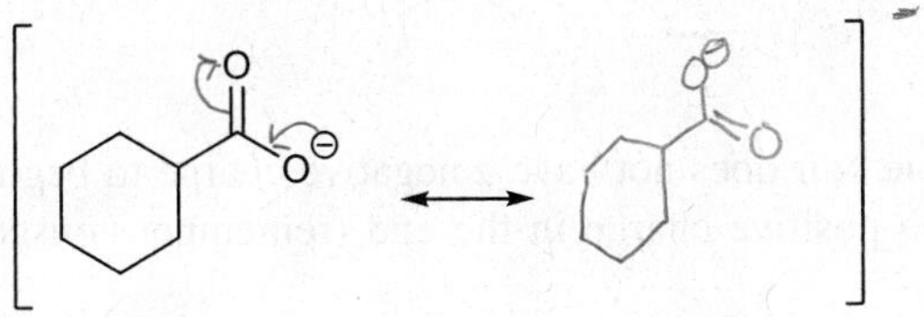

2.37

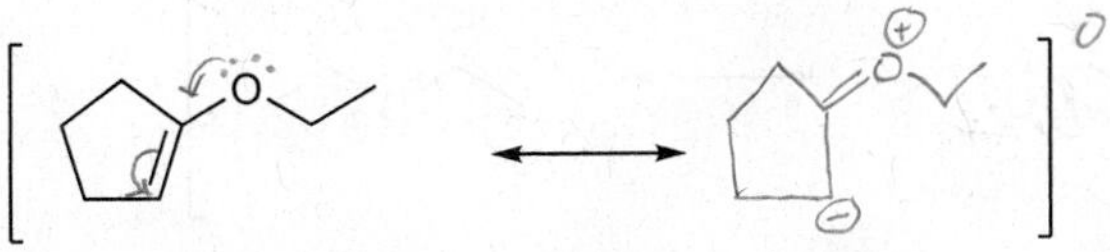

2.38

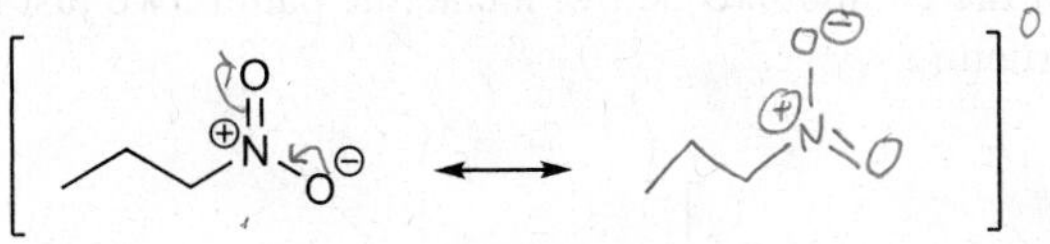

2.39

Notice that the lone pair needs to be directly next to the pi bond. If we move the lone pair one atom away, this does not work anymore:

GOOD   BAD

A Lone Pair Next to a Positive Charge

Consider the following two examples:

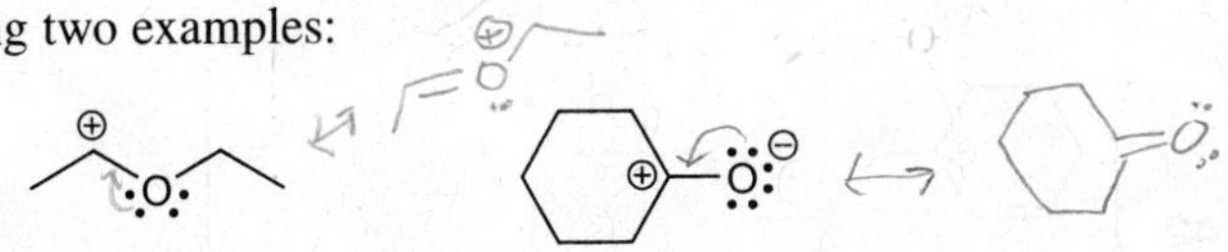

Both examples exhibit a lone pair next to a positive charge. In each case, we can bring down the lone pair to form a pi bond:

Notice what happens with the formal charges. When the atom with the lone pair has a negative charge, then the charges end up canceling each other:

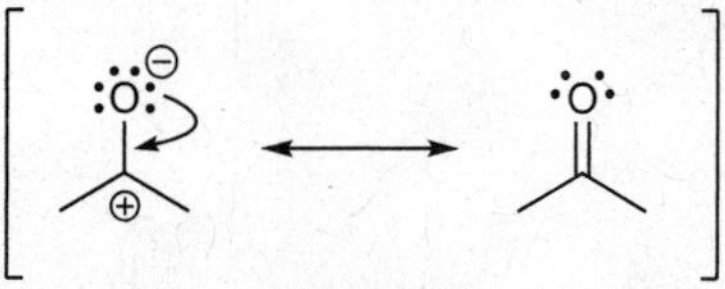

When the atom with the lone pair does not have a negative charge to begin with, then it will end up with the positive charge in the end (remember conservation of charge):

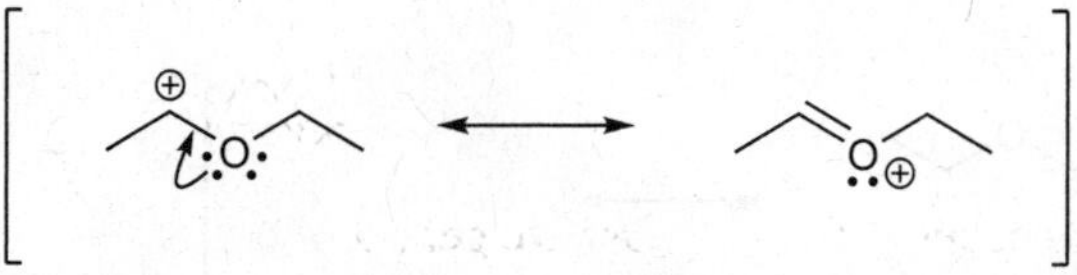

PROBLEMS   For each of the compounds below, locate the pattern we just learned and draw the resonance structure.

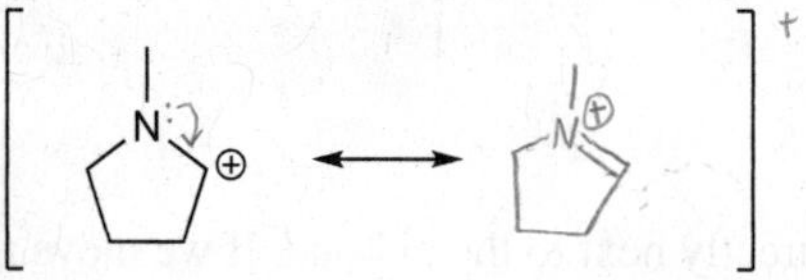

2.40

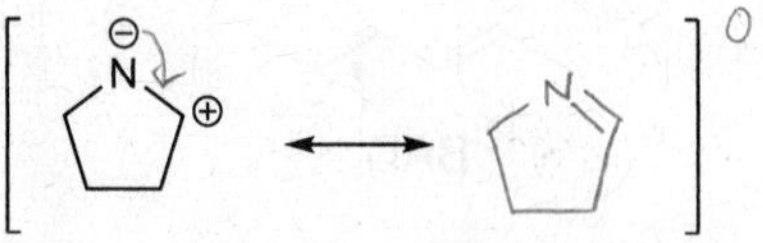

2.41

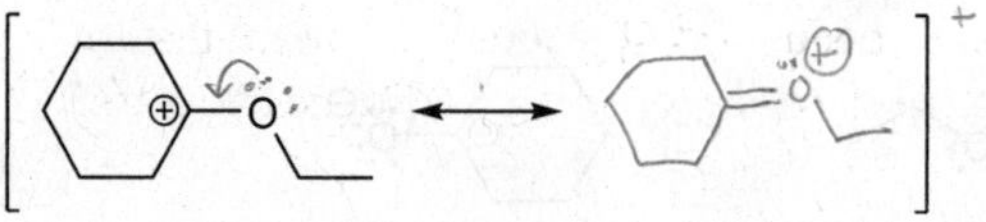

2.42

2.43

Notice that in this problem, a negative and positive charge cancel each other to become a double bond. There is one situation when we cannot combine charges to give a double bond: the nitro group. The structure of the nitro group looks like this:

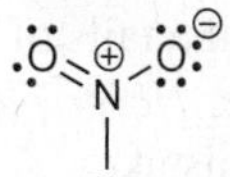

We cannot draw a resonance structure where there are no charges:

Violates
octet rule

This might seem better at first, because we get rid of the charges, but our two commandments show us why it cannot be drawn like this: the nitrogen atom would have five bonds, which would violate the octet rule.

A Pi Bond Next to a Positive Charge

These cases are very easy to see:

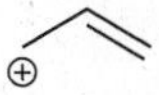

We need only one arrow going from the pi bond to form a new pi bond:

Notice what happens to the formal charge in the process. It gets moved to the other end:

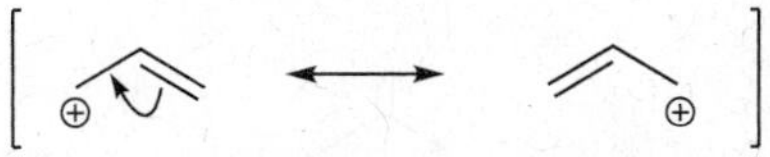

It is possible to have many double bonds in conjugation (this means that we have many double bonds that are each separated by only one single bond) next to a positive charge:

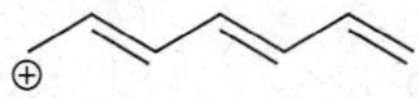

When this happens, we push each of the double bonds over, one at a time:

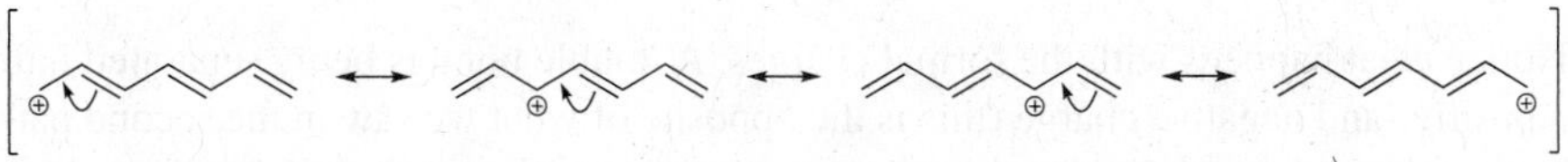

It is not necessary to waste time recalculating formal charges for each resonance structure, because the arrows indicate what is happening. Think of a positive charge as a hole (a place that is missing an electron). When we push electrons to plug up the hole, a new hole is created nearby. In this way, the hole is simply moved from one location to another. Notice that the tails of the curved arrows are placed on the pi bonds, not on the positive charge. *Never place the tail of a curved arrow on a positive charge* (that is a common mistake).

PROBLEMS For each of the compounds below, locate the pattern we just learned and draw the resonance structure.

2.44

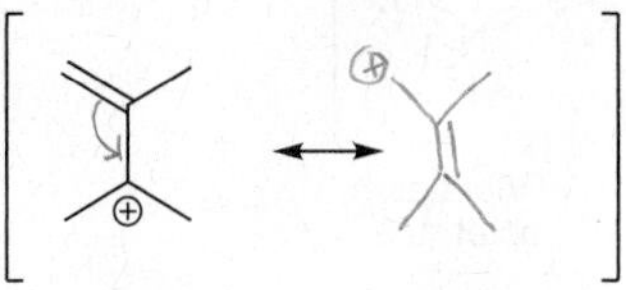

2.45

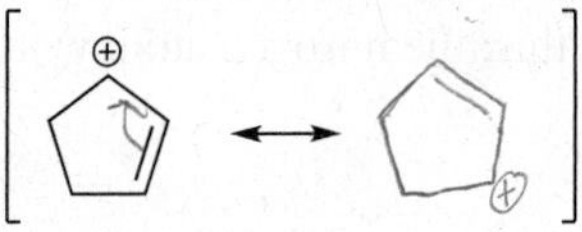

2.46

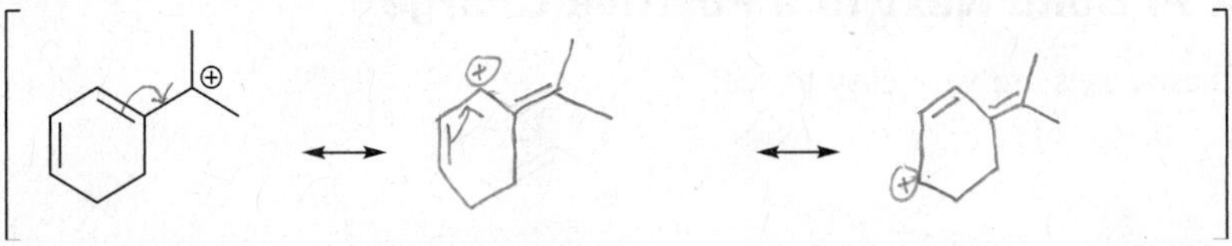

A Pi Bond Between Two Atoms, Where One of Those Atoms Is Electronegative (N, O, etc.)

Let's see an example:

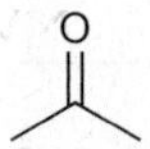

In cases like this, we move the pi bond up onto the electronegative atom to become a lone pair:

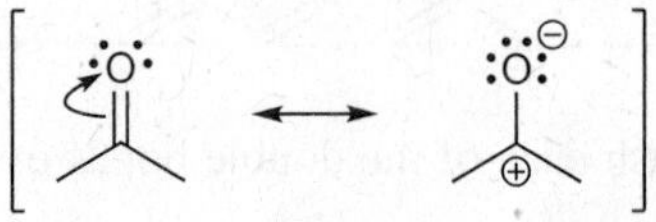

Notice what happens with the formal charges. A double bond is being separated into a positive and negative charge (this is the opposite of what we saw in the second pattern we looked at, where the charges came together to form a double bond).

PROBLEMS   For each of the compounds below, locate the pattern we just learned and draw the resonance structure:

2.47

2.48

2.49

Pi Bonds Going All the Way Around a Ring

Whenever we have alternating double and single bonds, we refer to the alternating bond system as *conjugated:*

Conjugated double bonds

When we have a conjugated system that wraps around in a circle, then we can always move the electrons around in a circle:

It does not matter whether we push our arrows clockwise or counterclockwise (either way gives us the same result, and remember that the electrons are not really moving anyway).

In summary, we have seen the following five arrow-pushing patterns:

1. A lone pair next to a pi bond.
2. A lone pair next to a positive charge.
3. A pi bond next to a positive charge.
4. A pi bond between two atoms, where one of those atoms is electronegative.
5. Pi bonds going all the way around a ring.

Let's get some practice with all five patterns.

PROBLEMS For each of the following compounds, draw the resonance structures.

2.50

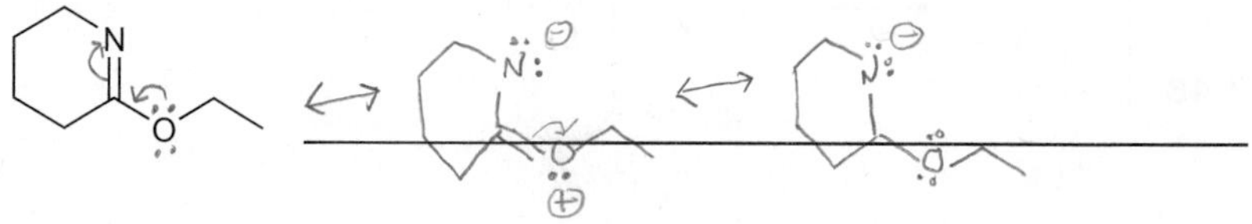

2.51

2.52

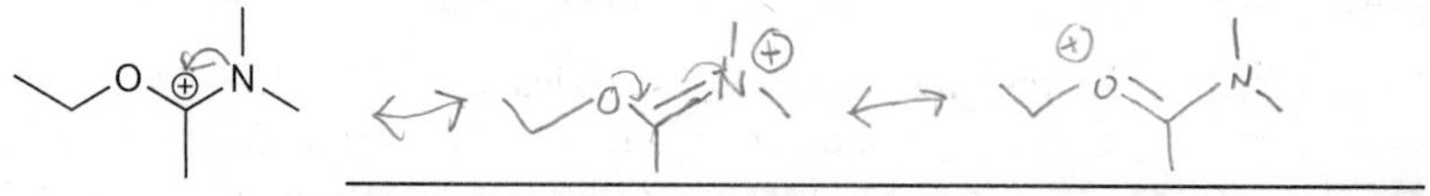

2.53

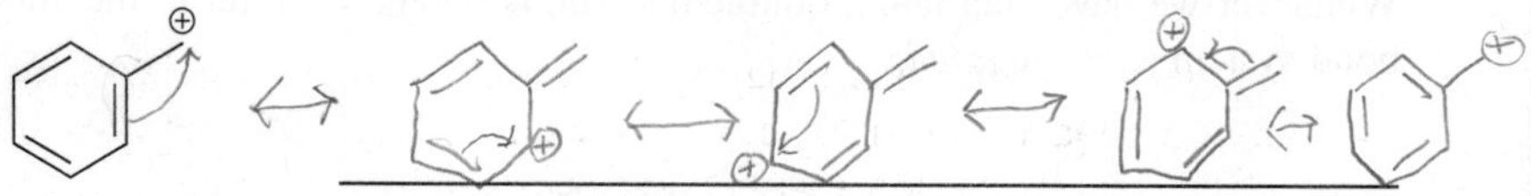

2.54

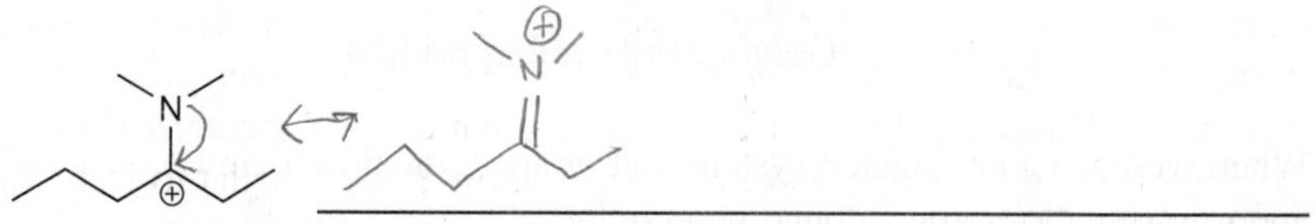

2.55

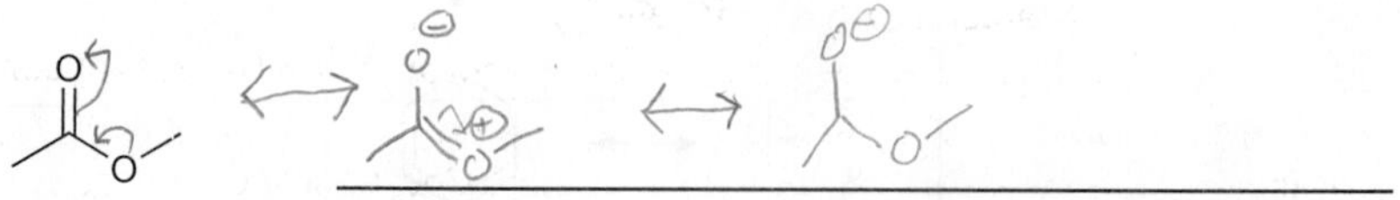

2.56

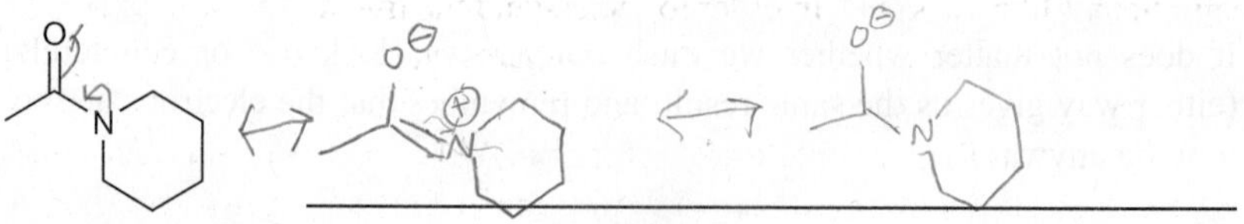

2.57

2.58

2.59

2.60

2.8   ASSESSING THE RELATIVE IMPORTANCE OF RESONANCE STRUCTURES

Not all resonance structures are equally significant. A compound might have *many* valid resonance structures (that do not violate the two commandments), but it is possible that one or more resonance structures might be insignificant. To understand what we mean when we say "insignificant," let's revisit the analogy we used in the beginning of the chapter.

Recall that we used the analogy of a nectarine (being a hybrid between a peach and plum) to explain the concept of resonance. Now, imagine that we create a new type of fruit that is a hybrid between *three* fruits: a peach, a plum, and a kiwi. Suppose that the hybrid fruit that we produce has the following character: 65% peach character, 34% plum character, and 1% kiwi character. This hybrid fruit will look almost exactly like a nectarine, because the amount of kiwi character is too small to have an effect on the nature of the resulting hybrid. Even though this fruit is actually a hybrid of three fruits, nevertheless it will look like a hybrid of only two fruits— because the kiwi character is "insignificant."

A similar concept exists when comparing resonance structures. One compound might have three resonance structures, but all three resonance structures might not contribute equally to the overall resonance hybrid. One resonance structure might be the major contributor (like the peach), while another resonance structure might be insignificant (like the kiwi). In order to understand the true nature of the compound, we must be able to compare the resonance structures and determine which structures are major contributors and which structures are not significant.

There are three simple rules to follow when comparing resonance structures. At this point, you are probably thinking that it is hard enough to keep track of everything we have seen so far—there are two commandments for how not to push arrows, there are three steps for determining valid resonance structures, there are five patterns, and now there are three rules for determining which resonance structures are significant. The good news is that this is the end of the line. There are no more rules or steps. We are almost done with resonance structures. More good news—drawing resonance structures really is very much like riding a bike. When you first learn to ride a bike, you need to concentrate on every movement to avoid from falling. And you have to remember a lot of rules, such as which way to lean your body and which way to turn the handlebars when you feel you are falling to the left. But eventually, you get the hang of it, and then you can ride the bike with no hands. The same is true here. It will take a lot of practice. But before you know it, you will be the resonance guru, and that is where you need to be to do well in this class.

Let's see the three rules for determining which resonance structures are significant:

Rule 1 *Minimize charges.* The best kind of structure is one without any charges. It is OK to have two charges, but you should try to avoid structures that have more than two charges. Compare the following two cases:

Both compounds have a pi bond between a carbon atom and an electronegative atom (C=O), and both compounds have a lone pair next to the pi bond. So we would expect their resonance structures to be similar, and we would expect these compounds to have the same number of significant resonance structures. But they do not. Let's see why. Let's start by drawing the resonance structures of the first compound:

The first resonance structure is the major contributor to the overall resonance hybrid, because it has no charge separation. The other two drawings have charge separation, but there are only two charges in each drawing, so they are both significant resonance structures. They might not contribute as much character as the first resonance structure does, but they are still significant. Therefore, this compound has three significant resonance structures.

Now, let's do the same thing for the other compound:

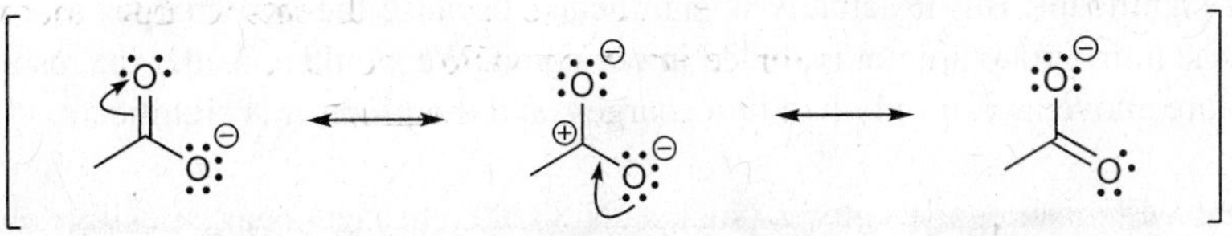

OK   Bad   OK

The first and last structures are OK, but the second resonance structure is bad because there are too many charges. This resonance structure is not significant, because it will not contribute much character to the overall resonance hybrid. It is like the kiwi in our analogy above. Therefore, we would say that this compound has only two significant resonance structures.

There is one notable exception to this rule: compounds containing the nitro group ($-NO_2$) will often have resonance structures with more than two charges. Why? The nitro group looks like this:

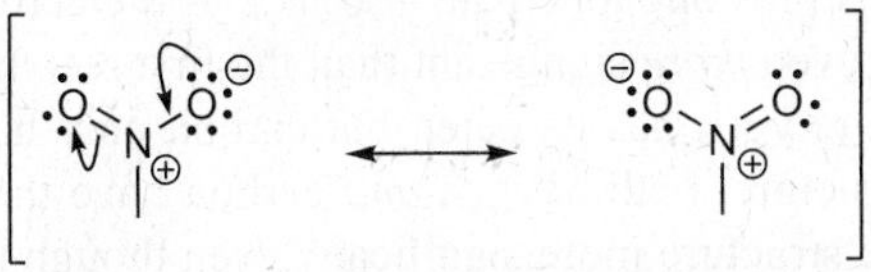

Notice that we have two resonance structures, each of which has charge separation. Even though the molecule has no net charge, nevertheless, we cannot draw a single resonance structure that is free of charge. If we try to do so, we will end up with a structure that violates the second commandment:

Violates
octet rule

Nitrogen cannot have five bonds, so we cannot draw the nitro group without charges. We must draw the nitro group with charge separation. Therefore, the two charges of a nitro group don't really count when we are counting charges. Consider the following case as an example:

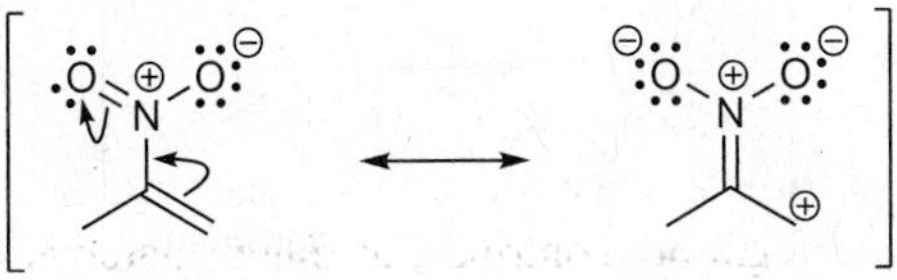

If we apply our rule (about limiting charge separation to no more than two charges), then we might say that the second resonance structure above has too many charges to be significant. But it actually is significant, because the two charges associated with the nitro group are not included in our count. We would consider the resonance structure above as if it only had two charges, and therefore, it is significant.

Rule 2   Electronegative atoms (such as N, O, Cl, etc.) can bear a positive charge, but only if they possess an octet of electrons. Consider the following as an example:

The second resonance structure above is significant, even though it has a positive charge on oxygen. Why? Because the positively charged oxygen has an octet of electrons (three bonds plus one lone pair = 6 + 2 = 8 electrons). In fact, it is not only significant, it is even *more* significant than the first resonance structure. Why? In the first structure, oxygen has its octet, but carbon only has 6 electrons. In the second resonance structure, both oxygen *and* carbon have their octet. This makes the second resonance structure more significant, even though the positive charge is on oxygen.

Here is another example, this time with the positive charge on nitrogen:

Once again, the second structure is significant (in fact, more significant than the first resonance structure).

When a resonance structure has a positive charge on an electronegative atom, that resonance structure will only be significant if the electronegative atom has an octet. If it does not have an octet, the resonance structure will not be significant. For example, consider the following:

Not significant

In the example above, the second resonance structure has an oxygen with a positive charge. But this oxygen does *not* have its octet, and therefore, this resonance structure is not significant.

Rule 3   Avoid drawing a resonance structure in which two carbon atoms bear opposite charges. Such resonance structures are generally insignificant, for example:

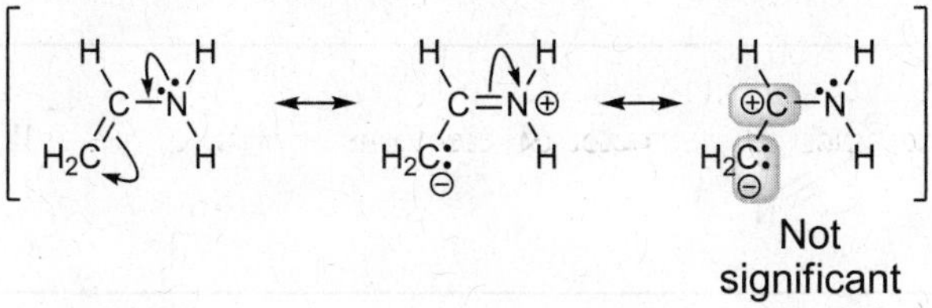

Not
significant

In this case, the third resonance structure is insignificant because it has both a C+ and a C−. The presence of carbon atoms with opposite charges, whether close to each other (as in the example above) or far apart, renders the structure insignificant.

PROBLEMS   For each of the following compounds, draw all of the *significant* resonance structures.

2.61

2.62

2.63

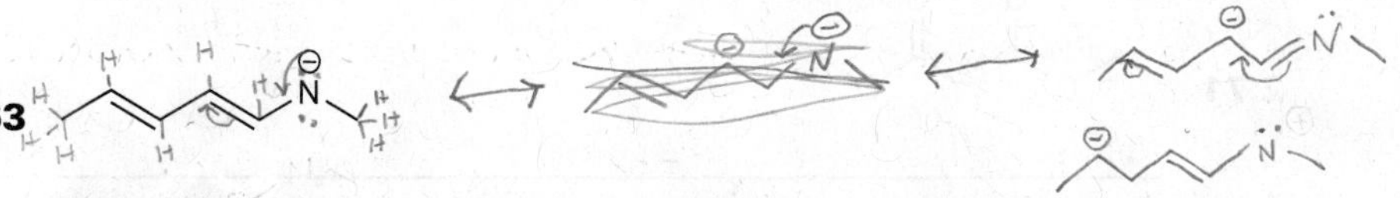

2.64

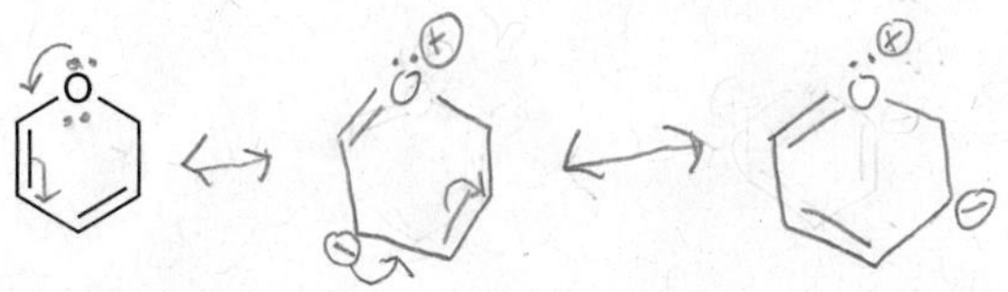

2.65

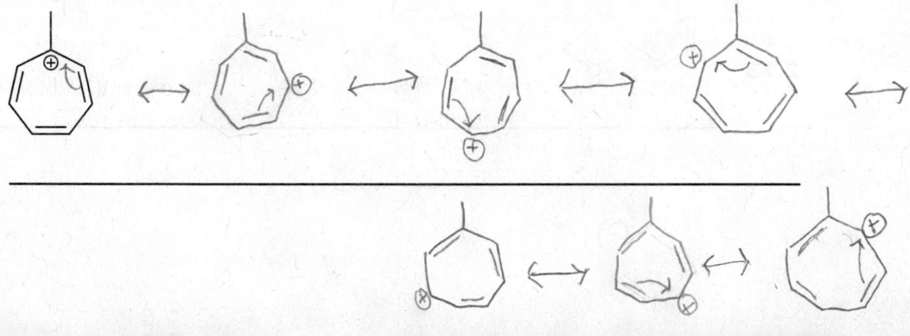

2.66

2.67

2.68

2.69

2.70

2.71

2.72

2.73

ACID–BASE REACTIONS

The first several chapters of any organic chemistry textbook focus on the structure of molecules: how atoms connect to form bonds, how we draw those connections, the problems with our drawing methods, how we name molecules, what molecules look like in 3D, how molecules twist and bend in space, and so on. Only after gaining a clear understanding of structure do we move on to reactions. But there seems to be one exception: acid–base chemistry.

Acid–base chemistry is typically covered in one of the first few chapters of an organic chemistry textbook, yet it might seem to belong better in the later chapters on reactions. There is an important reason why acid–base chemistry is taught so early on in your course. By understanding this reason, you will have a better perspective of why acid–base chemistry is so incredibly important.

To appreciate the reason for teaching acid–base chemistry early in the course, we need to first have a very simple understanding of what acid–base chemistry is all about. Let's summarize with a simple equation:

$$H-A \; \underset{}{\overset{-H^+}{\rightleftharpoons}} \; A^{\ominus}$$

In the equation above, we see an acid (HA) on the left side of the equilibrium, and the conjugate base (A^-) on the right side. HA is an acid by virtue of the fact that it has a proton (H^+) to give. A^- is a base by virtue of the fact that it wants to take its proton back (acids give protons and bases take protons). Since A^- is the base that we get when we deprotonate HA, we call A^- the *conjugate base* of HA.

So the question is: how much is HA willing to give up its proton? If HA is very willing to give up the proton, then HA is a strong acid. However, if HA is not willing to give up its proton, then HA is a weak acid. So, how can we tell whether or not HA is willing to give up its proton? *We can figure it out by looking at the conjugate base.*

Notice that the conjugate base has a negative charge. The real question is: how stable is that negative charge? If that charge is stable, then HA will be willing to give up the proton, and therefore HA will be a strong acid. If that charge is not stable, then HA will not be willing to give up its proton, and HA will be a weak acid.

So you only need one skill to completely master acid–base chemistry: you need to be able to look at a negative charge and determine how stable that negative charge is. If you can do that, then acid–base chemistry will be a breeze for you. If you cannot determine charge stability, then you will have problems even after you finish acid–base chemistry. To predict reactions, you need to know what kind of charges are stable and what kind of charges are not stable.

Now you can understand why acid–base chemistry is taught so early in the course. Charge stability is a vital part of understanding the structure of molecules. It is so incredibly important because reactions are all about how charges interact with one another. You cannot begin to discuss reactions until you have an excellent understanding of what factors stabilize charges and what factors destabilize charges. This chapter will focus on the four most important factors, one at a time.

3.1 FACTOR 1—WHAT ATOM IS THE CHARGE ON?

The most important factor for determining charge stability is to ask what atom the charge is on. For example, consider the following two structures:

The one on the left has a negative charge on oxygen, and the one on the right has the charge on sulfur. How do we compare these? We look at the periodic table, and we need to consider two trends: comparing atoms *in the same row* and comparing atoms *in the same column:*

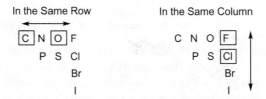

Let's start with comparing atoms *in the same row*. For example, let's compare carbon and oxygen:

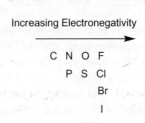

The structure on the left has the charge on carbon, and the structure on the right has the charge on oxygen. Which one is more stable? Recall that electronegativity increases as we move to the right on the periodic table:

Increasing Electronegativity

C N O F
P S Cl
Br
I

Since electronegativity is the measure of an element's affinity for electrons (how willing the atom will be to accept a new electron), we can say that a negative charge on oxygen will be more stable than a negative charge on carbon.

Now let's compare atoms *in the same column,* for example, iodide (I^-) and fluoride (F^-). Here is where it gets a little bit tricky, because the trend is the opposite of the electronegativity trend:

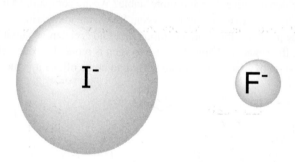

It is true that fluorine is more electronegative than iodine, but there is another more important trend when comparing atoms in the same column: the *size* of the atom. Iodine is *huge* compared to fluorine. So when a charge is placed on iodine, the charge is spread out over a very large volume. When a charge is placed on fluorine, the charge is stuck in a very small volume of space:

Even though fluorine is more electronegative than iodine, nevertheless, iodine can better stabilize a negative charge. If I^- is more stable than F^-, then HI must be a stronger acid than HF, because HI will be more willing to give up its proton than HF.

To summarize, there are two important trends: *electronegativity* (for comparing atoms in the same row) and *size* (for comparing atoms in the same column). The first factor (comparing atoms in the same row) is a much stronger effect. In other words, the difference in stability between C^- and F^- is much greater than the difference in stability between I^- and F^-.

Now we have all of the information we need to answer the question presented earlier in this section: Which charge below is more stable?

When comparing these two ions, we see an oxygen atom bearing the negative charge (on the left) and a sulfur atom bearing the negative charge (on the right). Oxygen and sulfur are in the same column of the periodic table, so size is the important trend to look at. Sulfur is larger than oxygen, so sulfur can better stabilize the negative charge.

EXERCISE 3.1 Compare the two protons highlighted below and determine which one is more acidic.

Answer For each compound, we remove the highlighted proton and draw the resulting conjugate base.

Now we need to compare these conjugate bases and ask which one is more stable. In other words, which negative charge is more stable? We are comparing a negative charge on nitrogen with a negative charge on oxygen. So we are comparing two atoms in the same row of the periodic table, and the important trend is electronegativity. Oxygen can better stabilize the negative charge, because oxygen is more electronegative than nitrogen. The proton on oxygen will be more willing to come off, so it is more acidic:

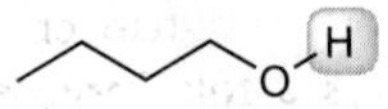

PROBLEMS

3.2 Compare the two highlighted protons in the following compound and determine which is more acidic. Remember to begin by drawing the two conjugate bases, and then compare them.

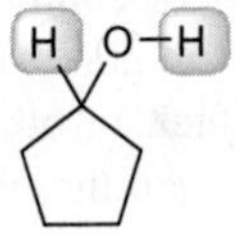

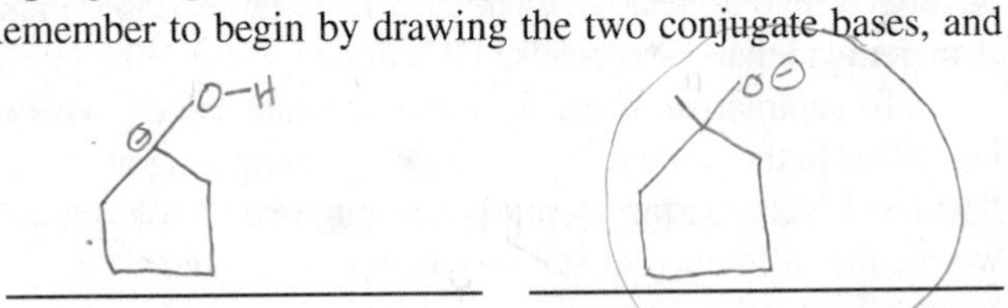

<table>
<tr><td>Conjugate base 1</td><td>Conjugate base 2</td></tr>
</table>

3.3 Compare the two highlighted protons in the following compound and determine which is more acidic.

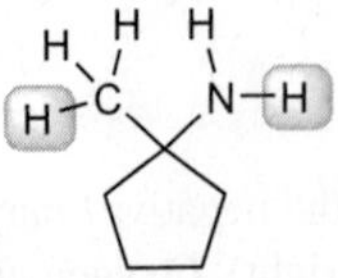

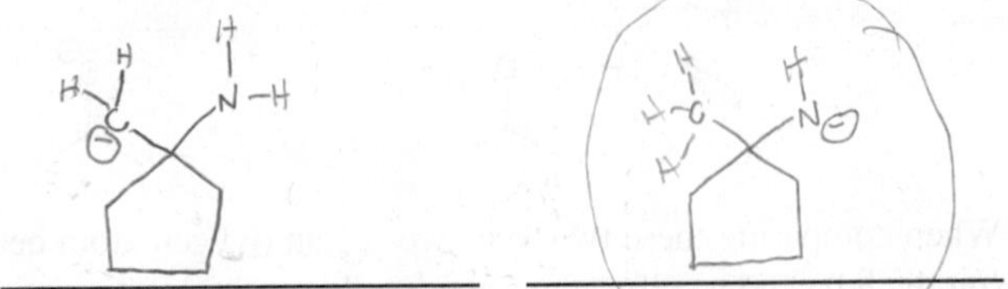

<table>
<tr><td>Conjugate base 1</td><td>Conjugate base 2</td></tr>
</table>

3.4   Compare the two highlighted protons in the following compound and determine which is more acidic.

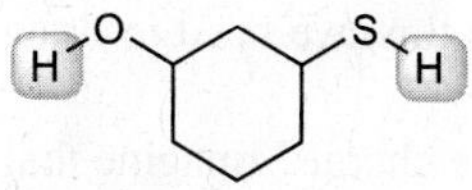

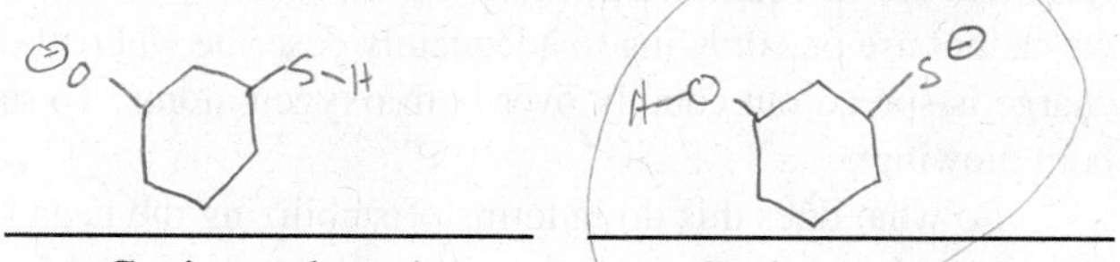

_____   _____
Conjugate base 1   Conjugate base 2

3.5   Compare the two highlighted protons in the following compound and determine which is more acidic.

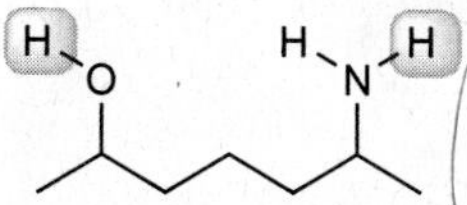

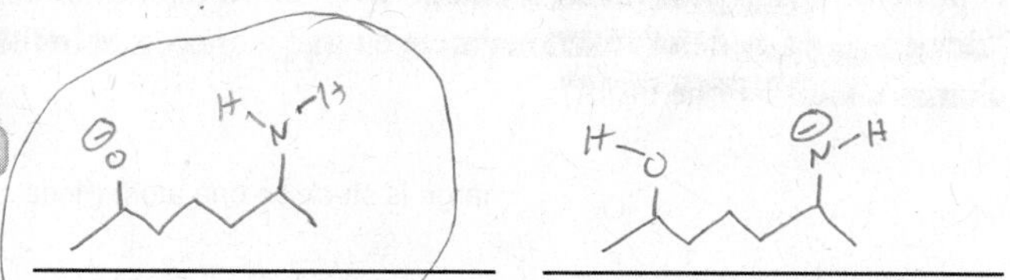

_____   _____
Conjugate base 1   Conjugate base 2

3.2   FACTOR 2—RESONANCE

The previous chapter was devoted solely to drawing resonance structures. If you have not yet completed that chapter, do so before you begin this section. We said in the previous chapter that resonance would find its way into every single topic in organic chemistry. And here it is in acid–base chemistry.

To see how resonance plays a role here, let's compare the following two compounds:

In both cases, we remove a proton, resulting in a charge on oxygen:

So we cannot use factor 1 (what atom is the charge on) to determine which proton is more acidic. In both cases, we are dealing with a negative charge on oxygen. But there is a critical difference between these two negative charges. The first one is stabilized by resonance, as shown here:

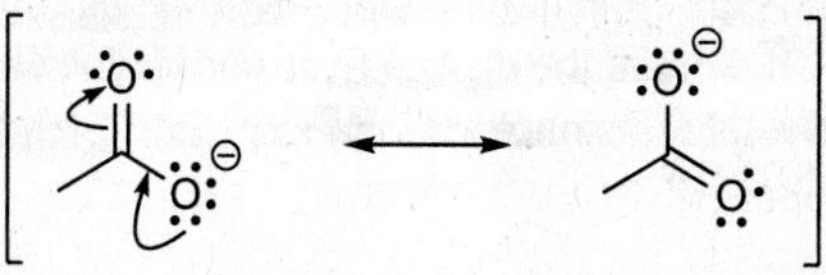

Remember what resonance means. It does not mean that we have two structures that are in equilibrium. Rather, it means that there is only one compound, and we cannot use one drawing to adequately describe where the charge is. In reality, the charge is spread out equally over both oxygen atoms. To see this, we need to draw both drawings.

So what does this do in terms of stabilizing the negative charge? Imagine that you have a hot potato in your hand (too hot to hold for long). If you could grab another potato that is cold and transfer half of the warmth to the second potato, then you would have two potatoes, each of which is not too hot to hold. It's the same concept here. When we spread a charge over more than one atom, we call the charge "delocalized." A delocalized negative charge is more stable than a localized negative charge (stuck on one atom):

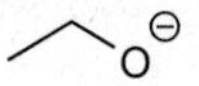

charge is stuck on one atom ("localized")

This factor is very important and explains why carboxylic acids are acidic:

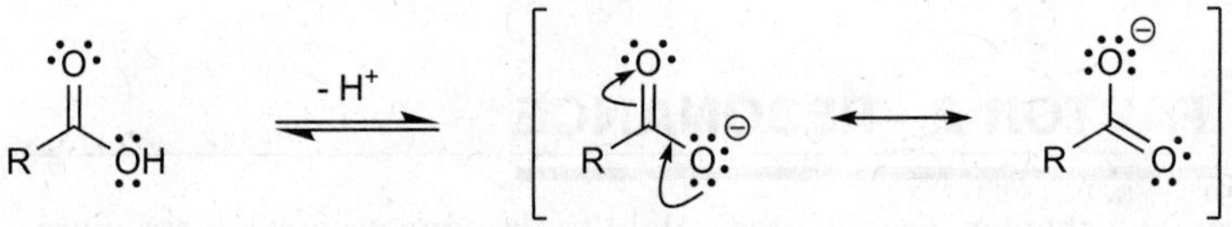

They are acidic because the conjugate base is stabilized by resonance. It is worth noting that carboxylic acids are not terribly acidic. They are acidic when compared with other organic compounds, such as alcohols and amines, but not very acidic when compared with inorganic acids, such as sulfuric acid or nitric acid. In the equilibrium above showing a carboxylic acid losing a proton, we have one molecule losing its proton for every 10,000 molecules that do not give up their proton. In the world of acidity, this is not very acidic, but everything is relative.

So we have learned that resonance (which delocalizes a negative charge) is a stabilizing factor. The question now is how to roughly determine how stabilizing this factor is. Consider, for example, the following case:

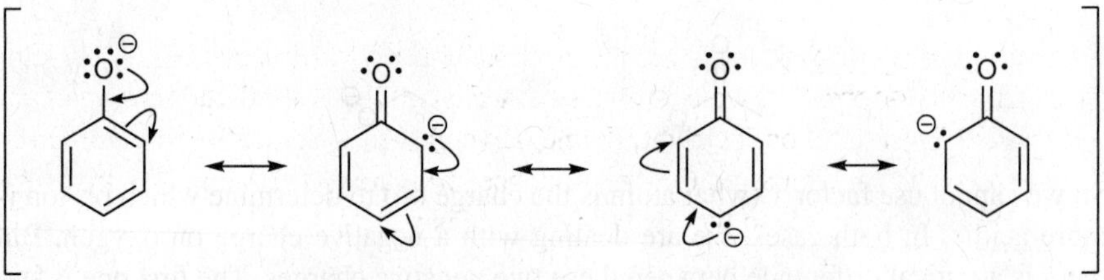

The negative charge is stabilized over four atoms: one oxygen atom and three carbon atoms. Even though carbon is not as happy with a negative charge as oxygen is, nevertheless, it is better to spread the charge over one oxygen and three carbon atoms than to leave the negative charge stuck on one oxygen. Spreading the charge around helps to stabilize that charge.

But the number of atoms sharing the charge isn't everything. For example, it is better to have the charge spread over two oxygen atoms than to have the charge spread over one oxygen and three carbon atoms:

More Stable

So now we have the basic framework to compare two compounds that are both resonance stabilized. We need to compare the compounds, keeping in mind the rules we just learned:

1. The more delocalized the better. A charge spread over four atoms is more stable than a charge spread over two atoms, *but*

2. One oxygen is better than many carbon atoms.

Now let's do some problems.

EXERCISE 3.6   Compare the two protons highlighted below and determine which one is more acidic.

Answer   We begin by pulling off one proton and drawing the conjugate base that we get. Then we do the same thing for the other proton:

Now we need to compare these conjugate bases and ask which one is more stable. In the structure on the left, we are looking at a charge that is localized on a nitrogen atom. For the structure on the right, the negative charge is delocalized over a nitrogen atom and an oxygen atom (draw resonance structures). It is more stable for the charge to be delocalized, so the second structure is more stable.

The more acidic proton is that one that leaves to give the more stable conjugate base.

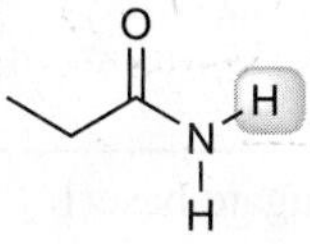

PROBLEMS

3.7 Compare the two highlighted protons and determine which one is more acidic.

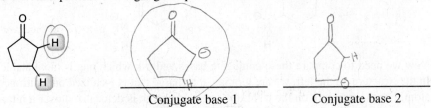

Conjugate base 1   Conjugate base 2

3.8 Compare the two highlighted protons and determine which one is more acidic.

Conjugate base 1   Conjugate base 2

3.9 Compare the two highlighted protons and determine which one is more acidic.

Conjugate base 1   Conjugate base 2

3.10 Compare the two highlighted protons and determine which one is more acidic.

Conjugate base 1   Conjugate base 2

3.11 Compare the two highlighted protons and determine which one is more acidic.

Conjugate base 1   Conjugate base 2

3.12 Compare the two highlighted protons and determine which one is more acidic.

Conjugate base 1	Conjugate base 2

3.3 FACTOR 3—INDUCTION

Let's compare the following compounds:

Which compound is more acidic? The only way to answer that question is to pull off the protons and draw the conjugate bases:

Let's go through the factors we learned so far. Factor 1 does not answer the problem: in both cases, the negative charge is on oxygen. Factor 2 also does not answer the problem: in both cases, there is resonance that delocalizes the charge over two oxygen atoms. Now we need factor 3.

The difference between the compounds is clearly the placement of the chlorine atoms. What effect will this have? For this, we need to understand a concept called induction.

We know that electronegativity measures the affinity of an atom for electrons, so what happens when you have two atoms of different electronegativity connected to each other? For example, consider a carbon–oxygen bond (C–O). Oxygen is more electronegative, so the two electrons that are shared between carbon and oxygen (the two electrons that form the bond between them) are pulled more strongly by the oxygen atom. This creates a difference in the electron density on the two atoms—the oxygen atom becomes electron rich and the carbon atom becomes electron poor. This is usually shown with the symbols $\delta+$ and $\delta-$, which indicate "partial" positive and "partial" negative charges:

This "pulling" of electron density is called induction. Going back to our first example, the three chlorine atoms withdraw electron density from the carbon atom to which they are attached, rendering the carbon atom electron poor ($\delta+$). This carbon atom can then withdraw electron density from the region that has the negative charge, and this effect will stabilize the negative charge:

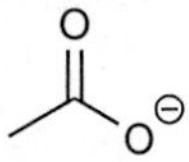

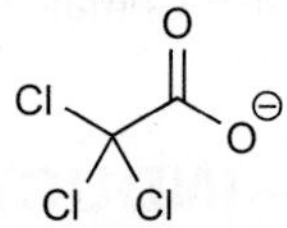

More stable

Inductive effects fall off rapidly with distance, so there is a large difference between the following structures:

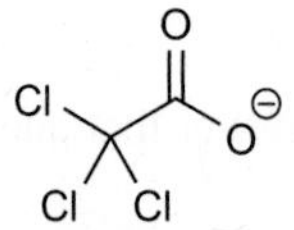

More stable

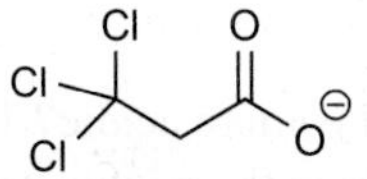

Now we know the effect that electronegative atoms (N, O, Cl, Br, etc.) can have if they are near a negative charge. It is a stabilizing effect. But what effect do carbon atoms have (alkyl groups)? For example, is there a difference in acidity between the following two compounds?

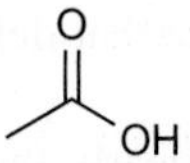

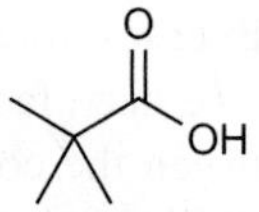

Yes, there is, because *alkyl groups are electron donating.* This is so because of a concept called *hyperconjugation,* which we will not get into here; if you are interested, you can look it up in your textbook. But the bottom-line, take-home message is that alkyl groups are electron donating. So, what effect will this have on a negative charge? If electron density is being given to an area where there is a negative charge, then this area becomes less stable. It would be as if you are holding a hot potato, and someone with a hot iron heats up your potato even more.

So the comparison goes like this:

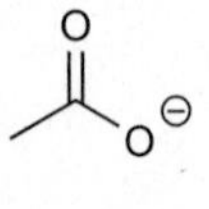

More stable 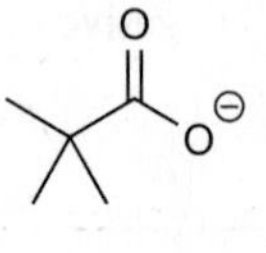   Less stable and therefore,

More acidic   Less acidic

EXERCISE 3.13 Compare the two protons highlighted below and determine which one is more acidic.

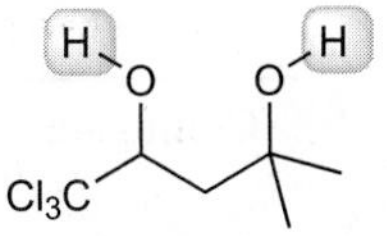

Answer Begin by drawing the conjugate bases:

In the structure on the left, the charge is somewhat stabilized by the inductive effects of the neighboring chlorine atoms. In contrast, the structure on the right is destabilized by the presence of methyl groups. Therefore, the structure on the left is more stable.

The more acidic proton is the one that will leave to give the more stable negative charge. So the following proton is more acidic:

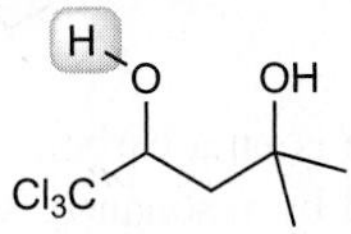

PROBLEMS

3.14 Compare the two highlighted protons and determine which one is more acidic.

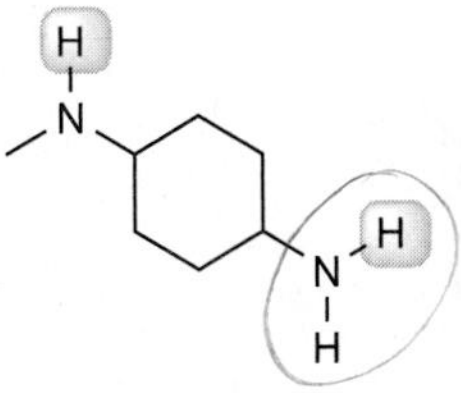

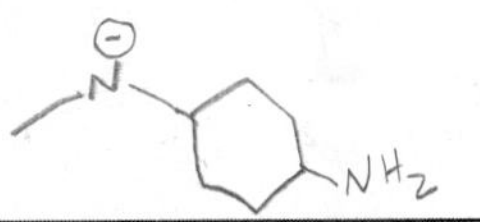

Conjugate base 1 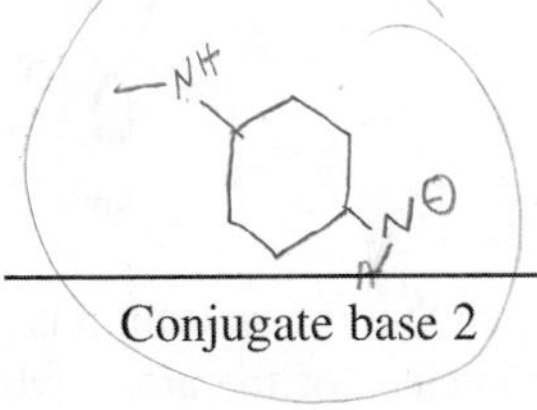   Conjugate base 2

3.15 Compare the two highlighted protons and determine which one is more acidic.

Conjugate base 1	Conjugate base 2

3.16 Compare the two highlighted protons and determine which one is more acidic.

Conjugate base 1	Conjugate base 2

3.4 FACTOR 4—ORBITALS

The three factors we have learned so far will not explain the difference in acidity between the two highlighted protons in the compound below:

In order to determine which proton is more acidic, we remove each proton and compare the resulting conjugate bases:

In both cases, the negative charge is on a carbon, so factor 1 does not help. In both cases, the charge is not stabilized by resonance, so factor 2 does not help. In both cases, there are no inductive effects to consider, so factor 3 does not help. The answer here comes from looking at the type of orbital that is accommodating the charge.

Let's quickly review the shape of hybridized orbitals. sp^3, sp^2, and sp orbitals all have roughly the same shape, but they are different in size:

sp^3 sp^2 sp

Notice that the sp orbital is smaller and tighter than the other orbitals. It is closer to the nucleus of the atom, which is located at the point where the front lobe (white) meets the back lobe (gray). Therefore, a lone pair of electrons residing in an *sp* orbital will be held closer to the positively charged nucleus and will be *stabilized* by being close to the nucleus.

So a negative charge on an *sp* hybridized carbon is more stable than a negative charge on an *sp*³ or *sp*² hybridized carbon:

More stable

Determining which carbon atoms are *sp*, *sp*², or *sp*³ is very simple: a carbon with a triple bond is *sp*, a carbon with a double bond is *sp*², and a carbon with all single bonds is *sp*³. For more on this topic, turn to the next chapter (covering geometry).

EXERCISE 3.17   Locate the most acidic proton in the following compound:

Answer   It is important to recognize where all of the protons (hydrogen atoms) are. If you cannot do this, then you should review Chapter 1, which covers bond-line drawings. Only one proton can leave behind a negative charge in an *sp* orbital. All of the other protons would leave behind a negative charge on either *sp*² or *sp*³ hybridized orbitals. So the most acidic proton is

3.5   RANKING THE FOUR FACTORS

Now that we have seen each of the four factors individually, we need to consider what order of importance to place them in. In other words, what should we look for first? And what should we do if two factors are competing with each other?

In general, the order of importance is the order in which the factors were presented in this chapter.

1. What atom is the charge on? (Remember the difference between comparing atoms in the same row and comparing atoms in the same column.)

2. Are there any resonance effects making one conjugate base more stable than the others?

3. Are there any inductive effects (electronegative atoms or alkyl groups) that stabilize or destabilize any of the conjugate bases?

4. In what orbital do we find the negative charge for each conjugate base that we are comparing?

There is an important exception to this order. Compare the following two compounds:

$$H-\!\!\!\equiv\!\!\!-H \qquad\qquad NH_3$$

In order to predict which compound is more acidic, we remove a proton from each compound and compare the conjugate bases:

$$H-\!\!\!\equiv:^{\ominus} \qquad\qquad {}^{\ominus}\!:\!\!\overset{..}{N}H_2$$

When comparing these two negative charges, we find two competing factors: the first factor (what atom is the charge on?) and the fourth factor (what orbital is the charge in?). The first factor says that a negative charge on nitrogen is more stable than a negative charge on carbon. However, the fourth factor says that a negative charge in an sp orbital is more stable than a negative charge in an sp^3 orbital (the negative charge on the nitrogen atom is an sp^3 orbital). In general, we would say that factor 1 wins over the others. But this case is an exception, and factor 4 (orbitals) actually wins here, so the negative charge on carbon is more stable in this case:

$$H-\!\!\!\equiv:^{\ominus} \qquad\qquad {}^{\ominus}\!:\!\!\overset{..}{N}H_2$$

More stable

This example illustrates that a negative charge on an sp hybridized carbon atom is more stable than a negative charge on an sp^3 hybridized nitrogen atom. For this reason, NH_2^- can be used as a base to deprotonate a triple bond.

There are, of course, other exceptions, but the one explained above is the most common. In most cases, you should be able to apply the four factors and provide a qualitative assessment of acidity.

EXERCISE 3.18 Compare the two protons highlighted below and determine which one is more acidic.

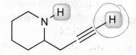

Answer   The first thing we need to do is draw the conjugate bases:

Now we can compare them and ask which negative charge is more stable, using our four factors:

1. *Atom*   The first conjugate base has a negative charge on a nitrogen atom, while the second conjugate base has a negative charge on a carbon atom. Based on this factor alone, we predict that the first conjugate base should be more stable.

2. *Resonance*   Neither of the conjugate bases is stabilized by resonance.

3. *Induction*   Neither of the conjugate bases is stabilized by induction.

4. *Orbital*   The first conjugate base has a negative charge on an sp^3 hybridized atom, while the second conjugate base has a negative charge on an sp hybridized atom. Based on this factor alone, we predict that the second conjugate base should be more stable.

So, we have a competition between the first factor (atom) and the fourth factor (orbital). In general, the first factor takes precedence over the fouth factor. But this is the one exception that we saw. In this case, the fourth factor is the dominating factor because a negative charge on an sp hybridized cabon atom is more stable than a negative charge on an sp^3 hybridized nitrogen atom. Therefore, the proton highlighted below is the more acidic proton:

Remember the four factors, and what order they come in:

1. Atom

2. Resonance

3. Induction

4. Orbital

If you have trouble remembering the order, try remembering this acronym: ARIO.

PROBLEMS   For each of the following compounds, two protons have been highlighted. In each case, determine which of the two protons is more acidic.

3.19

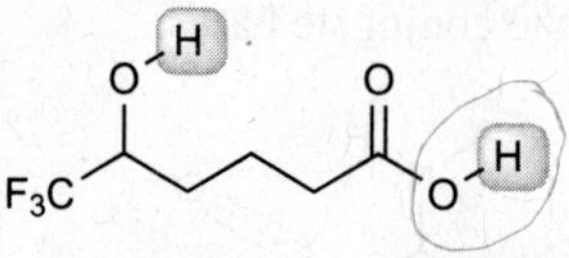

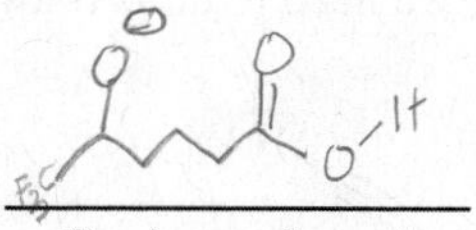

Conjugate base 1 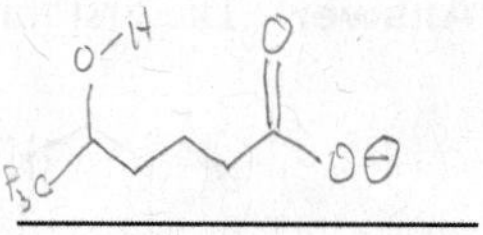   Conjugate base 2

3.20

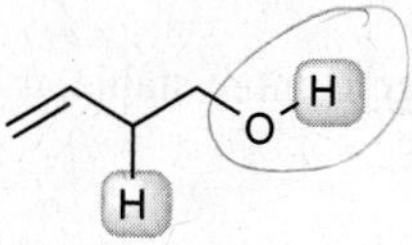

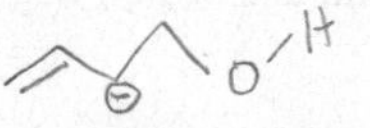

Conjugate base 1 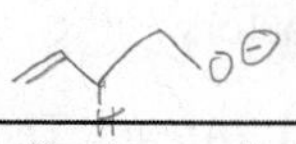   Conjugate base 2

3.21

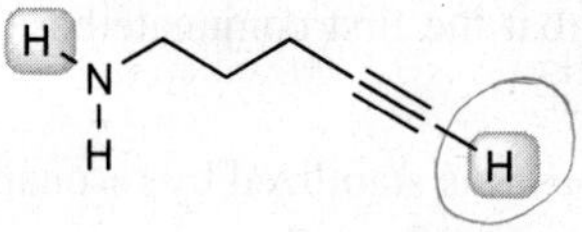

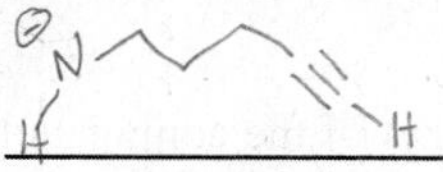

Conjugate base 1 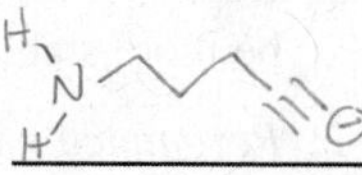   Conjugate base 2

3.22

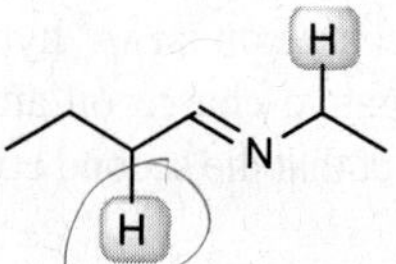

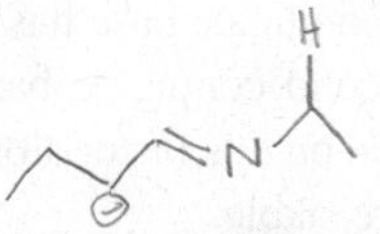

Conjugate base 1 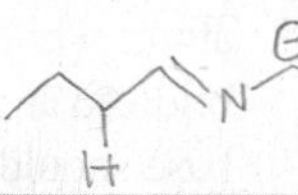   Conjugate base 2

3.23

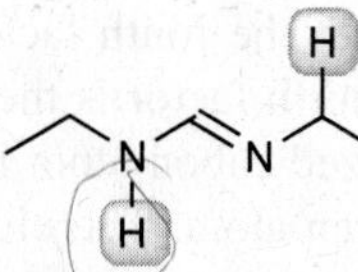

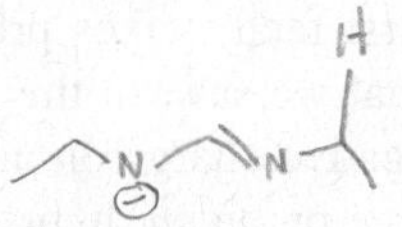

Conjugate base 1 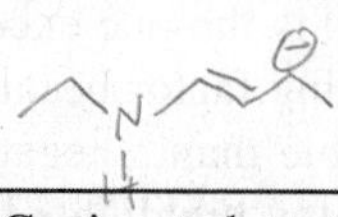   Conjugate base 2

3.24

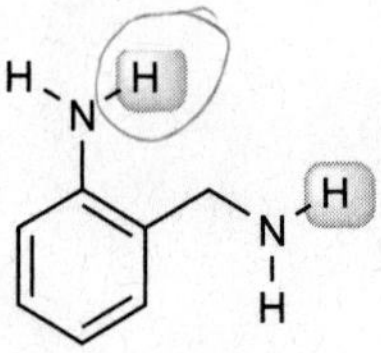

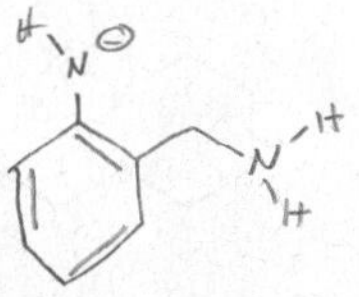

Conjugate base 1 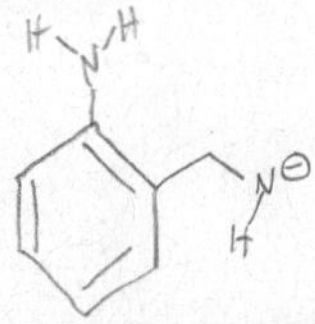   Conjugate base 2

3.25

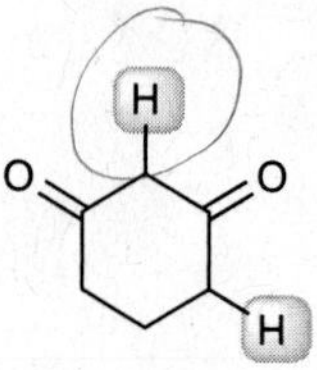

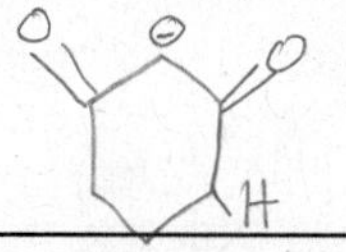

Conjugate base 1 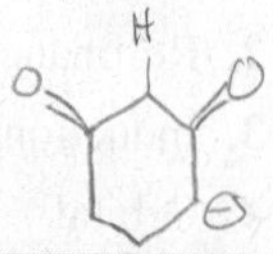   Conjugate base 2

3.26

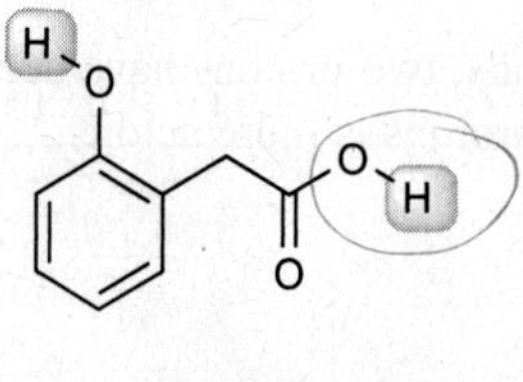

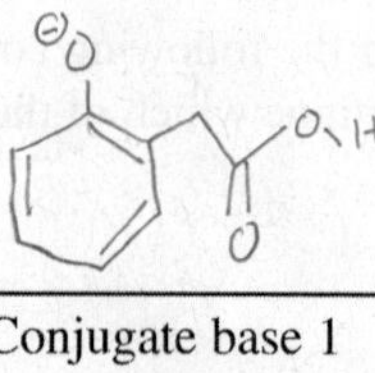

Conjugate base 1 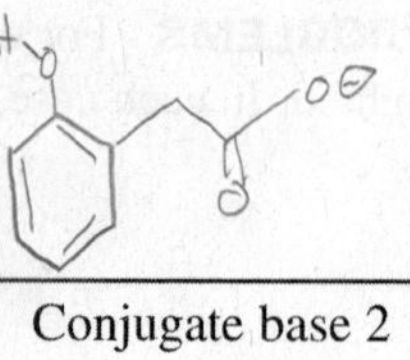   Conjugate base 2

3.27

Conjugate base 1	Conjugate base 2

PROBLEMS For each pair of compounds below, predict which will be more acidic.

3.28 HI   HBr

3.29 CH$_3$SH   CH$_3$OH

3.30 NH$_3$ H$_2$O

3.31

3.32

3.33 Cl$_3$C—OH—CCl$_3$

3.6 QUANTITATIVE MEASUREMENT (pK_a VALUES)

Everything we have mentioned so far has been the *qualitative* method for comparing acidity of different protons. In other words, we never said how *much more* acidic one proton is over another, and we never said *exactly* how acidic each proton is. We have talked only about relative acidities: which proton is *more* acidic?

There is also a *quantitative* method of measuring acidities. All protons can be given a number that quantifies exactly how acidic they are. This value is called pK_a. It is impossible to figure out the exact pK_a by just looking at a structure. The pK_a must be determined empirically through experimentation. Many professors require that you know some general pK_a's for certain classes of compounds (for instance, all alcoholic protons, RO–H, will have the same ballpark pK_a). Most textbooks will have a chart that you can memorize. Your instructor will tell you if you are expected to memorize this chart. Either way, you should know what the numbers mean.

The smaller the pK_a, the more acidic the proton is. This probably seems strange, but that's the way it is. A compound with a pK_a of 4 is more acidic than a compound with a pK_a of 7. Next, we need to know what the difference is between 4 and 7. These numbers measure orders of magnitude. So the compound with a pK_a of 4 is 10^3 times more acidic (1000 times more acidic) than a compound with a pK_a of 7. If we compare a compound with a pK_a of 10 to a compound with a pK_a of 25, we find that the first compound is 10^{15} times more acidic than the second compound (1,000,000,000,000,000 times more acidic).

3.7 PREDICTING THE POSITION OF EQUILIBRIUM

Now that we know how to compare stability of charge, we can begin to predict which side of an equilibrium will be favored. Consider the following scenario:

$$H-A \quad + \quad B^{\ominus} \quad \rightleftharpoons \quad A^{\ominus} \quad + \quad H-B$$

This equilibrium represents the struggle between two bases competing for H^+. A^- and B^- are competing with each other. Sometimes A^- gets the proton and sometimes B^- gets the proton. If we have a very large amount of A^- and B^- and not enough H^+ to protonate both of them, then at any given moment in time, there will be a certain number of A's that have a proton (HA) and a certain number of B's that have a proton (HB). These numbers are controlled by the equilibrium, which is controlled by (you guessed it) *stability of the negative charges*. If A^- is more stable than B^-, then A will be happy to have the negative charge and B^- will grab most of the protons. However, if B^- is more stable than A^-, then we will have the reverse effect.

Another way of looking at this is the following. In the equilibrium above, we see an A^- on one side and a B^- on the other side. The equilibrium will favor whichever side has the more stable negative charge. If A^- is more stable, then the equilibrium will lean so as to favor the formation of A^-:

$$H-A \quad + \quad B^{\ominus} \quad \rightleftharpoons \quad \boxed{A^{\ominus}} \quad + \quad H-B$$

If B^- is more stable, then the equilibrium will lean so as to favor the formation of B^-:

$$H-A \quad + \quad \boxed{B^{\ominus}} \quad \rightleftharpoons \quad A^{\ominus} \quad + \quad H-B$$

The position of equilibrium can be easily predicted by comparing the relative stability of negative charges.

EXERCISE 3.34 Predict the position of equilibrium for the following reaction:

$$H_2O \quad + \quad CH_3O^{\ominus} \quad \rightleftharpoons \quad HO^{\ominus} \quad + \quad CH_3OH$$

Answer We look at both sides of the equilibrium and compare the negative charge on either side. Then we ask which one is more stable. We use the four factors:

1. *Atom* The negative charge on the left is on oxygen, and negative charge on the right is also on oxygen. So this factor does not help us.

2. *Resonance* Neither one is resonance stabilized.

3. *Induction* The negative charge on the left is destabilized by an electron-donating alkyl group. The one on the right is not destabilized in this way.

4. *Orbital* No difference between the right and left.

Based on factor 3, we conclude that the one on the right is more stable, and therefore the equilibrium lies to the right. We show this in the following way:

$$H_2O \quad + \quad CH_3O^{\ominus} \quad \rightleftharpoons \quad HO^{\ominus} \quad + \quad CH_3OH$$

PROBLEMS

3.35 Predict the position of equilibrium for the following reaction:

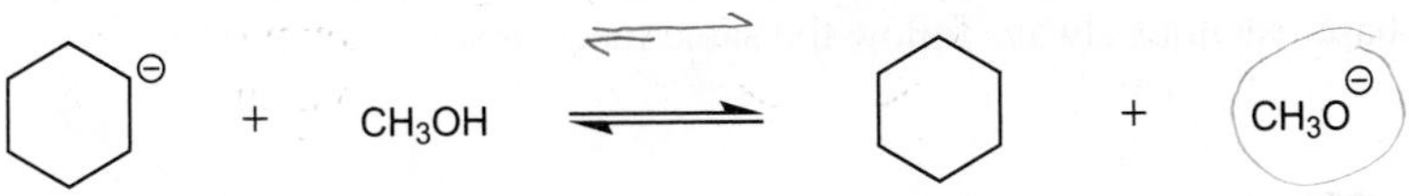

3.36 Predict the position of equilibrium for the following reaction:

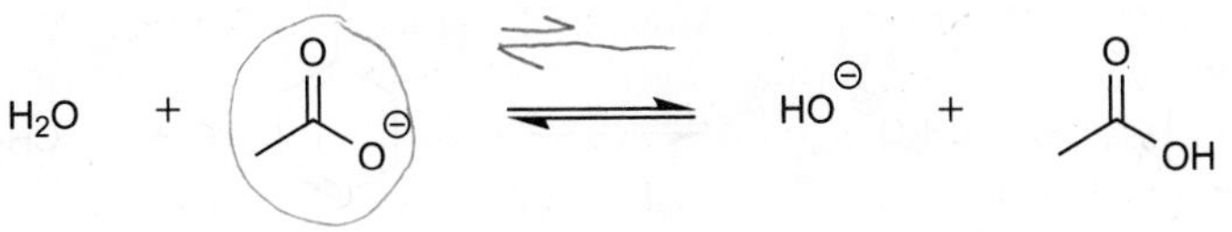

3.37 Predict the position of equilibrium for the following reaction:

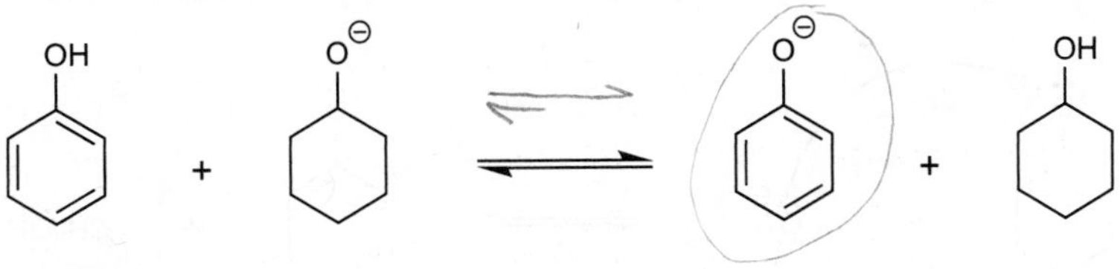

3.8 SHOWING A MECHANISM

Later on in the course, you will spend a lot of time drawing mechanisms of reactions. A mechanism shows how the electrons move during a reaction to form the products. Sometimes many steps are required, and sometimes only one step is required. In acid–base reactions, mechanisms are very straightforward because there is only one step. We use curved arrows (just like we did when drawing resonance structures) to show how the electrons flow. The only difference is that here we are allowed to break single bonds, because we are using arrows to show how a reaction happened (a reaction that involved the breaking of a single bond). With resonance drawings, we can never break a single bond (remember the first commandment). The second commandment—never violate the octet rule—is still true, even when we are drawing mechanisms. Second-row elements can never have more than four bonds, as described in Section 2.3.

From an arrow-pushing point of view, all acid–base reactions are the same. It goes like this:

There are always two arrows. One is drawn coming from the base and grabbing the proton. The second arrow is drawn coming from the bond (between the proton and whatever atom is connected to the proton) and going to the atom currently connected to the proton. That's it. There are always two arrows. Each arrow has a head and a tail, so there are four possible mistakes you can make. You might accidentally draw either of the heads incorrectly, or you might draw either of the tails incorrectly. With a little bit of practice you will see just how easy it is, and you will realize that acid–base reactions always follow the same mechanism.

EXERCISE 3.38   Show the mechanism for the following acid–base reaction:

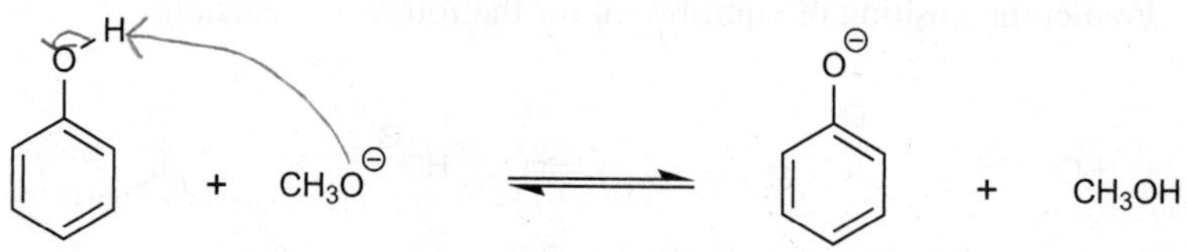

Answer   Remember—2 arrows. One from the base to the proton and the other from the bond (that is losing the proton) to the atom (currently connected to the proton):

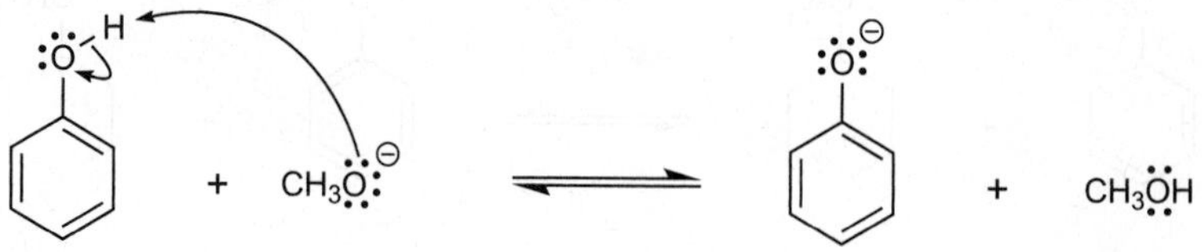

PROBLEMS

3.39   Show the mechanism for the following acid–base reaction:

3.40 Show the mechanism for the following acid–base reaction:

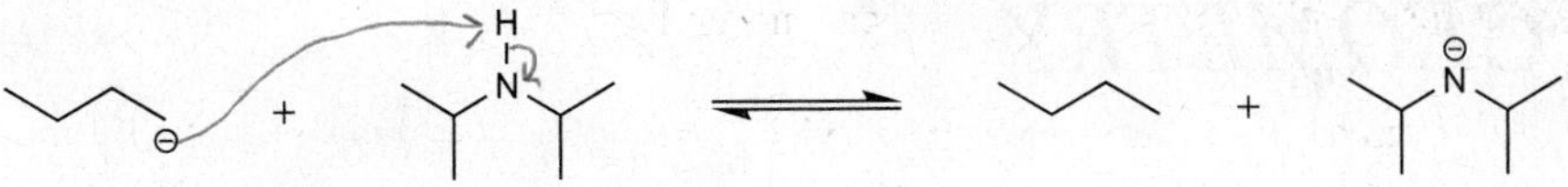

PROBLEMS Show the mechanism for the reaction that takes place when you mix hydroxide (HO⁻) with each of the following compounds (remember that you need to look for the most acidic proton in each case).

3.41

3.42

3.43

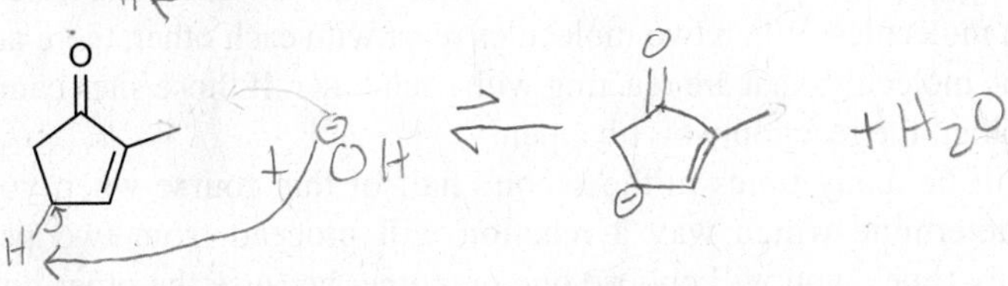

PROBLEMS Show the mechanism for the reaction that takes place when you mix the amide ion (H₂N⁻) with each of the following compounds (remember that you need to look for the most acidic proton in each case).

3.44

3.45

CHAPTER 4

GEOMETRY

In this chapter, we will see how to predict the 3D shape of molecules. This is important because it limits much of the reactivity that you will see in the second half of this course. For molecules to react with each other, the reactive parts of the molecules must be able to get close in space. If the geometry of the molecules prevents them from getting close, then there cannot be a reaction. This concept is called *sterics*.

Let's use an analogy to help us see the importance of geometry. Imagine that you are stuffing a turkey for Thanksgiving dinner and your hand gets stuck inside the turkey. Just at that moment, someone wants to shake your hand. You can't shake the person's hand because your hand is unavailable at the moment. It's kind of the same way with molecules. When two molecules react with each other, there are specific sites on the molecules that are reacting with each other. If those sites cannot get close to each other, the reaction won't happen.

There will be many times in the second half of this course when you will be trying to determine which way a reaction will proceed from two possible outcomes. Many times, you will choose one outcome, because the other outcome has steric problems to overcome (the geometry of the molecules does not permit the reactive sites to get close together). In fact, you will learn to make decisions like this as soon as you learn your first reactions: S_N2 versus S_N1 reactions. Now that we know why geometry is so important, we need to brush up on some basic concepts.

To determine the geometry of an entire molecule, we need to be able to determine the geometry of each atom. This is accomplished by analyzing how it is connected to the atoms around it. After all, that is what defines the geometry—how the atoms are connected in 3D space. Since atoms are connected to each other with bonds, it makes sense that we need to take a close look at bonds. In particular, we need to know the exact locations and angles of every bond to every atom. This might sound difficult, but it is actually straightforward, and with a little bit of practice, you can get to the point where you know the geometry of a molecule as soon as you look at it (without even needing to think about it). That is the point that we need to get to, and that is what this chapter is all about.

4.1 ORBITALS AND HYBRIDIZATION STATES

To determine the geometry of a molecule, we need to know how atoms bond with each other three dimensionally, so it makes sense for our discussion to start with orbitals. After all, bonds come from overlapping atomic orbitals.

A bond is formed when an electron of one atom overlaps with an electron of another atom. The two electrons are shared between both atoms, and we call that a bond. Since electrons exist in regions of space called orbitals, then what we really need to know is; what are the locations and angles of the atomic orbitals around every atom? It is not so complicated, because the number of possible arrangements of atomic orbitals is very small. You need to learn the possibilities, and how to identify them when you see them. So, we need to talk about orbitals.

There are two simple atomic orbitals: *s* and *p* orbitals (we don't really deal with *d* and *f* orbitals in organic chemistry). *s* orbitals are spherical and *p* orbitals have two lobes (one front lobe and one back lobe):

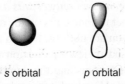

s orbital   *p* orbital

Atoms in the second row (such as C, N, O, and F) have one *s* orbital and three *p* orbitals in the valence shell. These orbitals are usually mixed together to give us hybridized orbitals (*sp³*, *sp²*, and *sp*). We get these orbitals by mixing the *properties* of *s* and *p* orbitals. What do we mean by mixing?

Imagine one swimming pool shaped like a triangle and another shaped like a pentagon; now we put them next to each other. We wave a magic wand and they magically turn into two rectangular pools. That would be a neat trick. That's what *sp* orbitals are: we take one *s* orbital and one *p* orbital, then wave a magic wand, and poof—we now have two equivalent orbitals that look the same. The two new orbitals have a different shape from the original two orbitals. This new shape is somewhat of an average of the two original shapes.

If we mix two *p* orbitals and one *s* orbital, then we get three equivalent *sp²* orbitals. Let's go back to the pool analogy. Imagine two pools shaped like octagons and one shaped like a triangle. We wave our magic wand and get three pools shaped like hexagons. We started with three pools and we ended with three pools. But the three pools in the end all look the same. The same thing is true here with orbitals. We start with three orbitals (two *p* orbitals and one *s* orbital). Then we mix them together and end up with three orbitals that all look the same. The three new orbitals (since they came from one *s* orbital and two *p* orbitals) are called *sp²* orbitals. Similarly, when you combine three *p* orbitals and one *s* orbital, you get four equivalent *sp³* orbitals.

To truly understand the geometry of bonds, we need to understand the geometry of these three different hybridization states. The hybridization state of an atom describes the type of hybridized atomic orbitals (*sp³*, *sp²*, or *sp*) that contain the valence electrons. Each hybridized orbital can be used either to form a bond with another atom or to hold a lone pair.

It is not difficult to determine hybridization states. If you can add, then you should have no trouble determining the hybridization state of an atom. Just count how many other atoms are bonded to your atom, and count how many lone pairs your atom has. Add these numbers. Now you have the total number of hybridized orbitals that contain the valence electrons. This number is all you need to determine the hybridization state of the atom. Let's look at an example to clear it up.

Consider the molecule below:

Let's try to determine the hybridization state of the carbon atom in the center. We begin by counting the number of atoms connected to this carbon atom. There are 3 atoms (O, H, and H). *The oxygen atom only counts as one.*

Next we count the number of lone pairs on the carbon atom. There are no lone pairs on the carbon atom. (If you are not sure how to tell that there are no lone pairs there, go back to Chapter 1 and review the section on counting lone pairs.) Now we take the sum of the attached atoms and the number of lone pairs—in this case, $3 + 0 = 3$. Therefore, three hybridized orbitals are being used here. That means that we have mixed two p orbitals and one s orbital (a total of three orbitals) to get three equivalent sp^2 orbitals. Thus, the hybridization is sp^2. Let's take a closer look at how this works.

Recall that the second row elements have three p orbitals and one s orbital that can be hybridized in one of three ways: sp^3, sp^2, or sp. If we are using three hybridized orbitals, then we must have mixed two p orbitals with one s orbital:

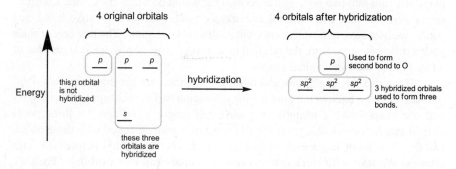

So here's the rule: Just add the number of bonded atoms to the number of lone pairs. The number you get tells you how many hybridized orbitals you need according to the following:

If the sum is 4, then you have 4 sp^3 orbitals.

If the sum is 3, then you have 3 sp^2 orbitals and one p orbital (as in our example).

If the sum is 2, then you have 2 sp orbitals and two p orbitals.

There are several exceptions. For now, don't worry. We will focus on simple cases.

Once you get used to looking at drawings of molecules, you should not have to count anymore. There are certain arrangements that are always sp^3 hybridized, and the same is true for sp^2 and sp. Here are some common examples:

sp^3	![structures: —C—, N, O]
sp^2	![structures: C=, C⊕, N, O]
sp	![structures: —C≡, =C=]

If you can determine the hybridization state of any atom, you will be able to easily determine the geometry of that atom. Let's do another example.

EXERCISE 4.1   Identify the hybridization state for the nitrogen atom in ammonia (NH₃).

Answer   First we need to ask how many atoms are connected to this nitrogen atom. There are three hydrogen atoms. Next we need to ask how many lone pairs the nitrogen atom has. It has 1 lone pair. Now, we take the sum. $3 + 1 = 4$. If we need to have four hybridized orbitals, then the hybridization state must be sp^3.

PROBLEMS   For each compound below, identify the hybridization state for the central carbon atom.

4.2 H–C(=O)–OH sp^2

4.3 O=C=O sp

4.4 CCl₄ (C with four Cl) sp^3

4.5 ⊕ carbon sp^2

4.6 ⊖ carbon sp^3

4.7 H–C≡C–CH₃ sp

4.8   For each carbon atom in the following molecule, identify the hybridization state. Do not forget to count the hydrogen atoms (they are not shown). Use the following simple method: A carbon with 4 single bonds is sp^3 hybridized. A carbon with a double bond is sp^2 hybridized, and a carbon with a triple bond is sp hybridized.

Once you get used to it, you do not need to count anymore—just look at the number of bonds. If carbon has only single bonds, then it is sp^3 hybridized. If the carbon atom has a double bond, then it is sp^2 hybridized. If the carbon atom has a triple bond, then it is sp hybridized. Consult the chart of common examples on the previous page.

4.2   GEOMETRY

Now that we know how to determine hybridization states, we need to know the geometry of each of the three hybridization states. One simple theory explains it all. This theory is called the *valence shell electron pair repulsion theory* (VSEPR). Stated simply, all orbitals containing electrons in the outermost shell (the valence shell) want to get as far apart from each other as possible. This one simple idea is all you need to predict the geometry around an atom. First, let's apply the theory to the three types of hybridized orbitals.

1. Four equivalent sp^3-hybridized orbitals achieve maximum distance from one another when they arrange in a tetrahedral structure:

Think of this as a tripod with an additional leg sticking straight up in the air. In this arrangement, each of the four orbitals is exactly 109.5° from each of the other three orbitals.

2. Three equivalent sp^2-hybridized orbitals achieve maximum distance from one another when they arrange in a trigonal planar structure:

All three orbitals are in the same plane, and each one is 120° from each of the other orbitals. The remaining p orbital is orthogonal to (perpendicular to the plane of) the three hybridized orbitals.

3. Two equivalent sp-hybridized orbitals achieve maximum distance from one another when they arrange in a linear structure:

Both orbitals are 180° from each other. The remaining two p orbitals are 90° from each other and from each of the hybridized orbitals.

So far its very simple:

1. sp^3 = tetrahedral
2. sp^2 = trigonal planar
3. sp = linear

But here's where students usually get confused. What happens when a hybridized orbital holds a lone pair? What does that do to the geometry? The answer is that the geometry of the orbitals does not change, but the geometry of the molecule is affected. Why?

Let's look at an example. In ammonia (NH_3), the nitrogen atom is sp^3 hybridized, so *all four orbitals arrange in a tetrahedral structure,* just as we would expect. But only three of the orbitals in this arrangement are responsible for bonds. So, if we look just at the atoms that are connected, we do not see a tetrahedron. Rather, we see a trigonal pyramidal arrangement:

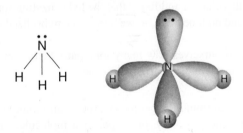

Trigonal, because there are three bonds pointing away from the central nitrogen atom, and pyramidal because it's shaped like a pyramid.

Similarly, in H_2O, the oxygen atom is sp^3 hybridized. So all four orbitals are in a tetrahedral arrangement, just as we would expect for an sp^3 hybridized atom. But only two of the orbitals are being used for bonds. So if we look just at the atoms that are connected, we do not see a tetrahedron. Rather, we see a bent arrangement:

Let's now put all of this information together:

sp^3 with 0 lone pairs = tetrahedral
sp^3 with 1 lone pair = trigonal pyramidal
sp^3 with 2 lone pairs = bent
sp^2 with 0 lone pairs = trigonal planar
sp^2 with 1 lone pair = bent
sp with 0 lone pairs = linear

That's it. There are only six different types of geometry that we need to know. First we determine the hybridization state. Then, using the number of lone pairs, we can figure out which of the six different types of geometry we are dealing with. Let's try it out on a problem.

EXERCISE 4.9 Identify the geometry of the carbon atom below:

Answer First, we need to determine the hybridization state. We did this for this molecule earlier in this chapter and found that the hybridization state is sp^2 (there are 3 atoms connected and no lone pairs, so we need three hybridized orbitals; therefore, it is sp^2).

Next we remind ourselves how many lone pairs there are; in this case, there are none. So the geometry must be trigonal planar.

Once you can determine the geometry around an atom, you should have no problem determining the geometry, or shape, of a molecule. Simply repeat your analysis for each and every atom in the molecule. This may seem like a large task at first, but once you get the hang of it, you will be able to determine the geometry of an atom immediately upon seeing it.

For the next set of problems, you should get to the point where you can do these problems very quickly. The first few will take you longer than the last ones. If the last problem is still taking you a long time, then you have not mastered the process and you will need more practice. If this is the case, open to any page in the second half of your textbook. You will probably see drawings of structures. Point to any atom in a structure and try to determine what the geometry is. Use the list above to help you. Go from one drawing to the next until you can do it without the list. That is the important part—doing it without needing the list.

PROBLEMS   Identify the hybridization state and geometry of each atom in the following compounds. Do not worry about the geometry of atoms connected to only one other atom. For example, do not worry about the geometry of any hydrogen atoms or about the geometry of the oxygen atoms in problems 4.12, 4.13, 4.15, and 4.17.

4.10   (benzene ring structure with handwritten annotations: *sp²*, *trigonal planar*)

4.11   (alkene structure with handwritten annotations: *sp³*, *sp²*, *sp³ tetrahedral*, *sp² trigonal planar*)

4.12   (bicyclic structure with handwritten annotations: *tetrahedral sp³*, *sp²*, *sp³*, *sp²*, *sp² tpl.*, *sp tpl*, *sp linear*, *linear*, *sp*, *sp²*, *sp³ sp³*, *tetra*)

4.13   (ketone/amine structure with handwritten annotations: *sp³*, *sp²*, *trig. pl.*, *tetra*, *sp²*, *sp³*, *tetra*, *sp³*, *sp³ trig. pyr.*, *tetra*, *N*, *sp³ tetra*)

4.14   (morpholine structure with handwritten annotations: *tetra sp³*, *tetra sp³*, *trig pyr.*, *N*, *sp³*, *sp³*, *tetra*, *O*, *sp³ bent*, *sp³ tetra*)

4.15   (ketone/alkyne structure with handwritten annotations: *sp³*, *sp²*, *sp³*, *sp³*, *sp²*, *sp*, *sp*, *sp³*, *sp²*, *sp³*)

4.16   (cyclopentene structure with handwritten annotations: *sp³*, *sp³*, *sp³*, *sp²*, *sp²*, *sp²*, *sp²*)

4.17   (cyclobutanone structure with handwritten annotations: *sp³*, *sp²*, *O*, *sp²*, *sp³*, *sp³*)

4.3 LONE PAIRS

In general, lone pairs occupy hybridized orbitals. For example, consider the lone pair on the nitrogen atom in the following compound:

(structure of an amine, N with lone pair)

This nitrogen atom has three bonds and one lone pair, so it is sp^3 hybridized, just as we would expect. The lone pair occupies an sp^3 hybridized orbital, and the nitrogen atom has trigonal pyramidal geometry, just as we saw in the previous section. But now consider the nitrogen atom in the following compound:

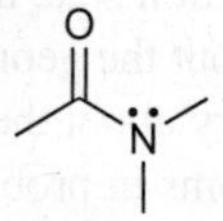

The lone pair on this nitrogen atom does NOT occupy a hybridized orbital. Why not? Because this lone pair is participating in resonance:

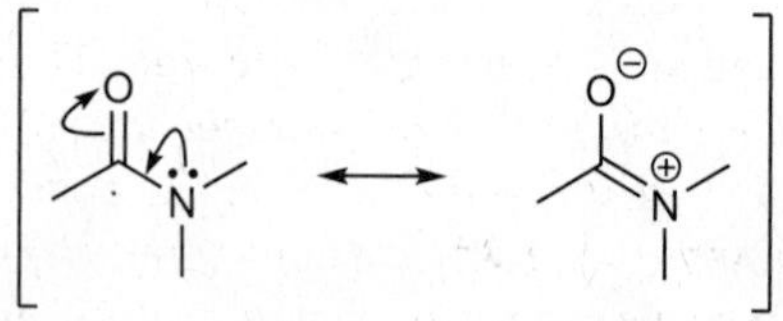

Inspection of the second resonance structure reveals that this nitrogen atom is actually sp^2 hybridized, not sp^3. It might look like it is sp^3 hybridized in the first resonance structure, but it isn't. Here is the general rule: a lone pair that participates in resonance must occupy a p orbital. In other words, the nitrogen atom in the compound above is sp^2 hybridized. And as a result, this nitrogen atom is trigonal planar rather than trigonal pyramidal.

Now let's get some practice identifying atoms that possess a lone pair participating in resonance:

PROBLEMS Identify the hybridization state and geometry of each nitrogen atom and each oxygen atom in the following compounds.

4.18

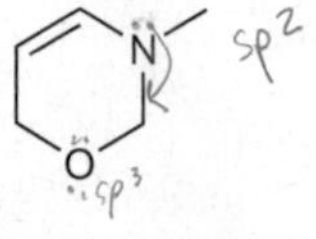

4.19

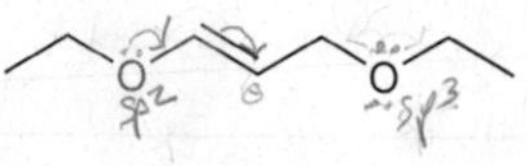

4.20

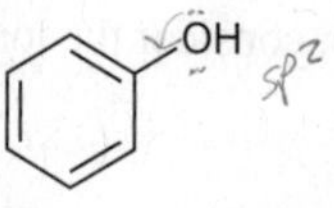

CHAPTER 5

NOMENCLATURE

All molecules have names, and we need to know their names to communicate. Consider the molecule below:

Clearly, it would be inadequate to refer to this compound as "you know, that thing with five carbons and an OH coming off the side with a chlorine on a double bond." First of all, there are too many other compounds that fit that fuzzy description. And even if we could come up with a very adequate description that could only be this one compound, it would take way too long (probably an entire paragraph) to describe. By following the rules of nomenclature, we can unambiguously describe this molecule with just a few letters and numbers: Z-2-chloropent-2-en-1-ol.

It would be impossible to memorize the names of every molecule, because there are too many to even count. Instead, we have a very systematic way of naming molecules. What you need to learn are the rules for naming molecules (these rules are referred to as IUPAC nomenclature). This is a much more manageable task than memorizing names, but even these rules can become challenging to master. There are so many of them, that you could study only these rules for an entire semester and still not finish all of them. The larger the molecules get, the more rules you need to account for every kind of possibility. In fact, the list of rules is regularly updated and refined.

Fortunately, you do not need to learn all of these rules, because we deal with very simple molecules in this course. You need to learn only the rules that allow you to name small molecules. This chapter focuses on most of the rules you need to name simple molecules.

There are five parts to every name:

Stereoisomerism	Substituents	Parent	Unsaturation	Functional group

1. *Stereoisomerism* indicates whether double bonds are *cis/trans*, and indicates stereocenters (*R, S*), which we will cover in the chapter on configuration.

2. *Substituents* are groups coming off of the main chain.

3. *Parent* is the main chain.

4. *Unsaturation*   identifies if there are any double or triple bonds.

5. *Functional group*   This is the group after which the compound is named.

Let's use the compound above as an example:

Stereoisomerism	Substituents	Parent	Unsaturation	Functional group
(Z)	2-chloro	pent	2-en	1-ol

We will systematically go through all five parts to every name, starting at the end (functional group) and working our way backward to the first part of the name (stereoisomerism). It is important to do it backward like this, because the position of the functional group affects which parent chain you choose.

5.1 FUNCTIONAL GROUP

Stereoisomerism	Substituents	Parent	Unsaturation	Functional group

The term *functional group* refers to specific arrangements of atoms that have certain characteristics for reactivity. For example, when an –OH is connected to a compound, we call the molecule an alcohol. Alcohols will display similar reactivity, because alcohols all have the same functional group, the –OH group. In fact, most textbooks have chapters arranged according to functional groups (one chapter on alcohols, one chapter on amines, etc.). Accordingly, many textbooks treat nomenclature as an ongoing learning process: As you work through the course, you slowly add to your list of functional group names. Here we focus on six common functional groups, because you will certainly learn at least these six throughout your course.

When a compound has one of these six groups, we show it in the name of the compound by adding a suffix to the name:

Functional group	Class of compound	Suffix
O $\parallel$ R—C—OH	Carboxylic acid	-oic acid
O $\parallel$ R—C—OR	Ester	-oate

(continued)

Functional group	Class of compound	Suffix
$\underset{R\quad H}{\overset{\displaystyle O}{\big\|}}$	Aldehyde	-al
$\underset{R\quad R}{\overset{\displaystyle O}{\big\|}}$	Ketone	-one
R—O—H	Alcohol	-ol
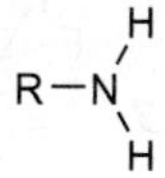 $\underset{H}{\overset{H}{R-N}}$	Amine	-amine

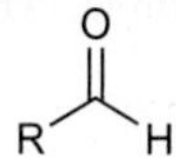

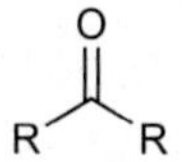

Halogens (F, Cl, Br, I) are usually not named in the suffix of a compound. They get named as substituents, which we will see later on.

Notice that the carboxylic acid is like a ketone and an alcohol placed next to each other. But be careful, because carboxylic acids are very different from ketones or alcohols. So don't make the mistake of thinking that a carboxylic acid is a ketone and an alcohol:

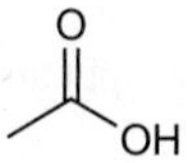

a carboxylic acid

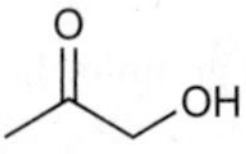

a ketone
and an alcohol

The second compound above raises an important issue: how do you name the functional group when you have two functional groups in a compound? One will go in the suffix of the name and the other will be a prefix in the substituent part of the name. But how do we choose which one goes as the suffix of the name? There is a hierarchy that needs to be followed. The six groups shown above are listed according to their hierarchy, so a carboxylic acid takes precedence over an alcohol. A compound with both of these groups is named as a ketone and we put the term "hydroxy" in the substituent part of the name.

EXERCISE 5.1 Identify what suffix you would use in naming the following compound:

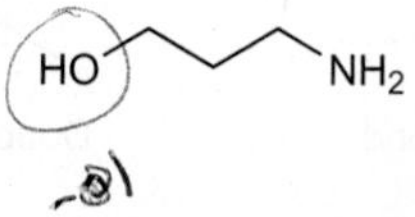

Answer There are two functional groups in this compound, so we have to decide between calling this compound an amine or calling it an alcohol. If we look at the hierarchy above, we see that an alcohol outranks an amine. Therefore, we use the suffix -ol in naming this compound.

PROBLEMS Identify what suffix you would use in naming each of the following compounds.

5.2

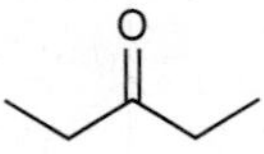

Suffix: _one_

5.3

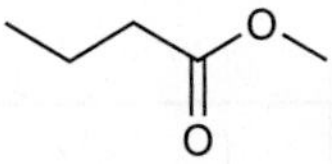

Suffix: _oate_

5.4

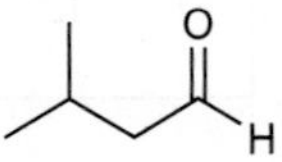

Suffix: _-al_

5.5 H₂N ⟶ Cl

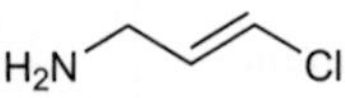

Suffix: _amine_

5.6 Br ⟶ OH

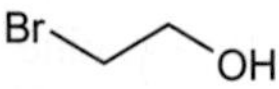

Suffix: _-ol_

5.7

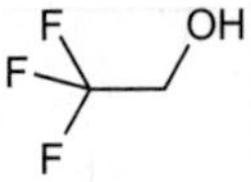

Suffix: _-ol_

5.8

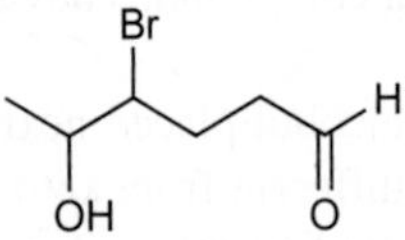

Suffix: _-al_

5.9

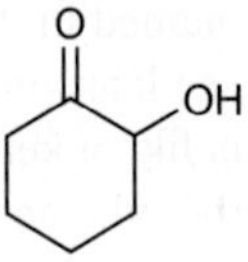

Suffix: _-one_

5.10

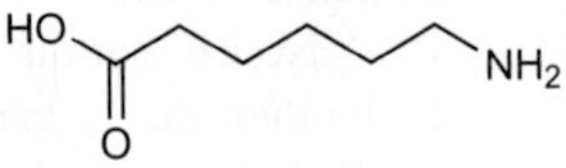

Suffix: _-oic acid_

If there is no functional group in the compound, then we put an "e" at the end of the name:

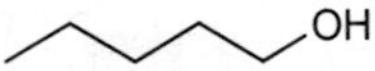

pentan**ol**

pentan**e**

5.2 UNSATURATION

Stereoisomerism	Substituents	Parent	Unsaturation	Functional group

Many compounds have double or triple bonds, and are said to be "unsaturated" because a compound with a double or triple bond has less hydrogen than it would have without the double or triple bond. These double and triple bonds are very easy to see in bond-line drawings:

Triple bond

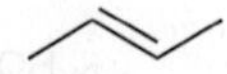

Double bond

The presence of a double bond or triple bond is indicated in the following way, where "en" (pronounced *een*) represents a double bond, and "yn" (pronounced *ine*) represents a triple bond:

pent**an**e	 pent**en**e	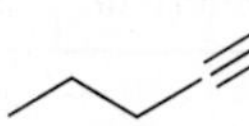 pent**yn**e

Notice that "an" is used to indicate the absence of a double bond or triple bond, such as in "pentane" above. If there are two double bonds in a compound, then the unsaturation is "-dien-." Three double bonds is "-trien-." Similarly, two triple bonds is "-diyn-," and three triple bonds is "-triyn-." For multiple double and triple bonds, we use the following terms:

di = 2   penta = 5

tri = 3   hexa = 6

tetra = 4

You will rarely ever see this many double or triple bonds in one compound, but it is possible to see both double and triple bonds in the same compound. For example,

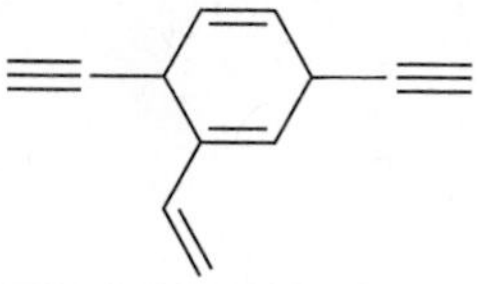

The compound shown here has three double bonds and two triple bonds. So it is a triendiyne. Double bonds always get listed first.

EXERCISE 5.11 Identify how you would describe the unsaturation in the name of the following compound:

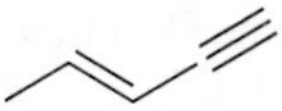

Answer This compound has one double bond and one triple bond. For the double bond, we use the term "en." For the triple bond, we use the term "yn." Double bonds get listed first, so this compound is –enyn-.

PROBLEMS Identify how you would describe the unsaturation in the name of the following compounds.

5.12   *en*

5.13   *yn*

5.14   *diene*

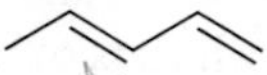

5.15   *trien*

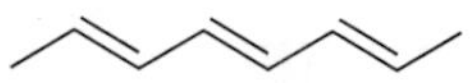

5.16   *trien*

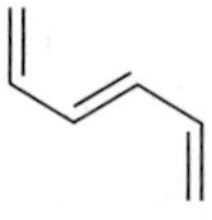

5.17   *endiyn*

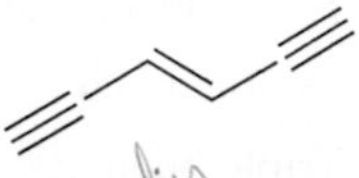

5.3 NAMING THE PARENT CHAIN

Stereoisomerism	Substituents	Parent	Unsaturation	Functional group

When naming the parent of the compound, we are looking for the chain of carbon atoms that is going to be the root of our name. Everything else in the compound is connected to that chain at a specific location, designated by numbers. So we need to know how to choose the parent carbon chain and how to number it correctly.

The first step is learning how to say "a chain of three carbon atoms" or "a chain of seven carbon atoms." Here is a table showing the appropriate names:

Number of carbon atoms in the chain	Parent
1	meth
2	eth
3	prop
4	but
5	pent
6	hex
7	hept
8	oct
9	non
10	dec

If we have carbon atoms in a ring, we add the term cyclo, so a ring of six carbon atoms is called cyclohex- as the parent and a ring of five carbon atoms is cyclopent-.

You must commit these terms to memory. I am not a big advocate of memorization, but for now, you must memorize these terms. After a while, it will become habitual, like a phone number that you dial all of the time, and you won't have to think about it anymore.

The tricky part comes when you need to figure out which carbon chain to use. Consider the following example, which has three different possibilities for the parent chain:

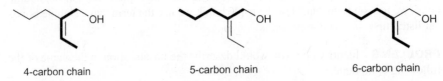

4-carbon chain   5-carbon chain   6-carbon chain

So how do we know whether to call this –but- (which is 4) or –pent- (which is 5) or –hex- (which is 6)? There is a hierarchy for this as well. The chain should be as long as possible, making sure to include the following groups, in this order:

Functional group

Double bond

Triple bond

First we need to find the functional group and make sure that the functional group is connected directly to the parent chain. Remember from the last section that if there are two functional groups, one of them gets priority. The functional group that gets priority is the one that needs to be connected to our parent chain. Of the three possibilities shown above, this rule eliminates the last possibility, because the functional group (OH) is not connected directly to the parent chain.

If there are still more choices of possible parent chains (as there are in this case), then we look for the chain that also includes the double bond (if there is one in the compound). In our case, there is a double bond, and this rule determines for us which chain to use:

CORRECT   INCORRECT

Of the remaining two possibilities, we must choose a parent that includes both carbon atoms of the double bond. Only one parent chain contains both the functional group and the two carbon atoms of the double bond. "Containing the functional group" means that the OH is connected to a carbon that is part of the chain. We do not count the oxygen atom itself as part of the chain. It is only attached to the chain. So the chain above is made up of four carbon atoms.

In cases where there is no functional group, then we look for the longest chain that includes the double bond. If there is no double bond, then we look for a triple bond, and choose the longest chain that has the triple bond in it.

If there are no functional groups, no double bonds, and no triple bonds, then we simply choose the longest chain possible.

Now you can see why we are moving our way backward through the naming process. We cannot name the parent correctly unless we can pick out the highest ranking functional group in the compound. So we start naming a compound by first asking which functional group has priority.

EXERCISE 5.18   Name the parent chain in the following compound:

Answer   First we look for a functional group. This compound is a carboxylic acid, so we know the parent chain must include the carboxylic acid group. Next we look for a double bond. The parent chain should include that as well. This gives us our answer. The triple bond will not be included in the parent chain, because the functional group and the double bond are a higher priority than a triple bond.

So we count the number of carbon atoms in this chain. There are six (notice that we include the carbon atom of the carboxylic acid group). Therefore, the parent will be called "hex."

PROBLEMS Name the parent chain in each of the following compounds.

5.19 Parent: hex

5.20 Parent: hept

5.21 Parent: hex

5.22 Parent: non

5.23 Parent: oct

5.24 Parent: hex

5.25 Parent: hex

5.26 Parent: hex

5.27 Parent: pent.

5.4 NAMING SUBSTITUENTS

Stereoisomerism	Substituents	Parent	Unsaturation	Functional group

Once we have identified the functional group and the parent chain, then everything else connected to the parent chain is called a substituent. In the following example, all of the highlighted groups are substituents, because they are not part of the parent chain:

We start by learning how to name the alkyl substituents. These groups are named with the same terminology that we used for naming parent chains, but we add "yl" to the end to indicate that it is a substituent:

Number of carbon atoms in the substituent	Substituent
1	methyl
2	ethyl
3	propyl
4	butyl
5	pentyl
6	hexyl
7	heptyl
8	octyl
9	nonyl
10	decyl

Methyl groups can be shown in a number of ways, and all of them are acceptable:

Ethyl groups can also be shown in a number of ways:

Propyl groups are usually just drawn, but sometimes you will see the term Pr (which stands for propyl):

Look at the propyl group above and you will notice that it is a small chain of 3 carbon atoms that is attached to the parent chain by the first carbon of the small chain. But what if it is attached by the middle carbon? Then it is not called propyl anymore:

Attached by the first carbon of the chain

propyl

Attached by the middle carbon of the chain

isopropyl

It is still a chain of three carbon atoms, but it is attached to the parent chain differently than a propyl group is attached, so we call it isopropyl. This is an example of a branched substituent (branched, because it is not connected in one straight line to the parent chain, like a propyl group is).

Another important branched substituent to be familiar with is the *tert*-butyl group:

butyl   *tert*-butyl

The *tert*-butyl group is made up of four carbon atoms, just like a butyl group, but the *tert*-butyl group is not a straight line connected to the parent. Rather, the group has three methyl groups attached to one carbon atom, which is itself attached to the parent chain. So, we call this group *tert*-butyl.

There are two other ways to attach four carbon atoms to a parent chain (other than butyl and *tert*-butyl). As a small assignment, see if you can find their names in your textbook.

There is another important type of substituent that we need to consider. When we learned about functional groups, we saw that some compounds can have two functional groups. When this happens, we choose one of the functional groups to be named as the suffix, and the other functional group must be named as a substituent. To choose the functional group that gets the priority, go back to the section on functional groups and you will see the list of functional groups. We need to know how to name these functional groups as substituents. The OH group is named –hydroxy- as a substituent. The NH_2 group is called –amino- if it is named as a substituent. A ketone is called –keto- as a substituent, and an aldehyde is called –aldo- as a substituent. Knowing how to name those four functional groups as substituents will probably cover you for anything you will see in your course.

Halogens are named as substituents in the following way: fluoro, chloro, bromo, and iodo. Essentially, we add the letter "o" at the end to say that they are substituents. If there are multiple substituents of the same kind (for example, if there are five chlorine atoms in the compound), we use the same prefixes that we used earlier when classifying the number of double and triple bonds:

di = 2   penta = 5

tri = 3   hexa = 6

tetra = 4

So a compound with five chlorine atoms would be "pentachloro." Each and every substituent needs to be numbered so that we know where it goes on the parent chain, but we will learn about this after we have finished going through the five parts of the name. At that time, we will also discuss in what order to place substituents in the name.

EXERCISE 5.28 In the following compound, identify all groups that would be considered substituents, and then indicate how you would name each substituent:

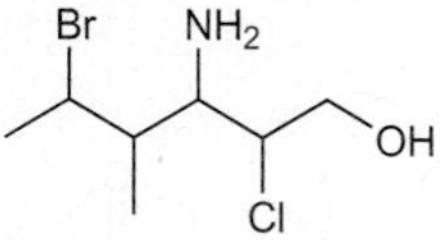

Answer First we must locate the functional group that gets the priority. Alcohols outrank amines, so the OH group is the priority functional group. Then, we need to locate the parent chain. There are no double or triple bonds, so we choose the longest chain containing the OH group:

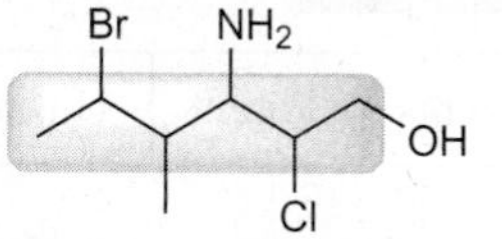

Now we know which groups must be substituents, and we name them accordingly:

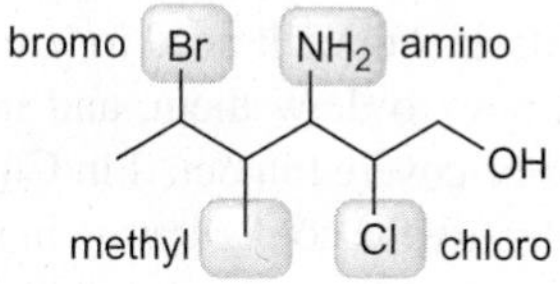

PROBLEMS For each of the following compounds, identify all groups that would be considered substituents, and then indicate how you would name each substituent.

5.29

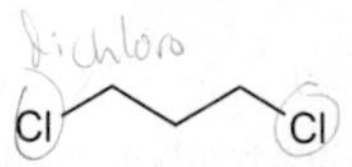

5.30

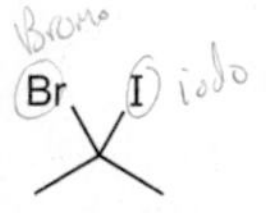

5.31

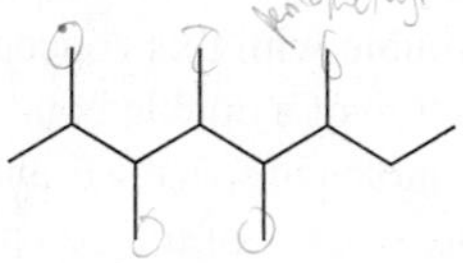

5.32

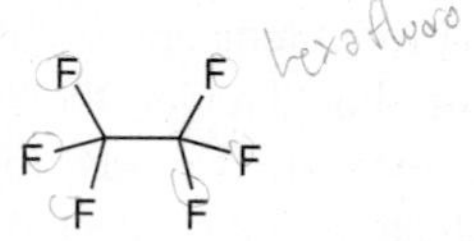

5.33

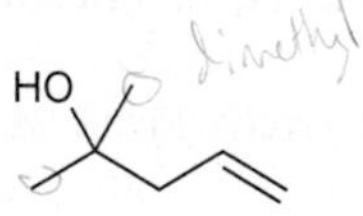

5.34

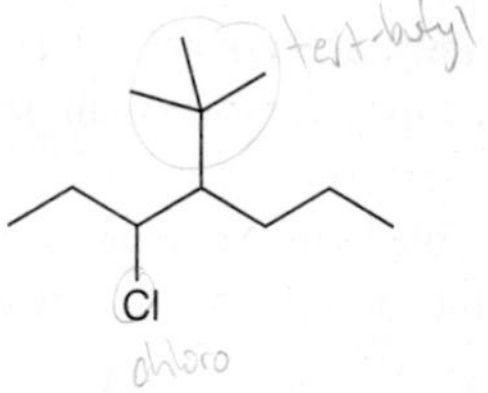

5.35

5.36

5.37

5.38

5.5   STEREOISOMERISM

| Stereoisomerism | Substituents | Parent | Unsaturation | Functional group |

Stereoisomerism is the first part of every name. It identifies the configuration of any double bonds or stereocenters. If there are no double bonds or stereocenters in the molecule, then you don't need to worry about this part of the name. If there are, you must learn how to identify the configuration of each. Identifying the configuration of a stereocenter requires a chapter to itself. You will need to learn what a stereocenter is, how to locate them in molecules, how to draw them, and how to assign a configuration (*R* or *S*). These topics will all be covered in detail in Chapter 7. At that time, we will revisit how to appropriately identify the configuration in the name of the molecule. For now, you should know that configurations are placed here in the first part of a name.

Here we will focus on double bonds, which can often be arranged in two ways:

cis   *trans*

This is very different from the case with single bonds, which are freely rotating all of the time. But a double bond is the result of overlapping *p* orbitals, and double bonds *cannot* freely rotate at room temperature (if you had trouble with this concept when you first learned it, you should review the bonding structure of a double bond in your textbook or notes). So there are two ways to arrange the atoms in space: *cis* and *trans*. If you compare which atoms are connected to each other in each of the two possibilities, you will notice that all of the atoms are connected in the same order. The difference is how they are connected *in 3D space*. This is why they are called stereoisomers (this type of isomerism stems from a difference of orientation in space—"stereo").

To name a double bond as being *cis* or *trans*, you need to have identical groups on *either side* of the double bond that can be compared to each other. If these identical groups are on the same side of the double bond, we call them *cis*. If they are on opposite sides, we call them *trans*:

two methyl groups
are *trans*

two fluorine atoms
are *cis*

two ethyl groups
are *trans*

two methyl groups
are *trans*

The two groups that we compare can even be hydrogen atoms. For example,

is *trans* because there are two hydrogen atoms not shown, and they are *trans* to each other:

But what do you do if you don't have two identical groups to compare? For example,

is not the same as

These compounds are clearly not the same. We cannot use *cis/trans* terminology to differentiate them, because we don't have two identical groups to compare. In situations when all four groups on the double bond are different, we have to use another method for naming them.

The other way of naming double bonds uses rules similar to those used in determining the configuration of a stereocenter (*R* versus *S*), so we will wait until the next chapter (when we learn about *R* and *S*), and then we will cover this alternative way of naming double bonds. The alternative method is far superior, because it can be used to name any double bond. In contrast, *cis/trans* nomenclature can be used only when we have two identical groups. The reason that we do not drop the *cis/trans* terminology altogether is probably based in deep-rooted tradition and usage of these terms.

There is one situation when we don't have to worry about *cis/trans* because there aren't two ways to arrange the double bond. If we have two identical groups connected *to the same atom,* then we cannot have stereoisomers. For example,

is the same as

because there are two chlorine atoms connected to one carbon atom on one side of the double bond. Why are the two drawings the same? Remember that the carbon atoms of the double bond are *sp*^{2} hybridized, and therefore trigonal *planar.* So, if we flip over the first drawing, we get the other drawing. They are the same thing. To see this, take two pieces of paper. Draw one of these compounds on one piece of paper, and draw the other compound on the second piece of paper. Then flip over one of the pieces of paper, and hold it up to the light so that you can see the drawing through the back side of the paper. Compare it to the other drawing and you will see that they are the same. If you try to do the same thing with some of the previous examples (that did have *cis* and *trans* stereoisomers), you will find that flipping the page over does not make the two drawings the same. This is a useful exercise, so take a few minutes and do it.

EXERCISE 5.39 Determine whether the double bond below is *cis* or *trans*:

Answer Begin by circling the four groups attached to the double bond and try to name them:

You should always use this technique, because it will help you see when you have two groups that are the same. There are always four groups on the double bond (even if some of them are just hydrogen atoms). In this case, it helps us see that there are two isopropyl groups on the same side of the double bond. Therefore, the double bond is *cis*.

PROBLEMS For each of the compounds below, determine whether the double bond is *cis* or *trans*.

5.40   trans

5.41   trans

5.42   trans

5.43   cis   D

5.44   cis

5.45   trans

5.6 NUMBERING

Stereoisomerism	Substituents	Parent	Unsaturation	Functional group

Numbering applies to all parts of the name

We're almost ready to start naming molecules. We finished learning about the individual parts of a name, and now we need to know how to identify how the pieces are connected. For example, let's say you determine that the functional group is OH (therefore, the suffix is –ol), there is one double bond (-en-), the parent chain is six carbon atoms long (hex), there are four methyl groups attached to the parent chain (tetramethyl), and the double bond is *cis*. Now you know all of the pieces, but we must find a way to identify where all of the pieces are on the parent chain. Where are all of those methyl groups? (and so on). This is where the numbering system comes in. First we will learn how to number the parent chain, and then we will learn the rules of how to apply those numbers in each part of the name.

Once we have chosen the parent chain, there are only two ways to number it: right to left or left to right. But how do we choose? To number the parent chain properly, we begin with the same hierarchy that we used when choosing the parent in the first place:

Functional group

Double bond

Triple bond

If there is a functional group, then number the parent chain so that the functional group gets the lower number:

OH gets the number 2 instead of 5

If there is no functional group, then number the chain so that the double bond gets the lower number:

the double bond is 1 instead of 5

If there is no functional group or double bond, then number the chain so that the triple bond gets the lower number:

the triple bond is 1 instead of 5

If there is no functional group, double bond, or triple bond, then we should number the chain so that the substituent has the lower number:

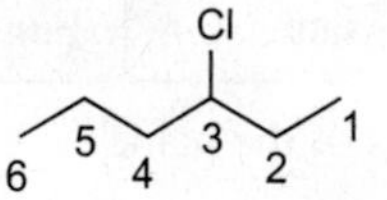

Cl gets the number 3 instead of 4

If there is more than one substituent on the parent chain, then we should number the chain so that the substituents get the lowest numbers possible:

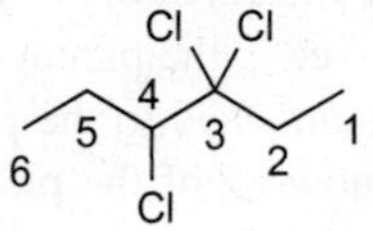

3,3,4-trichloro instead of 3,4,4-trichloro

EXERCISE 5.46 For the compound below, choose the parent chain and then number it correctly:

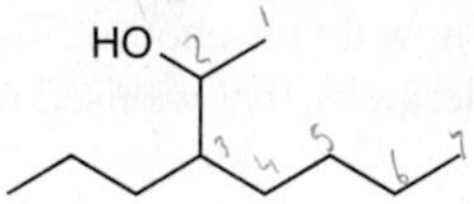

Answer To choose the parent chain, remember that we need to choose the longest chain that contains the functional group:

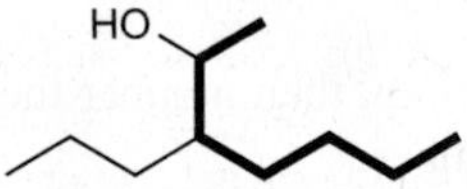

To number it correctly, we need to go in the direction that gives the functional group the lowest number:

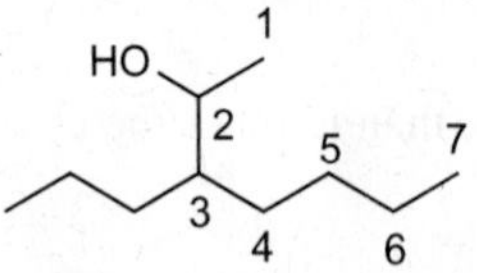

PROBLEMS For each of the compounds below, choose the parent chain and number it correctly.

5.47 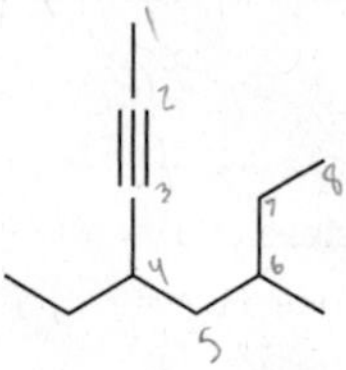   **5.48** 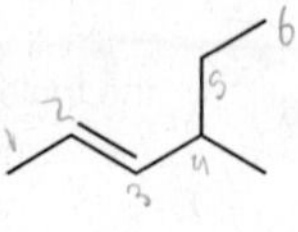   **5.49**

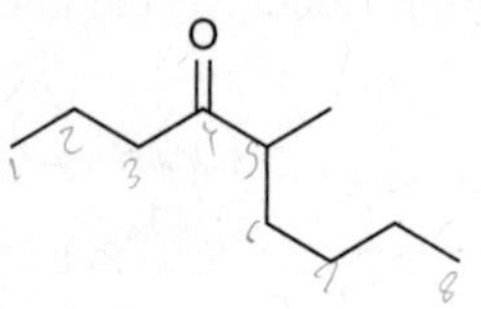

5.50 HO

5.51

5.52

5.53

5.54 OH

5.55

Now that we know how to number the parent chain, we need to see how to apply those numbers to the various parts of the name.

Functional Group   The number generally gets placed directly in front of the suffix (for example, hexan-2-ol). If the functional group appears at the number 1, then the number does not need to be placed in the name (for example, hexanol). It is assumed that the absence of a number means that the functional group is at the number 1 position. When placing a number, it is OK to place the number at the beginning of the parent. For example, 2-hexanol is the same as hexan-2-ol. Both names are acceptable.

Unsaturation   For double and triple bonds, the number indicates the lower number of the two carbon atoms. For example,

We use the number 2

The double bond is between C2 and C3, so we use C2 to number the double bond. So the example above is hex-2-ene (or 2-hexene). We treat triple bonds the same way.

If there are two double bonds in the molecule, then we must indicate both numbers; for example, hexa-2,4-diene, or 2,4-hexadiene. Every double and triple bond must be numbered.

Substituents   The number of the substituent goes immediately in front of the substituent. Examples:

CI

2-chlorohexane

3-methylpentane

If there are multiple substituents, then every substituent must be numbered:

CI

CI

2,3-dichlorohexane

2,2,4-trimethylpentane

If there are multiple substituents of different types, then we must alphabetize the substituents in the name. Consider the following example:

There are four types of substituents in the example above (chloro, fluoro, ethyl, and methyl). They must be alphabetized (c, e, f, m) (we do not count di, tri, tetra, etc. as part of the alphabetization system). So the compound above is called

2-chloro-3-ethyl-2,4-difluoro-4-methylnonane

Note that two numbers are always separated by commas (2,4 in example above) but letters and numbers are separated by dashes (2-chloro-3-ethyl . . .).

Stereoisomerism   If there are any double bonds, we place the term *cis* or *trans* at the beginning of the name. If there is more than one double bond, then we need to indicate *cis* or *trans* for each double bond, and we must number accordingly (for example, 2-*cis*-4-*trans* . . .). If there are any stereocenters, here is where we would indicate them; for example, (2*R*,4*S*). Stereocenters are placed in parentheses. We will see more of this when we learn about stereocenters in the upcoming chapters.

There it is. A lot of rules. No one ever said nomenclature would take 10 minutes to learn, but with enough practice, you should get the hang of it. Let's now take everything we have learned and practice naming compounds:

EXERCISE 5.56   Name the following compound:

Answer   We go through the five parts of the name backward. So we start by looking for the functional group. We see that this compound is a ketone. So we know the end of the name will be –one.

Next, we look for unsaturation. There is a double bond here, so there will be –en- in the name.

Next we need to name the parent. We locate the longest chain that includes the functional group and double bond. In this example, it is an obvious choice. The parent has 7 carbon atoms, so the parent is –hept-.

Next we look for substituents. There are two methyl groups and two chlorine atoms. We need to alphabetize, and c comes before m, so it will be dichlorodimethyl.

Next we look for stereoisomerism. The double bond in this molecule has two chlorine atoms on opposite sides, so it is *trans*. This part of the name (*trans*) is generally italicized and surrounded by parentheses. So far, we have

(*trans*)-dichlorodimethylheptenone

Now that we have figured out all of the pieces, we must assign numbers. We need to number the parent to give the functional group the lowest number, so the numbering, in this example, will go from left to right. This puts the functional group at the number 2 position, the double bond at the 4 position, the chlorine atoms at 4 and 5, and the methyl groups at 6 (both of them). So the name is

(*trans*)-4,5-dichloro-6,6-dimethylhept-4-en-2-one

PROBLEMS   Name each of the following compounds. (Ignore stereocenters for now. We will focus on stereocenters in the upcoming chapters.)

5.57   Name: _____

5.58   Name: _____

5.59   Name: _____

5.60   Name: _____

5.61   F   Cl   Name: _____

5.62   Name: _____

5.63   Name: _____

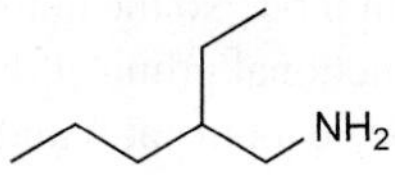

5.64   Name: _____

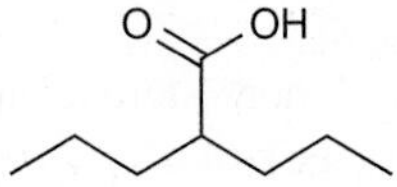

5.65   Name: _____

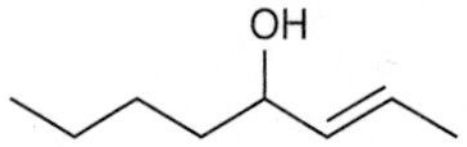

5.66   Name: _____

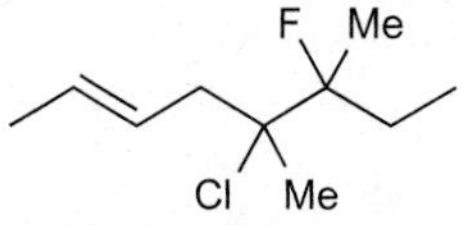

5.7   COMMON NAMES

In addition to the rules for naming compounds, there are also some common names for some simple and common organic compounds. You should be aware of these names to the extent that your course demands this of you. Each course will be different in terms of how many of these common names you should be familiar with. Here are some examples:

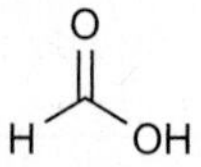

IUPAC Name = methanoic acid
Common Name = formic acid

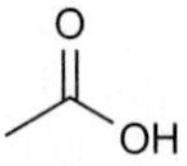

IUPAC Name = ethanoic acid
Common Name = acetic acid

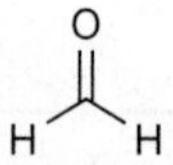

IUPAC Name = methanal
Common Name = formaldehyde

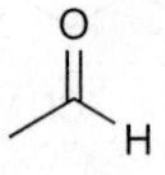

IUPAC Name = ethanal
Common Name = acetaldehyde

IUPAC Name = ethene
Common Name = ethylene

IUPAC Name = ethyne
Common Name = acetylene

Most of these examples are so common that it is quite rare to hear someone refer to these compounds by their IUPAC names. Their common names are much more "common," which is why we call them common names.

Ethers are typically called by their common names. The group on either side of the oxygen atom is named as a substituent before the term ether. Examples:

Diethyl ether

Dimethyl ether

The IUPAC method would be to treat the oxygen atom like a carbon atom and then indicate where the oxygen atom is with the term oxa-. So diethyl ether would be 3-oxapentane, but no one calls it that. Everyone calls it diethyl ether (or even just ethyl ether). It is not a bad idea to familiarize yourself with all of the common names listed in whatever textbook you are using.

5.8 GOING FROM A NAME TO A STRUCTURE

Once you have completed all of the problems in this chapter, you will find that it is much easier to draw a compound when you are given the name than it is to name a compound that is drawn in front of you. It is easier for the following reason: when naming a compound, there are a lot of decisions you need to make (which functional group has priority, what is the parent chain, how the chain should be numbered, in what order to put the substituents in the name, etc.). But when you have a name in front of you and you need to draw a structure, you do not need to make any of these decisions. Just draw the parent and then start adding everything else to it according to the numbering system provided in the name.

For practice, make a list of the answers to problems 5.57–5.66. This list should just be names. Wait a few days until you cannot remember what the structures looked like and then try to draw them based on the names. You can also use your textbook for more examples.

From this point on, I will assume that I can say names like 2-hexanol and you will know what I mean. That is what your textbook will do as well, so now is the time to master nomenclature.

CHAPTER 6

CONFORMATIONS

Molecules are not inanimate objects. Unlike rocks, they can twist and bend into all kinds of shapes, very much like people do. We have limbs and joints (elbows, knees, etc.) that give us flexibility. Although our bones might be very rigid, nevertheless, we can achieve a great range of movement by twisting our joints in different ways. Molecules behave the same way. Once you learn about the general types of joints that molecules have and the ranges of motion available to those joints, you will then be able to predict how individual molecules can twist around in space. Why is this important?

Let's stay with the people analogy. Think about how many common positions you assume every day. You can sit, you can stand, you can lean against something, you can lie down, you can even stand on your head, and so on. Some of these positions are comfortable (such as lying down), while other positions are very uncomfortable (such as standing on your head). There are some activities, like drinking a glass of water, which can be done only in certain positions. You cannot drink a glass of water while lying down or standing on your head.

Molecules are very similar. They have many positions into which they can twist and bend. There are some comfortable positions (low in energy), while other positions are uncomfortable (high in energy). The various positions available to a molecule are called *conformations*. It is important to be able to predict the conformations available to molecules, because there are certain activities that can be performed only in specific conformations. Just as a person standing on her head cannot drink a glass of water, so, too, molecules cannot undergo some reactions from certain conformations. Just as you can run only when you are in a standing position, molecules can often undergo certain reactions from only one specific conformation. If the molecule cannot twist into the conformation necessary for a reaction to take place, then the reaction will not happen. So you can understand why you will need to be able to predict what conformations are available to molecules—so that you can predict when reactions can and cannot happen.

There are two very important drawing styles that show conformations and give us the power to predict what conformations are available to different types of molecules: Newman projections and chair conformations:

Newman projection   Chair

We begin this chapter with Newman projections.

6.1 HOW TO DRAW A NEWMAN PROJECTION

Before we can talk about drawing Newman projections, we need to first review one aspect of drawing bond-line structures that we did not cover in Chapter 1. To show how groups are positioned in 3D space, we often use wedges and dashes:

In the bond-line drawing above, the fluorine atom is on a wedge and the chlorine atom is on a dash. The wedge means that the fluorine is coming out toward us in 3D space, and the dash means that the chlorine is going away from us in 3D space. Imagine that all four carbon atoms in the molecule above are positioned in the plane of the page. If you look at this page from the left side (so the page looks two-dimensional), you would see the fluorine sticking out of the page to your right and the chlorine sticking out of the page to the left.

Do not be confused by whether the dash is drawn on the right or left:

is the same as

In both drawings above, the chlorine is on a dash and the fluorine is on a wedge, so these drawings are the same. In reality, both the chlorine and the fluorine should be drawn straight up—the chlorine goes straight up and behind the page, while the fluorine goes straight up and in front of the page. If we drew it like that, we would not be able to see the chlorine because the fluorine would be blocking it (the way the moon blocks the sun in a solar eclipse). To clearly see both groups, we move one of the groups slightly to the left and the other slightly to the right. It does not matter which is on the right and which is on the left. All that matters is which one is on the wedge and which one is on the dash.

Now that we understand what the dash and wedge mean, let's consider what the molecule would look like from a slightly different angle:

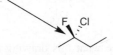

Imagine looking at the molecule from the angle shown by the arrow above. If you are not sure what angle we are talking about, try doing the following: Turn your book so that it is facing your stomach instead of your head. Now turn the page so you are looking at it from the side, like we did before. You should be looking down the path of this arrow at the molecule. In this view, you are looking directly down a carbon–carbon bond, where one carbon is in front and one is in back:

Look from here

F Cl

front
carbon back
carbon

In this view, you will see three groups connected to the front carbon atom. You should expect to see a fluorine atom sticking out of the page to the right and a chlorine atom sticking out of the page to the left. You would also see a methyl group pointing straight down. This is what it would look like from that view:

Cl F

Front
carbon

Me

You would see the three groups like this, and you would not see the back carbon atom, because the carbon atom in front would be covering it up (again, like the moon covers the sun during a solar eclipse). Let's try to draw that back carbon that we cannot see, and by convention, we will draw it as a big circle:

Back
carbon
Cl F

Me

Now we place the three groups that are on the back carbon atom into our drawing. There is one methyl group and two hydrogen atoms. If we put them into our drawing, then we get our Newman projection. It looks like this:

Me
Cl F

H H
Me

It is important that you can see what this drawing represents, because we cannot move forward until you see it very clearly. We are essentially looking at a carbon–carbon bond, focusing on the three groups attached to each carbon atom. The central point in our Newman projection (where the lines to the Cl, F, and Me meet each other) is the first carbon. The big circle in the back is our back carbon. All at once you can see all six groups (the three connected to the front carbon atom and the three connected to the back carbon atom). So a Newman projection is another way of drawing the compound we showed earlier:

is the same as

Let's use one more analogy to help us understand this. Imagine that you are looking at a fan that has three blades. Behind this fan, there is another fan that also has three blades. So you see a total of six blades. If both fans were spinning, and you started taking photographs, you might find some photos where you can clearly see all six blades, and other photos where the three blades in the front are blocking our view of the three blades in the back. In this last case, the three blades in front would be *eclipsing* the three blades in back.

This is where the analogy helps us understand why Newman projections are useful. The bond connecting the two carbon atoms is a single bond that can freely rotate. Sometimes, you can see all six groups because they are *staggered*. But other times, you can't see the groups in the back because they are being *eclipsed* by the three groups in the front:

Staggered

Eclipsed

Think of the front carbon atom and its three groups as one fan, and the back carbon atom and its three groups as a different fan. These two fans can spin independently of each other, which gives rise to many different possible conformations. This is why Newman projections are so incredibly powerful at showing conformations. They are drawn in a way that is perfect for showing the various conformations that arise as an individual single bond rotates.

It gets a little more complicated when we realize that every single bond in every molecule can freely rotate, giving rise to a very large number of conformations for molecules. We can avoid that kind of complexity by focusing just on one particular bond, and the various conformations that arise from free rotation of just that bond. If we can learn how to do that, then we can do that for the part of the molecule that is undergoing a reaction and we do not need to concern ourselves with the rest of the molecule.

EXERCISE 6.1 Draw a Newman projection of the following compound looking from the perspective of the arrow:

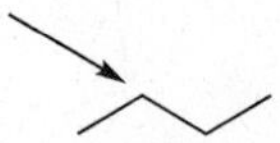

Answer The first thing to realize is that there are groups on dashes and wedges that have not been shown. They are the hydrogen atoms. We did not focus on drawing the dashes and wedges in Chapter 1, but the hydrogen atoms are, in fact, on dashes and wedges. We learned in the chapter on geometry that these carbon atoms would be classified as sp^3 hybridized, and, therefore, their geometry is tetrahedral. That means the hydrogen atoms are coming in and out of the page:

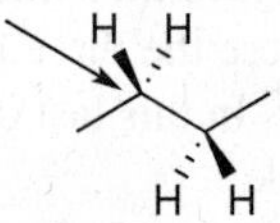

Now we draw the front carbon atom with its three groups. Looking along the direction of the arrow, we see a hydrogen atom that is up and to the left, another hydrogen atom that is up and to the right, and a methyl group pointing straight down. So we draw it like this:

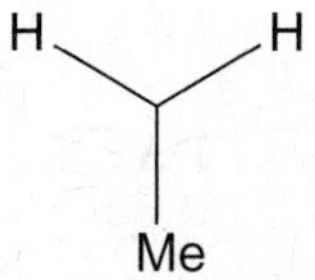

Next we draw the back carbon atom as a large circle and we look at all three groups attached to it. There is a methyl group pointing straight up and then two hydrogen atoms pointing left and right. So the answer is

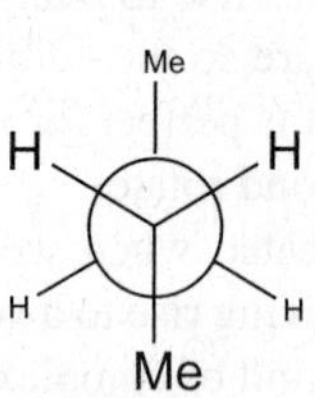

PROBLEMS Draw Newman projections for each of the following compounds. In each case the skeleton of the Newman projection is drawn for you. You just need to fill in the six groups in their proper places.

6.2 Me Me
        Me Me

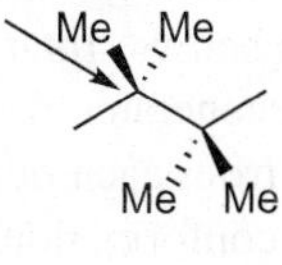

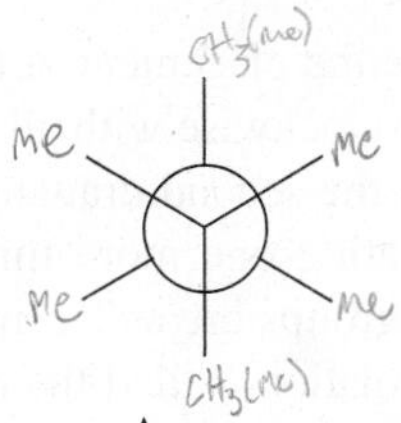

Answer

6.3 H Me
        H Me

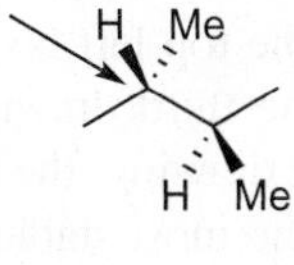

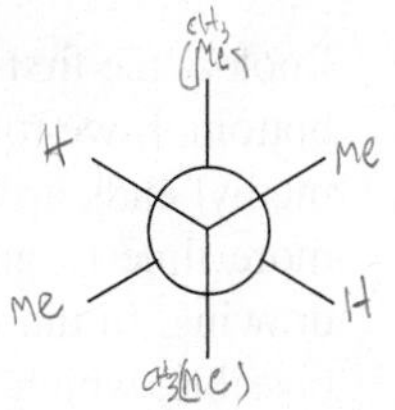

Answer

6.4 H H
        H H

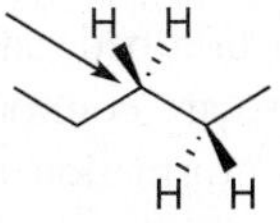

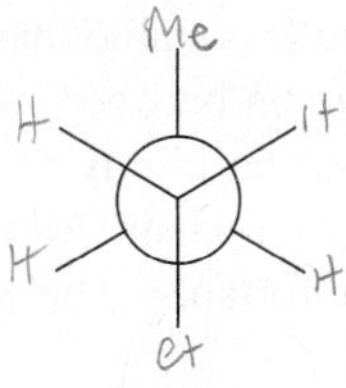

Answer

6.5 H H
        Me H

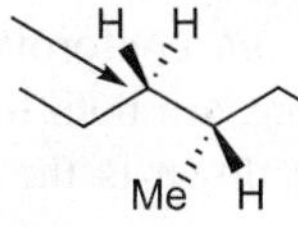

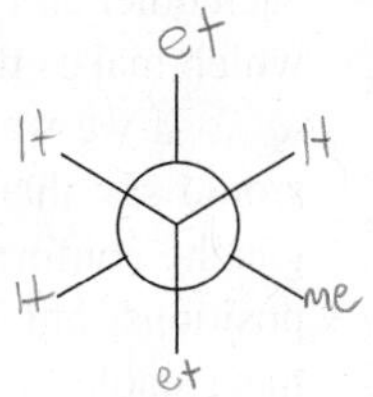

Answer

6.6 H H
        H H

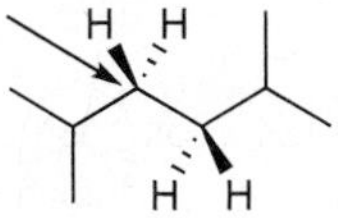

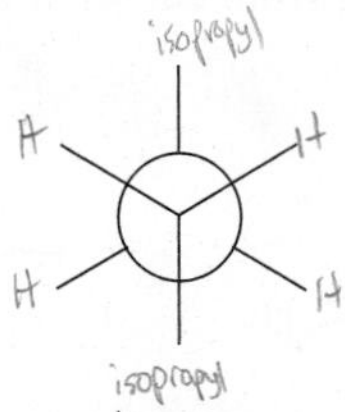

Answer

6.7 Me H
        H

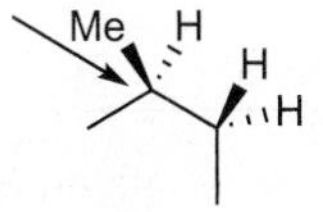

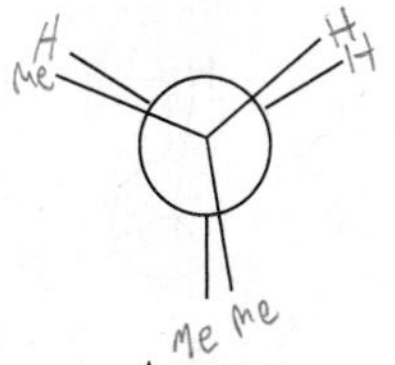

Answer

6.2 RANKING THE STABILITY OF NEWMAN PROJECTIONS

We have seen that Newman projections are a powerful way to show the different conformations of a molecule. We mentioned earlier that there are staggered conformations and eclipsed conformations. In fact, there are three staggered and three eclipsed conformations. Let's draw all three staggered conformations of butane. The best way to do this is to keep the back carbon atom motionless (so the fan in the back is not spinning), and let's slowly turn the groups in the front (only the front fan is spinning):

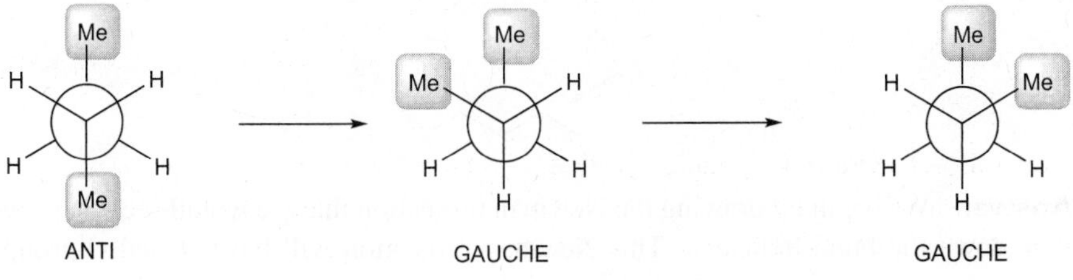

ANTI   GAUCHE   GAUCHE

Look at the first drawing above and notice the placement of the methyl group at the bottom. If we rotate the front carbon atom clockwise with all three of its groups, this methyl ends up in the top left (as seen in the second drawing). Then we rotate one more time to get the third drawing. Rotating one more time regenerates the first drawing. In the first drawing, the methyl groups are as far away from each other as possible, which is the most stable conformation, called the *anti* conformation. The other two drawings both have the methyl groups near each other in space. They feel each other and they are a bit crowded. This interaction is called a *gauche* interaction, which makes these conformations a little bit less stable than the *anti* conformation.

If we were to go back to our comparison between molecules and people, we would say that the *anti* conformation would be like lying down in a bed, and the gauche conformations are both like sitting in a chair. All of these are comfortable positions, but lying down is the most comfortable. The *anti* conformation is the most stable.

Now let's look at the three eclipsed conformations of butane. Again, let's keep the back where it is, and let's just rotate the front carbon atom with its three groups:

These conformations are all high in energy relative to the staggered conformations. All of the groups are eclipsing each other, so they are very crowded. All three of these would be like standing on your head, which is extremely uncomfortable. But the middle one is the most unstable, because the two methyl groups (the two largest groups) are eclipsing each other. This would be like standing on your head without using your hands for help—now that *really* hurts. So, all of these are high in energy, but the middle one is the most unstable.

To summarize, the most stable conformation will be the staggered conformation where the large groups are as far apart as possible (*anti*), and the least stable conformation will be the eclipsed conformation where the large groups are eclipsing each other.

EXERCISE 6.8 Draw the most stable and the least stable conformations of the following compound, using Newman projections looking down the following bond:

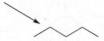

Answer We begin by drawing the Newman projection that we would see when we look down the bond indicated. This Newman projection will have a methyl group and two hydrogen atoms connected to the front carbon atom, and there will be an ethyl group and two hydrogen atoms connected to the back carbon atom, in the following way:

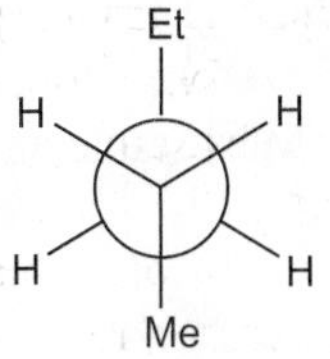

Now we decide how to rotate the front carbon atom so as to provide the most stable conformation. The most stable conformation will be a staggered conformation with the two largest groups *anti* to each other, so in this case, we do not need to rotate at all. The drawing that we just drew is already the most stable conformation, because the methyl and ethyl groups are *anti* to each other:

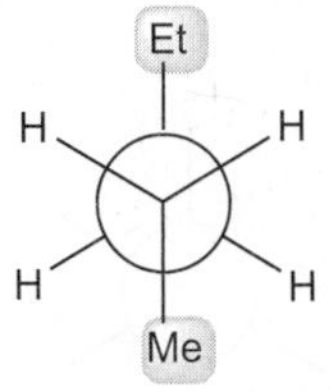

To find the least stable conformation, we need to rotate the front carbon atom and consider all three eclipsed conformations. The least stable conformation will be the one with the two largest groups eclipsing each other:

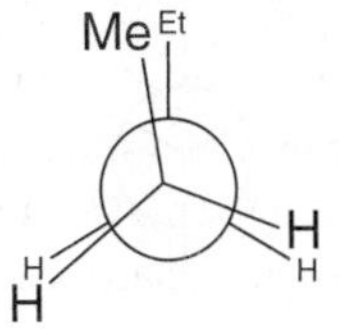

PROBLEMS   Draw the most stable and the least stable conformations for each of the following compounds. In each case, fill in the groups on the Newman projections below.

6.9

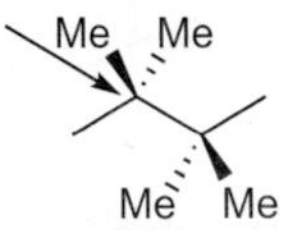

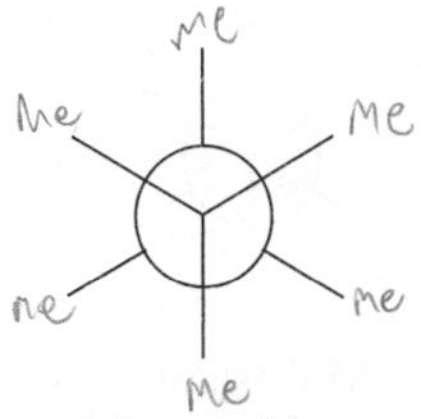

Most stable

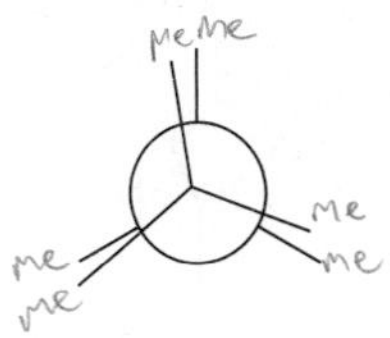

Least stable

6.10

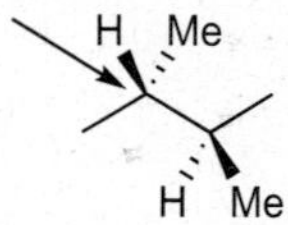

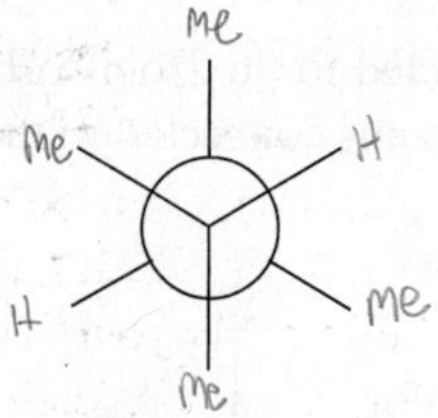

Most stable

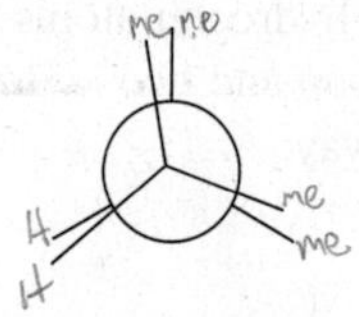

Least stable

6.11

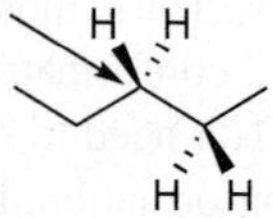

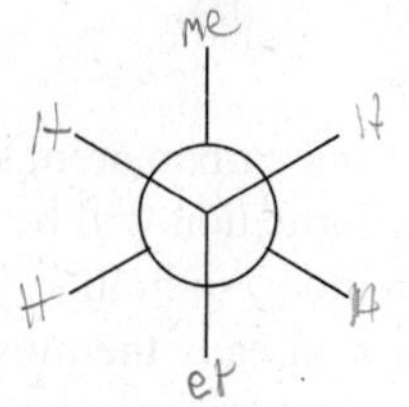

Most stable

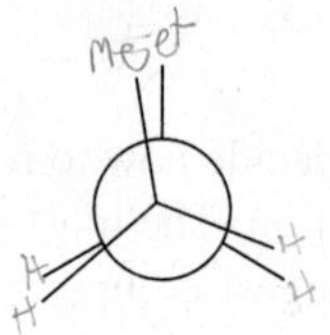

Least stable

6.12

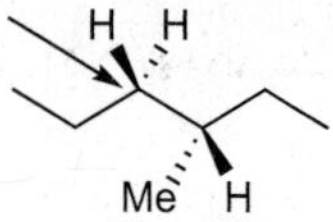

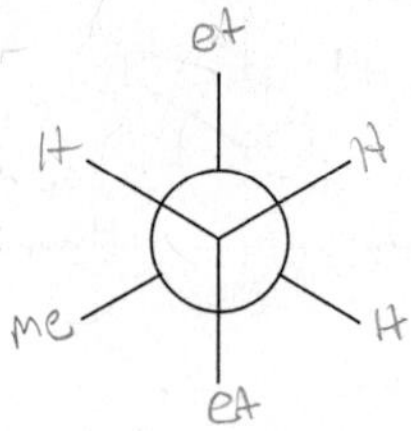

Most stable

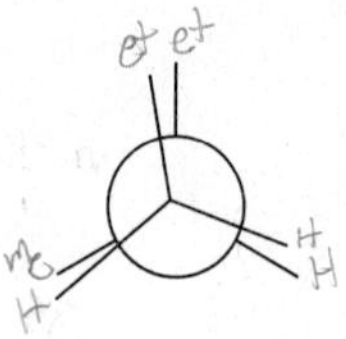

Least stable

6.13

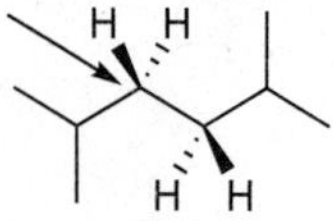

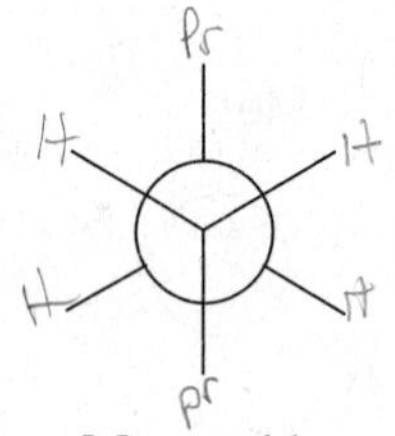

Most stable

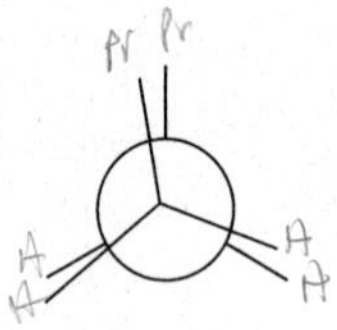

Least stable

6.14

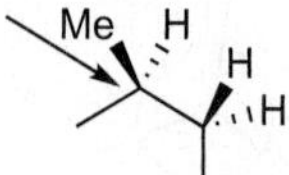

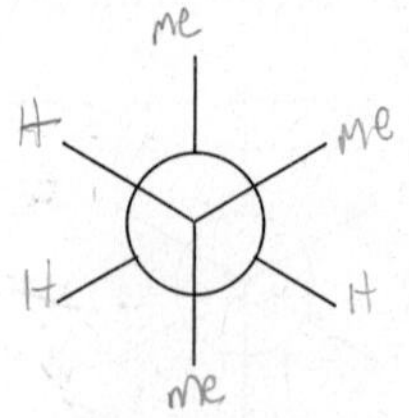

Most stable

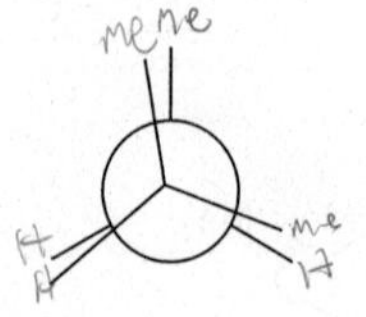

Least stable

6.3   **DRAWING CHAIR CONFORMATIONS**

An interesting case of conformational analysis comes to play when we consider a six-membered ring (cyclohexane). There are many conformations that this compound can adopt. You will see them all in your textbook: the chair, the boat, the twist-boat. The most stable conformation of cyclohexane is the chair. We call it a chair, because when you draw it, it looks like a chair:

You can almost imagine someone sitting on this structure, as if it were a beach chair. Most students have a difficult time drawing the chair and its substituents correctly. In this section, we will focus on learning how to draw the chair correctly. It is very important, because we cannot move on to see more about chairs until you know how to draw them.

You will need to practice this, step by step. Begin by drawing a *very wide* V:

Next, draw a line from the top right of the V, going down at a 60-degree angle, and stop a little bit to the right of an imaginary line coming straight down from the center of the V:

Next, draw a line parallel to the left side of the V and stop a little bit to the right of an imaginary line coming straight down from the top left of the V:

Then, start at the top left of the V and draw a line parallel to the line all the way on the right side, going down exactly as low as that line goes:

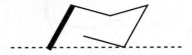

Finally, connect the last two lines:

Please don't *ever* draw a chair like this:

When you draw the chair sloppily (as so many students do), it makes it impossible to place the substituents correctly on the ring. And that's when you start losing silly points on your exam. So, take the time to practice and learn how to draw it properly. Practice in the space below:

Now we can start drawing in the substituents. Start with the top right corner and draw a line going straight up:

Then go around the ring and draw a straight line at each carbon atom, alternating between up and down:

These six substituents are called *axial* substituents. They go straight up and down, in the order shown above.

Next we need to learn how to draw the *equatorial* substituents. These are the substituents pointing out toward the sides. There are also six of them. Each one is drawn so that it is parallel to the two bonds from which it is once removed:

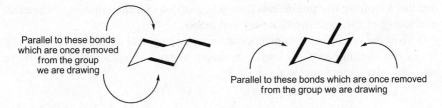

Parallel to these bonds which are once removed from the group we are drawing

Parallel to these bonds which are once removed from the group we are drawing

We go all the way around the ring like this, until we have drawn all of the equatorial groups:

Now we know how to draw all twelve substituents, but remember that if we draw a line and don't put anything at the end of that line, then this implies a methyl group. So, unless we are referring to dodecamethylcyclohexane (that's 12 methyl groups), we should really draw hydrogen atoms at the end of these lines:

Generally, we do not have to draw all 12 lines and place hydrogen atoms there. Remember how bond-line drawings work: if we don't draw any groups at all, it is assumed that there are hydrogen atoms. We are going through this exercise because it is important to know *how* to draw all 12 substituents. We will see in the next section that in most problems you will need to draw only a few of them. Which ones you draw will depend on the problem. So the only way to be sure that you can draw whatever the problem throws at you is to master drawing *all* of them. *Never* draw groups like this:

Very bad

Drawings like this will cause you to lose serious points on exams (not to mention the fact that a drawing like this defeats the whole purpose of a chair drawing—the exact positioning of the substituents is very important).

Use the space below to practice drawing a chair with all twelve substituents. When you are finished, label every substituent as axial or equatorial. Do not move on to the next section until you can do this.

6.4 PLACING GROUPS ON THE CHAIR

Now we need to see how to draw a chair with proper placement of the groups when we are given a regular hexagon-style drawing:

Br

is the same as

Br

Cl

Cl

Before we can get started, we need to remember what the dashes and wedges mean in the hexagon-style drawing. Remember that wedges are coming out toward you, and dashes are going back away from you. So, at each of the six carbon atoms of the ring, there are two groups—one coming out at you and one going away from you. If the groups are not drawn, then it means that there are two hydrogen atoms, one coming out and one going away.

Now let's introduce some new terminology. This is not scientific terminology, and you likely won't find it in your textbook, but it will help you master the task at hand. Anything coming off of the ring that is on a wedge, we will call *up*, because the group is coming up above the ring. Any group on a dash, we will call *down*, because the group is going down below the ring. So in the example above, Br is up and Cl is down.

Now let's apply the same terminology to the groups on a chair. Each carbon atom has two groups, one pointing above the ring (up) and one pointing below the ring (down):

You can do this for every carbon (and you should try on the drawing above), and you will see that each carbon has two groups (up and down). It is important to realize that there is *no correlation* between up/down and axial/equatorial. Look at the drawing above. For one of the carbon atoms, the up position is axial. For the other carbon atom showing its groups, the up position is equatorial. Take a close look at the two equatorial positions shown above. One of them is up and the other is down.

Now we are ready to draw a chair when we are given a hexagon. Let's work through the example that we started with:

We begin by placing numbers. These are not the same as the numbers that we used in naming compounds. These are just numbers that help us draw the chair with the groups in the right place. It does not matter where we start or which direction we go in, so let's just say that we will always start at the top and go clockwise:

Now, we draw a chair and we put numbers on the chair also. We can start our numbers anywhere on the chair, but we *must* go in the same direction that we did in the hexagon. If we went clockwise there, then we must go clockwise here. To avoid a mistake, let's just say that we will always go clockwise from now on and we will always start at the top right corner:

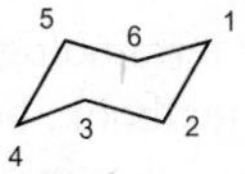

Now we know where to put in the groups. Br is on the carbon numbered 1, and Cl is on the carbon numbered 3. This brings us to the up/down system. Draw the chair, showing both positions (up and down) at each of the carbon atoms where we need to place a group:

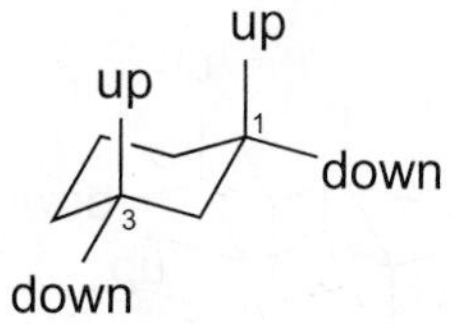

Look at the hexagon drawing again and ask whether each group is up or down. Br is on a wedge, so it is up. Cl is on a dash, so it is down. Now we are ready to put the groups into our chair drawing:

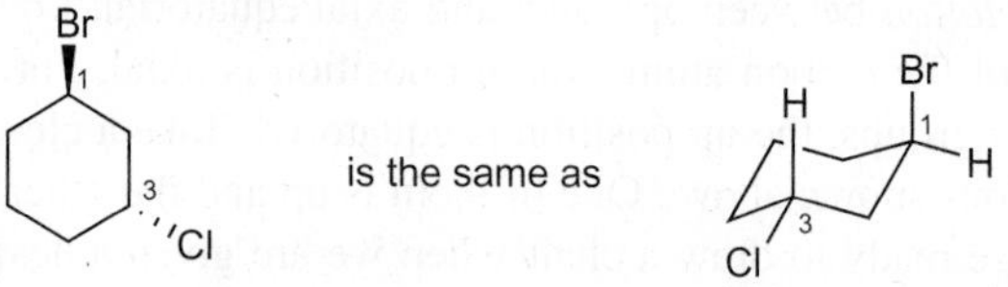

is the same as

That's all there is to it. To review, we need to draw the chair, number the chair and hexagon (both clockwise), determine where the groups go, determine whether they are up or down, and then draw them in. Let's get some practice now.

EXERCISE 6.15 Draw a chair conformation of the compound below:

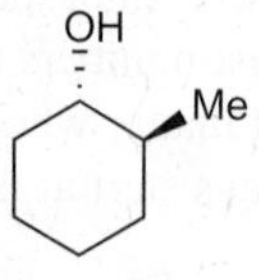

OH

Me

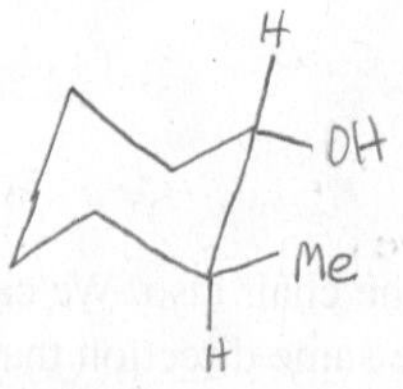

Answer   Begin by numbering the hexagon at the first group and then going clock-wise. This puts the OH at the position numbered 1 and the Me at the position numbered 2:

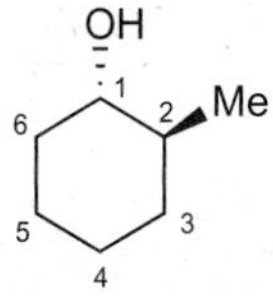

Next draw the chair, number it going clockwise, and then put in the up and down positions at the carbon atoms numbered 1 and 2:

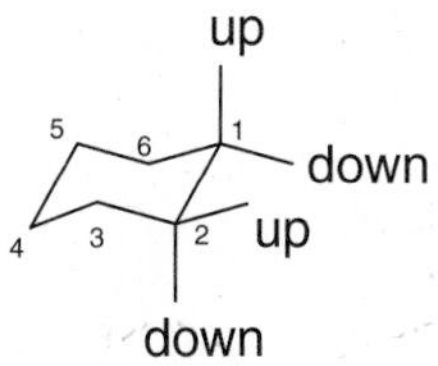

Finally, place the groups where they belong. The OH should be at the number 1 position in the down direction (because it was on a dash in the hexagon drawing), and the Me should be at the number 2 position in the up direction (because it was on a wedge):

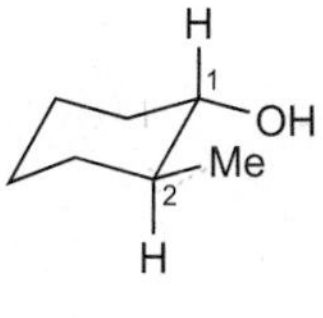

The example above illustrates an important point. Take a close look at the Me and the OH in the hexagon drawing. One is on a wedge and one is on a dash. We call this relationship *trans* (when you have two groups that are on opposite sides of the ring). If the two groups had been on the same side (either both on wedges, or both on dashes), then we would have called them *cis*. So in the example above, the groups are *trans* to each other. Now take a close look at the chair drawing we just drew above. The OH and Me don't look *trans* in this drawing. In fact, they look like they are *cis*, but they are *trans*. The OH is in the down position and the Me is in the up position.

It will become very clear that these groups are *trans* to each other when we learn how to draw the chair after it has "flipped" to give us a new chair drawing. We will see this in the next section. For now, let's get some practice drawing the first chair correctly.

PROBLEMS Draw a chair conformation for each of the compounds below.

6.16

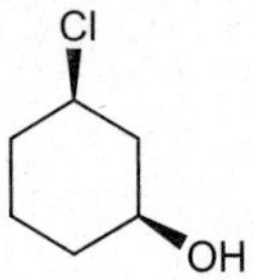

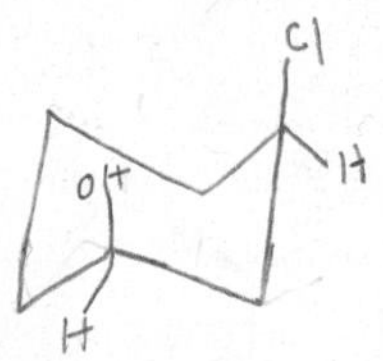

6.17

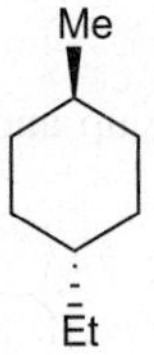

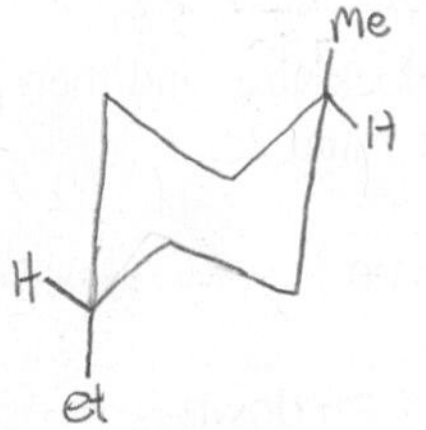

6.18

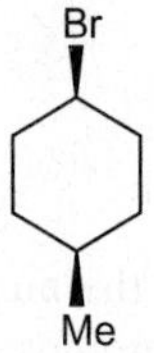

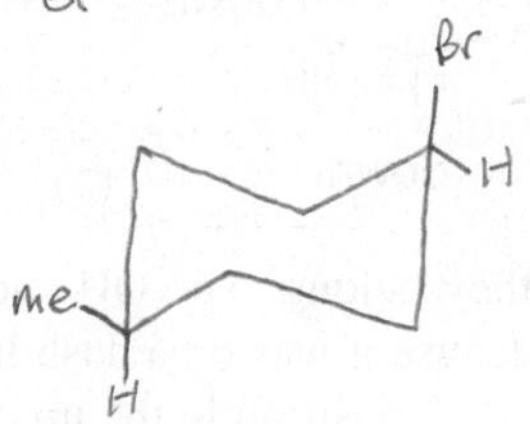

6.19

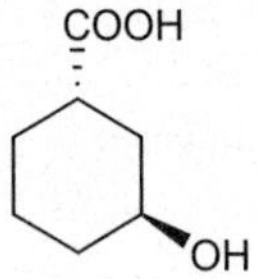

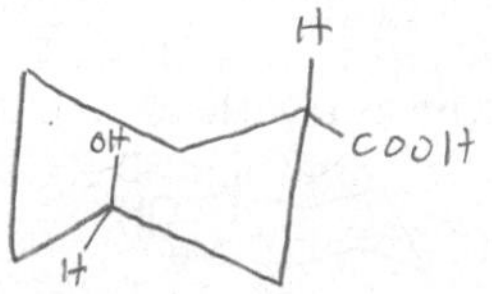

6.20

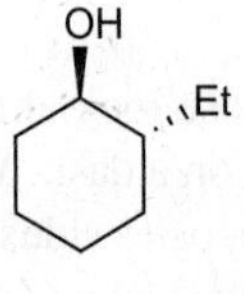

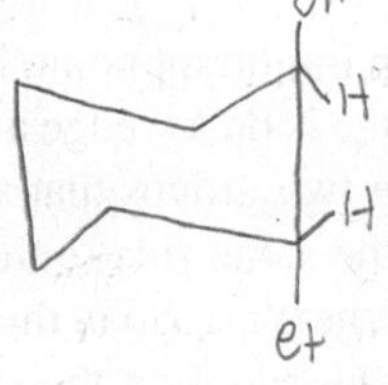

6.21

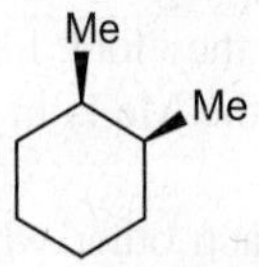

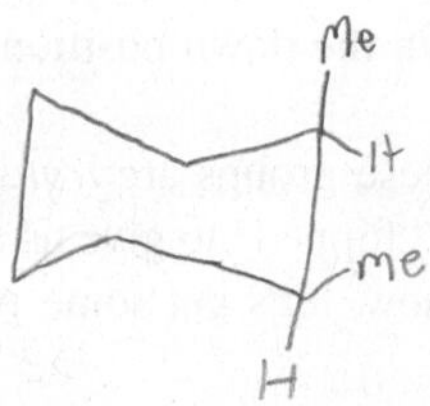

6.5 RING FLIPPING

Ring flipping is one of the most important aspects of understanding chair conformations, yet students commonly misunderstand this. Let's try to avoid the mistake by starting off with what ring flipping is not. It is *not* simply turning the ring over:

It makes sense why students think that this would be a flip—after all, this is the common meaning of the word "flipping." But we are talking about something very different when we say that rings can flip. Here is what we really mean:

Notice that in the drawing on the left, the left side of the chair is pointing down. In the drawing on the right, the left side of the chair is pointing up. This is a different chair. Also notice that the chlorine went from being in an equatorial position to being in an axial position. This is a critical feature of ring flips. When performing a ring flip, all axial positions become equatorial, and all equatorial positions become axial.

Let's consider an analogy to help us get through this. Imagine that you are walking down a long hallway. Your hands are swaying back and forth as you walk, as most people do with their hands when they walk. One second, your left hand is in front of you and your right hand is behind you; the next second, it switches. Your hands switch back and forth with every step you take. The cyclohexane ring is doing something similar. It is moving around all of the time, flipping back and forth between two different chair conformations. So all of the substituents are constantly flipping back and forth between being axial and being equatorial.

There is one more important feature to recognize. Let's go back to the example above with the chlorine. We said that the chair flip moves the chlorine from an equatorial position into an axial position. But what about the up/down terminology? Let's see:

Notice that the chlorine is down all of the time. In other words, up/down is *not* something that changes during a ring flip, but axial/equatorial does change during a ring flip. This illustrates that there is no relationship between up/down and axial/equatorial. If a substituent is up, then it will stay up all of the time, throughout the ring flipping process.

So now we can understand that a common hexagon-style drawing represents a molecule that is flipping back and forth between two chair conformations. The hexagon drawing shows us which substituents are up and which are down. That never changes. But whether those groups are axial or equatorial will depend on which chair you are drawing. So far, we have learned how to draw only one of these chairs. Now we will learn how to draw the other.

The process for drawing the skeleton of the chair is very similar to how we did it before. The only difference is that we draw our lines in the other direction. When we drew our first chair, we followed these steps:

Step 1   Step 2   Step 3   Step 4   Step 5

Now, to draw the other chair, we follow these steps:

Step 1   Step 2   Step 3   Step 4   Step 5

Compare the method for drawing the second chair to the method for drawing the first. The key is in step 2. If you compare step 2 for the first and second chair, everything else should flow from there. Use the following space to practice drawing the second chair:

Now, let's make sure you know how to draw the substituents. The rules are the same as before. All axial positions are drawn straight up and down, alternating:

and all equatorial positions are drawn parallel to the two bonds that are once removed:

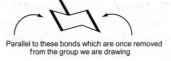

Parallel to these bonds which are once removed
from the group we are drawing

PROBLEM

6.22 In the space below, practice drawing the second chair, showing all 12 substituents.

Let's now go back and review, because it is important that you understand the following points. When we are given a hexagon-style drawing, the drawing shows us which positions are up and which positions are down. No matter which chair we draw, up will always be up, and down will always be down. There are two chair conformations for this compound, and the molecule is flipping back and forth between these two conformations. With each flip, axial positions become equatorial positions and vice versa. Let's see an example.

Consider the following compound:

Cl
Br

Notice that there are two groups on this ring. Cl is down (because it is on a dash), and Br is up (because it is on a wedge). There are two chair conformations that we can draw for this compound:

In both chair conformations, Cl is down and Br is up. The difference between these drawings is the axial/equatorial positions. In the conformation on the left, both groups are equatorial. In the conformation on the right, both groups are axial.

So any hexagon-style drawing will have two chair conformations. Now let's focus on making sure you can draw both conformations for any compound. We already saw in the last section how to draw the first one. We used a numbering system to determine where to put the groups, and we used the up/down terminology to figure exactly how to draw them (whether to draw them as equatorial or axial). To draw the second chair, we simply follow the same procedure. We begin by drawing the skeleton for the second chair (this is where you begin to see the difference between the chairs):

Skeleton for
First Chair

Skeleton for
Second Chair

Once we have drawn the skeleton, we number the carbon atoms going clockwise. Then we place the groups in the correct positions, making sure to draw them in the correct direction (up or down). So we can really use this method to draw both chairs at the same time. Let's do an example.

EXERCISE 6.23 Draw both chair conformations for the following compound:

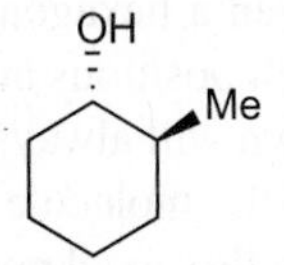

Answer Begin by numbering the hexagon clockwise. This puts the OH at the position numbered 1 and the Me at the position numbered 2.

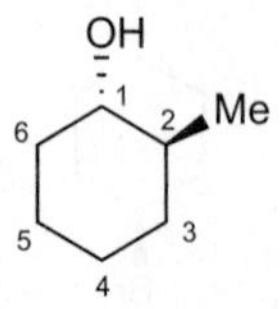

Next draw both chair skeletons and number them clockwise. Then put in the up and down positions at the carbon atoms numbered 1 and 2:

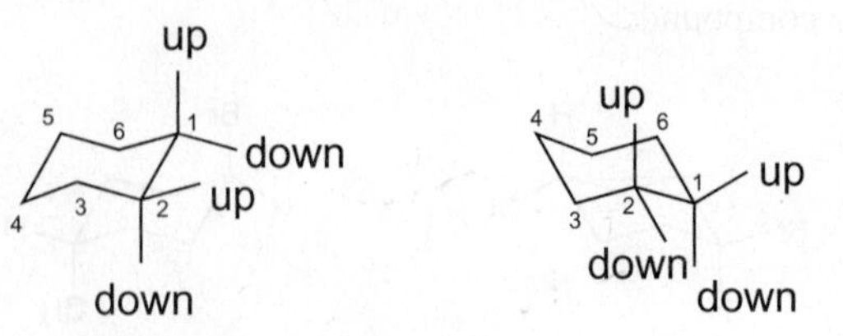

Finally, place the groups where they belong in both chairs. The OH should be at the number 1 position in the down direction (because it was on a dash in the hexagon drawing), and the Me should be at the number 2 position in the up direction (because it was on a wedge):

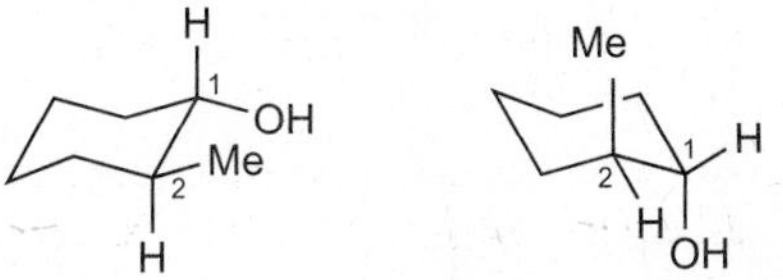

When we redraw these compounds without showing any numbers or hydrogen atoms, it is clear to see that we need to go through these steps methodically because the relationship between these two conformations is not so obvious:

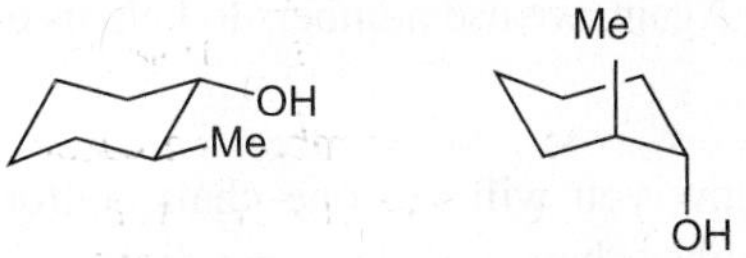

PROBLEMS For each of the compounds below, draw both chair conformations.

6.24

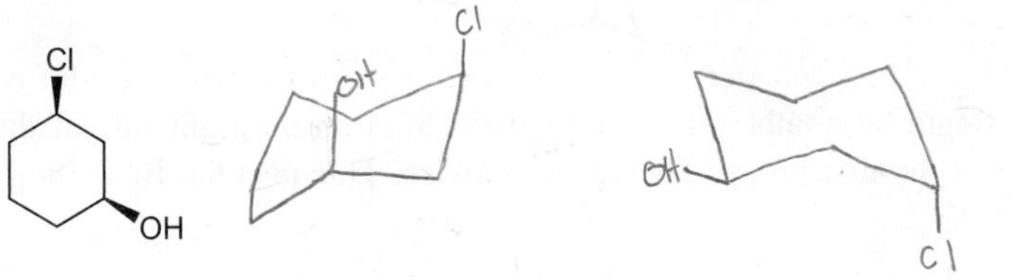

6.25

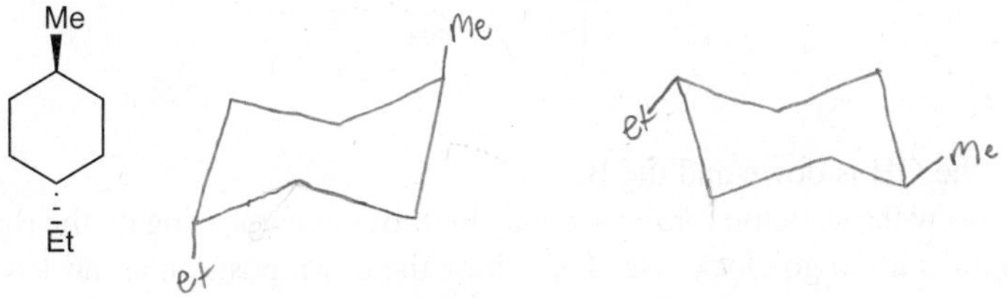

6.26

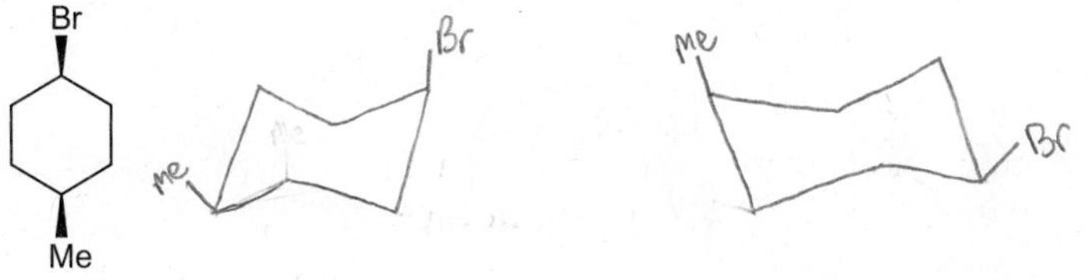

6.27

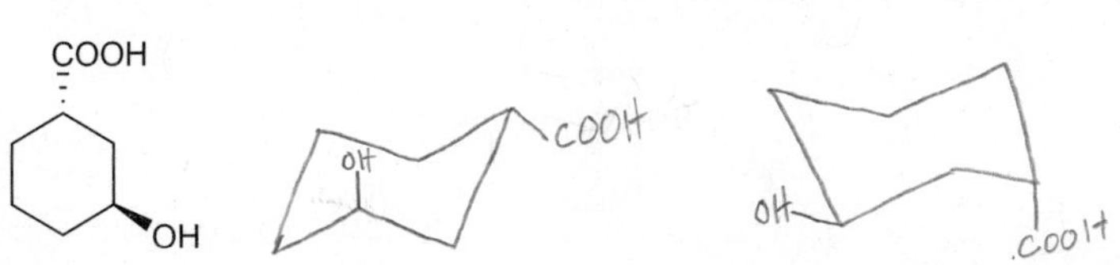

6.28

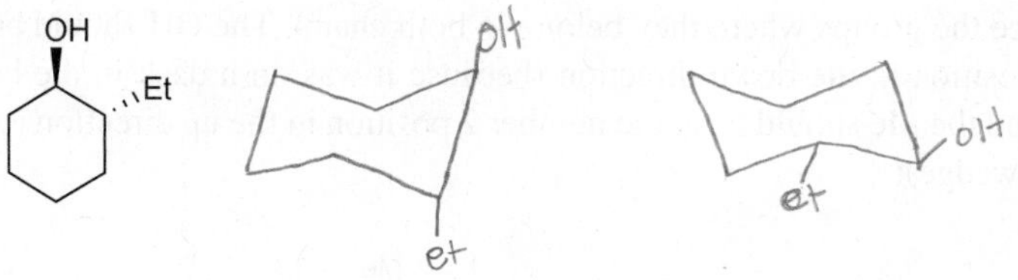

6.29

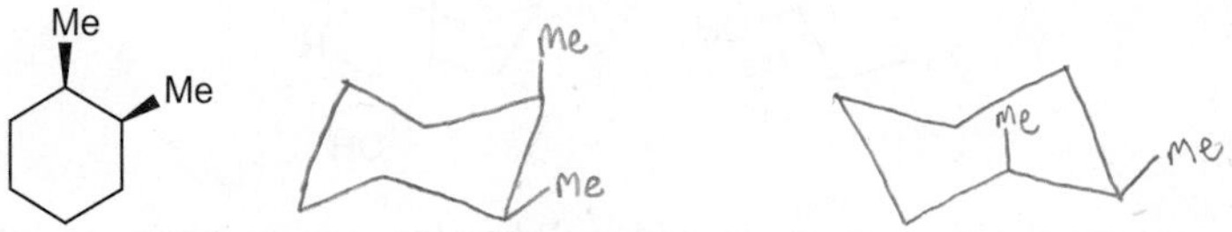

Sometimes, we might be given one chair conformation and be asked to draw the second chair conformation. Again, we use numbers to help us out. Let's see an example:

EXERCISE 6.30   Below you will see one chair conformation of a substituted cyclohexane. Draw the other chair (i.e., do a ring flip):

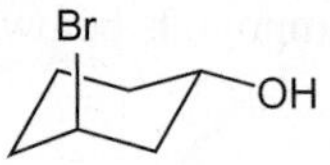

Answer   Begin by numbering the first chair. Start on the right side of the chair, and put a 1 at the first group. Then go clockwise. This puts the Br at the position numbered 3.

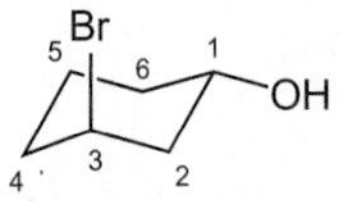

Notice that the OH is down and the Br is up.

Next draw the skeleton for the second chair. Begin numbering on the right side again, making sure to go clockwise. Then draw the down position at the 1 position, and draw the up position at the 3 position:

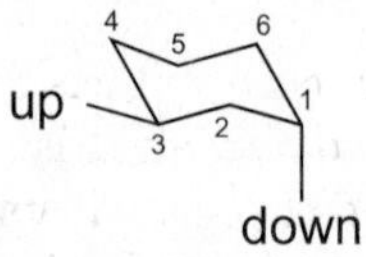

up

down

Finally, place the groups where they belong:

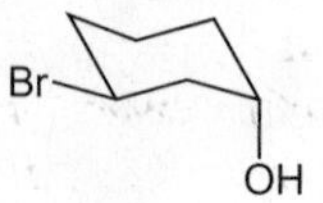

PROBLEMS For each chair conformation shown below, do a chair flip and draw the other chair conformation.

6.31

6.32

6.33

6.34

6.35

6.36

6.6 COMPARING THE STABILITY OF CHAIRS

Once you have drawn both chair conformations for a substituted cyclohexane, you should be able to predict which conformation is more stable. This is where it gets important for reactivity. Imagine that you are learning about a reaction that can proceed only if a certain group is in an axial position (you will learn about a reaction like this very soon—it is called E2). You already know that the groups are flipping back and forth between axial and equatorial positions (as it goes back and forth from one chair to the other). But what if one of the chairs is so unstable that the ring is spending 99% of its time in the other chair conformation? Then the question becomes, where is the important group in the stable chair conformation? Is it axial or equatorial? If it is equatorial, then the reaction can't happen. It could happen only during the 1% of the time that the group is in the axial position, so the reaction would be very slow. However, if the group is in an axial position 99% of the time, then the reaction will happen very quickly.

So you can see that it is important to understand what makes chair conformations unstable. There is really only one rule you need to worry about: a chair will be more stable with a group in an equatorial position, because it is not bumping into anything (this bumping is called *steric hindrance*). Axial positions are bumping into other axial positions, but equatorial positions are not:

The larger the group, the more it will prefer to be equatorial. So, a *tert*-butyl group will spend almost all of its time in an equatorial position. This essentially "locks" the ring in one conformation and prevents the ring from flipping (the truth is that the ring is still flipping, but the ring is spending more than 99% of its time in the more stable chair conformation):

So what happens if you also have a chlorine atom on the ring that is axial while the *tert*-butyl group is equatorial?

This will essentially lock the ring in the chair conformation that puts the chlorine in an axial position. If we are trying to run a reaction where the Cl needs to be axial, then this effect will speed up the reaction. However, if the Cl is locked in an equatorial position, then the reaction will be too slow:

Now we understand why this can be important. So let's go step by step in determining which of two chair conformations is more stable.

If you have only one group on the ring, then the more stable chair will be the one with the group in an equatorial position:

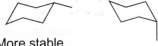

More stable

If you have two groups, then it is best if both can occupy equatorial positions:

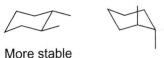

More stable

If only one can be equatorial in either chair, then the more stable chair will be the one with the larger group in the equatorial position:

More stable

In the example above, we have a choice to put the *tert*-butyl group in an equatorial position or the methyl group in an equatorial position. We can't have both. So we put the *tert*-butyl in an equatorial position.

If the ring has more than two groups, you just use the same logic that we used above to choose the more stable chair. Just try to put the largest groups in equatorial positions.

EXERCISE 6.37 For the following compound, draw the most stable chair conformation:

Answer We begin by drawing both chair conformations (if you have trouble with this, review the two previous sections in this chapter):

Now we select the chair that has the larger group in the equatorial position. In this case, both groups can be equatorial, so we choose this one:

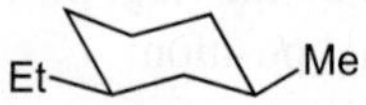

PROBLEMS For each compound below, draw the most stable chair conformation.

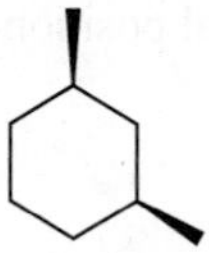

6.38

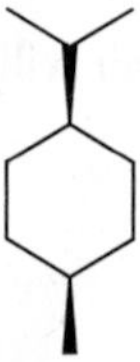

6.39

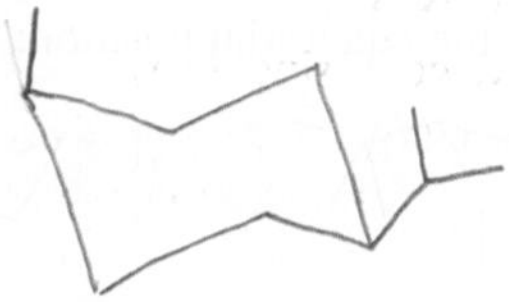

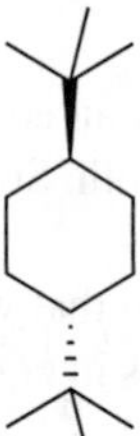

6.40

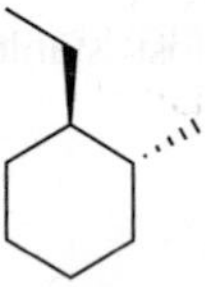

6.41

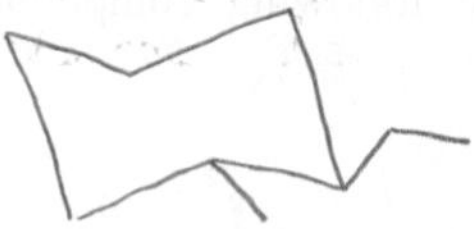

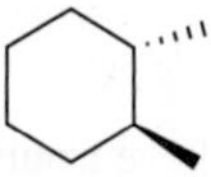

6.42

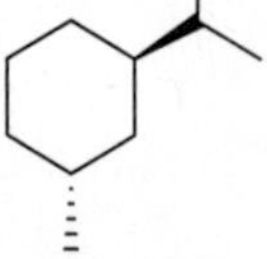

6.43

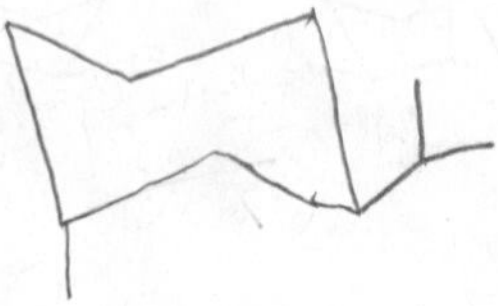

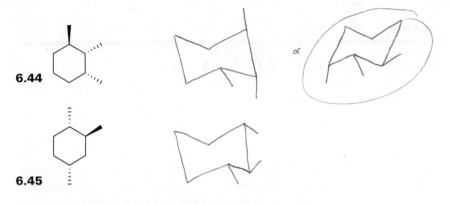

6.44

6.45

6.7 DON'T BE CONFUSED BY THE NOMENCLATURE

Some nomenclature is always confusing to students, so it is worth a couple of para-graphs to clear things up. When two groups are both up or both down, we call them *cis* to each other; when one group is up and one group is down, we call them *trans* to each other:

cis

trans

Do not confuse this with *cis* and *trans* double bonds. There are no double bonds here. Don't draw any double bonds. It is amazing how many students will draw a double bond when you ask them to draw *cis*-1,2-dimethylcyclohexane. Remember that the ending –ane– means that there are no double bonds anywhere in the mole-cule. The only comparison between double bonds and disubstituted cylohexanes is that, in both cases, *cis* means "on the same side" and *trans* means "on opposite sides."

CHAPTER 7

CONFIGURATIONS

In the previous chapter, we saw that molecules can assume many different conformations, much like a person. You can move your hands all around: hold them straight up in the air, out to the sides, straight down, and so on. In all of these positions, your right hand is still your right hand, regardless of how you move it around. There is no way to twist your right hand to turn it into a left hand. The reason it is always a right hand has nothing to do with the fact that it is connected to your right shoulder. If you were to chop off your arms and sew them on backward (*don't try this at home!*), you would not look normal. You would look like someone with his right hand attached to his left shoulder and vice versa. You would look very strange, to say the least.

Your right hand is a right hand because it fits into a right-handed glove, and it does not fit into a left-handed glove. This will always be true no matter how you move your hand. Molecules can have this property also.

It is possible for a molecule to have a region where there are two possibilities for how the atoms can be connected in 3D space, much like the difference between a right hand and a left hand. Instead of "right hand" and "left hand," we call the two possibilities *R* and *S*. When we talk about the configuration of a compound, we are talking about whether it is *R* or *S*. If the arrangement is *S*, then it will always be *S* no matter how the arms of the molecule twist about as the molecule moves. In other words, the molecule can move into any conformation it wants, but the configuration will never change. The only way for a configuration to change would be to undergo a chemical reaction.

This explains something we saw in the previous chapter: when drawing chair conformations, we saw that up is always up regardless of which chair you draw. That is because up and down are issues of *configuration,* which does not change when the molecule twists into another conformation.

Don't confuse *conformation* with *configuration*. Students confuse these terms all of the time. *Conformations* are the different positions that a molecule can twist into, but *configuration* is a matter of right-handedness or left-handedness (*R* or *S*).

In molecules, the regions that can be *R* or *S* are called *stereocenters* (or chiral centers—chiral is Greek for "hand," and we can understand the symbolism there). In this chapter, we will learn how to locate stereocenters, how to draw them properly, how to label them as *R* or *S*, and what happens when you have more than one stereocenter in a compound.

This is all *extremely* important stuff. You will understand this as soon as you begin learning reactions. You will see that some reactions convert a stereocenter from *R* into *S* (and vice versa), while other reactions will not. To predict the products of a reaction, you absolutely *must* know how to show these stereocenters, and you must understand what is happening to them in a reaction.

7.1 LOCATING STEREOCENTERS

For purposes of this course, we will define a stereocenter as a carbon atom with four different groups on it. For example,

Br   Cl
Et   Me

This drawing has a carbon in the center with four different groups: ethyl, methyl, bromine, and chlorine. Therefore, we have a stereocenter. Anytime you have four different groups connected to a carbon atom, there will be two ways to arrange the groups in space (always two; never more and never less). These two arrangements are different configurations:

Br   Cl   Cl   Br
Et   Me   and   Et   Me

These two compounds are different from each other even though the atoms are connected in the same way. The difference between them comes from their positions in 3D space. Therefore, they are called stereoisomers ("stereo" for space). More specifically, they are called *enantiomers,* because the two compounds are mirror images of each other and they are not superimposable. If we construct models of these two compounds, we see that they are not the same—i.e., they cannot be superimposed.

Notice that we are not looking at just the four *atoms* that are connected to the carbon atom in the middle (which would be Br, Cl, C, and C—and we might think that two of these are the same), but we are looking at the entire groups. In other words, whenever we look at the four groups connected to an atom, we are looking at the entire molecule, no matter how big those groups are. Consider the following example:

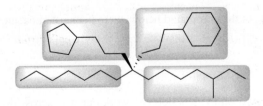

All four of these groups are different.

You must learn how to recognize when an atom has four different groups attached to it. To help you with this, let's begin with examples that are *not* stereocenters:

Not a stereocenter

The carbon atom indicated above is not a stereocenter because there are two groups that are the same (there are two ethyl groups). The same is true in the following case:

Not a stereocenter

Whether you go around the ring clockwise or counterclockwise, you see the same thing, so this is not a stereocenter. If we wanted to make it a stereocenter, we could do so by putting a group on the ring:

Stereocenter

Usually, stereocenters are drawn with dashes and wedges to show the configuration. If the dashes and wedges are not drawn, then we assume that there is a mixture of equal amounts of both configurations (which we call a *racemic* mixture). In fact, in the compound above, there is a second stereocenter. Can you find it? Each of the two stereocenters in the compound above can be either *R* or *S*. Since there are two stereocenters, there will be four possibilities: *R,R* and *R,S* and *S,R* and *S,S*. Since neither stereocenter has been drawn with dashes and wedges, we must assume that we have all four possible stereoisomers.

EXERCISE 7.1   In the compound below, there is one stereocenter. Find it.

OH

Answer   Let's start on the left side and work our way across the compound. The methyl group has three hydrogen atoms, so that can't be it. Then there is a CH_2 group, which also cannot be it, because two groups are the same (two H's). Then you have a carbon atom with four different groups: ethyl (on the left), methyl (on the right), OH sticking up, and H (don't forget about the H's that are not shown). This is our stereocenter.

PROBLEMS In each of the compounds below, there is one stereocenter. Find it.

7.2

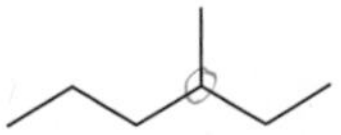

7.3

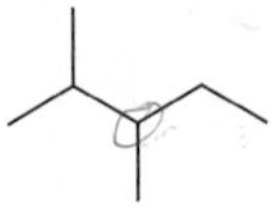

7.4

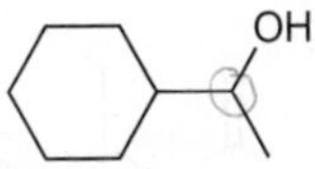

7.5

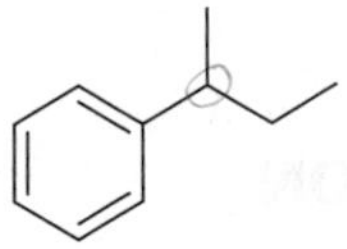

7.6

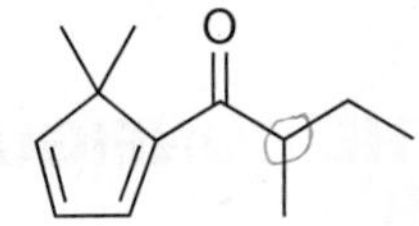

7.7

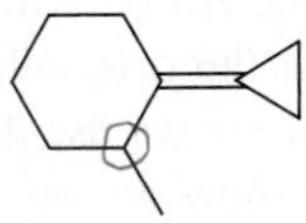

   In the previous problems, you knew that you were looking for just one stereo-center. Hopefully, you started to realize some tricks that make it faster to find the stereocenter (for example, ignore CH$_2$ groups). So, now, we will move on to examples where you don't know how many stereocenters there are. There could be five stereocenters or there could be none.

EXERCISE 7.8 In the following compound, find all of the stereocenters, if any:

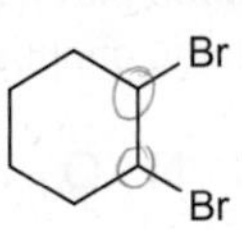

Answer If we go around the ring, we find that there are only six carbon atoms in this compound. Four of them are CH$_2$ groups, so we know that they are not stereo-centers. If we look at the remaining two carbon atoms, we see that each of them is connected to four different groups. They are both stereocenters.

PROBLEMS For each of the compounds below, find all of the stereocenters, if any.

7.9

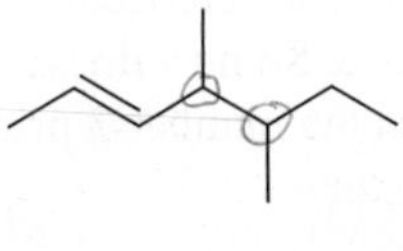

7.10

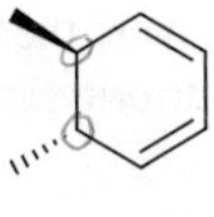

7.11

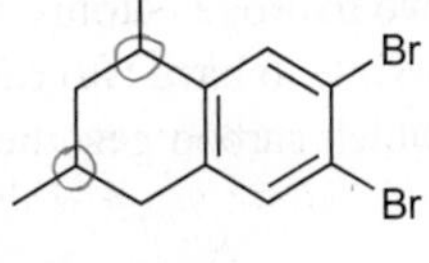

7.12

7.13

7.14

7.15

7.2 DETERMINING THE CONFIGURATION OF A STEREOCENTER

Now that we can find stereocenters, we must now learn how to determine whether a stereocenter is *R* or *S*. There are two steps involved in making the determination. First, we give each of the four groups a number (from 1 to 4). Then we use the orientation of these numbers to determine the configuration. So, how do we assign numbers to each of the groups?

We start by making a list of the four *atoms* attached to the stereocenter. Let's look at the following example:

The four atoms attached to the stereocenter are C, C, O, and H. We rank them from 1 to 4 based on atomic number. To do this, we must either consult a periodic table every time or commit to memory a small part of the periodic table—just those atoms that are most commonly used in organic chemistry:

$$C \quad N \quad O \quad F$$
$$P \quad S \quad Cl$$
$$Br$$
$$I$$

When comparing the four atoms in the example above, we see that oxygen has the highest atomic number, so we give it the first priority—we give it the number 1. Hydrogen is the smallest atom, so it will always get the number 4 (lowest priority) when a stereocenter has a hydrogen atom. We don't have to worry about what to do if there are two hydrogen atoms, because if there were, it would not be a stereocenter. But it is possible to have two carbon atoms, as in the example above. So how do we figure out which carbon gets the number 2 priority and which gets the number 3 priority?

This is how we rank the two carbon atoms: for each carbon atom, we write a list of three atoms it is connected to (other than the stereocenter). Let's do the example above to see how this works. The carbon atom on the left side of the stereocenter has four bonds: one to the stereocenter, one to another carbon atom, and then two hydrogen atoms. So, other than the stereocenter, it has three bonds (C, H, and H). Now let's look at the carbon atom on the right side of the stereocenter. It has four bonds: one to the stereocenter and then three hydrogen atoms. So, other than the stereocenter, it has three bonds (H, H, and H). We compare the two lists and look for the first point of difference:

$$
\begin{array}{cc}
C & H \\
H & H \\
H & H
\end{array}
$$

We see the first point of difference immediately: carbon beats hydrogen. So the left side of the stereocenter gets priority over the right side, and the numbering turns out like this:

EXERCISE 7.16 In the compound below, find the stereocenter, and label the four groups from 1 to 4 using the system of priorities based on atomic number.

Answer The four atoms attached to the stereocenter are C, C, Cl, and F. Of these, Cl has the highest atomic number, so it gets the first priority. Then comes F as number 2. We need to decide which carbon atom gets the number 2 and which carbon atom gets the number 3. We do this by listing the three atoms attached to each of them:

Left Side	*Right Side*
C	C
H	C
H	H

So the right side wins. Therefore, the numbering goes like this:

PROBLEMS   In each of the compounds below, find the stereocenter, and label the four groups from 1 to 4 using the system of priorities based on atomic number.

7.17   7.18 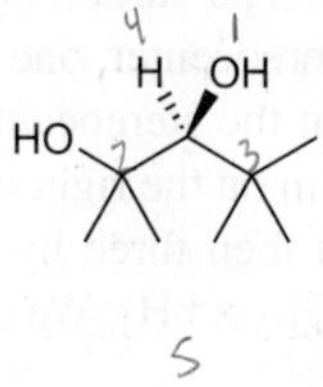   7.19

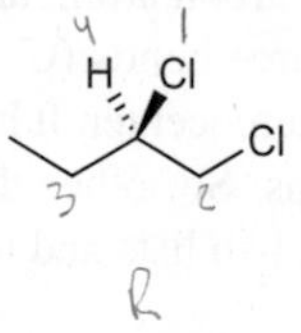

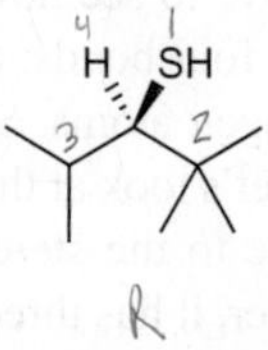

There are a few more situations you should know how to deal with when numbering the four groups. If you are comparing two carbon atoms and you find that the three atoms on one side are the same as the three atoms on the other, then keep going farther out until you find the first difference:

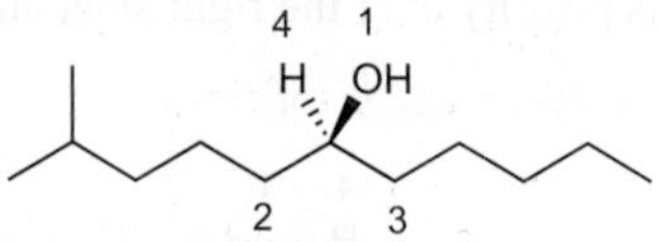

Also, you should know that we are looking for the first point of difference as we travel out, and we don't add the atomic numbers. This is best explained with an example:

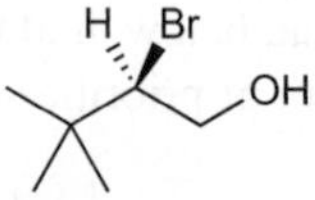

In this example, we know that the Br gets the first priority, and the H gets the number 4. When comparing the two carbon atoms, we find the following situation:

Left Side	Right Side
C	O
C	H
C	H

In this case, we do not add the atomic numbers and say that the left side wins. Rather we go down the list and compare each row. In the first row above, we have C versus O. That's it, end of story—the O wins. It doesn't matter what comes in the next two rows. Always look for the *first* point of difference. So the priorities go like this:

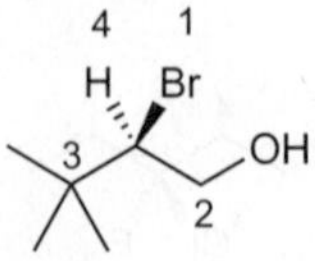

Finally, you should count a double bond as if the atom were connected to two carbon atoms. For example,

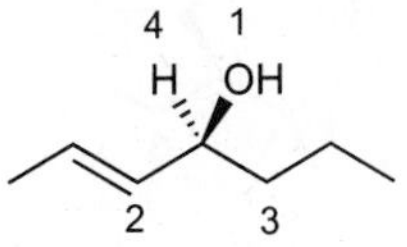

The group on the left gets the number 2, because we counted the following way:

Left Side	Right Side
C	C
C	H
H	H

EXERCISE 7.20   In the compound below, find the stereocenter, and label the four groups from 1 to 4 using the system of priorities based on atomic number.

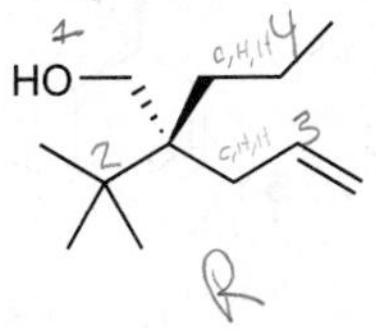

Answer   All four atoms connected to the stereocenter are carbon atoms, so we need to compare the three atoms connected in all four cases:

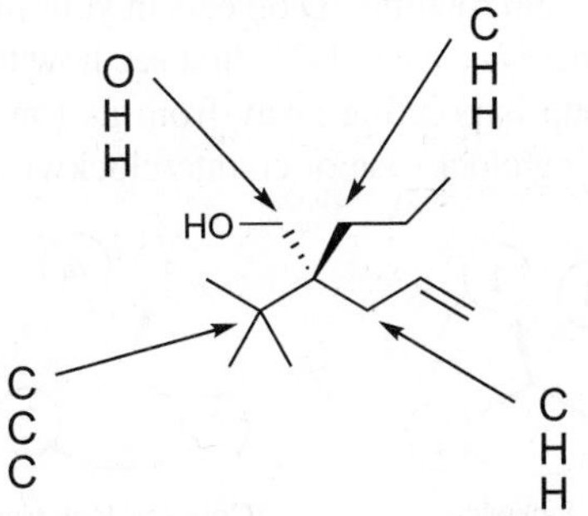

The oxygen atom wins. Next comes the one with three carbon atoms. The remaining two are the same, so we need to move out one farther on the chain and compare again. Remember to count the double bond like two carbon atoms:

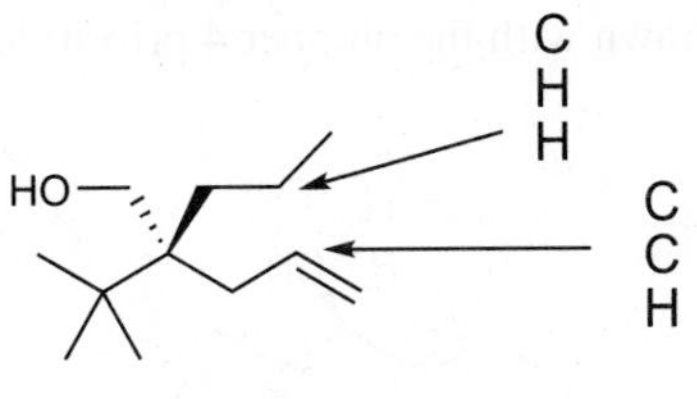

So the order of priorities goes like this:

PROBLEMS   In each of the compounds below, find the stereocenter and label the four groups from 1 to 4 using the system of priorities based on atomic number.

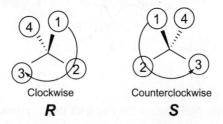

Now we need to learn how to use this numbering system to determine the configuration of a stereocenter. The idea is simple, but it is difficult to do if you have a hard time closing your eyes and rotating 3D objects in your mind. For those who cannot do this, don't worry. There is a trick. Let's first see how to do it without the trick.

If the number 4 group is pointing away from us (on the dash), then we ask whether 1, 2, and 3 are going clockwise or counterclockwise:

Clockwise   Counterclockwise
R   **S**

In the example on the left, we see that 1, 2, 3 go clockwise, which is called *R*. In the example on the right, we see that 1, 2, 3 go counterclockwise, which is called *S*. If the molecule is already drawn with the number 4 priority on the dash, then your life is very simple:

The 4 is already on the dash, so you just look at 1, 2, and 3. In this case, they go counterclockwise, so it is *S*.

It gets a little more difficult when the number 4 is not on a dash, because then you must rotate the molecule in your mind. For example,

Let's redraw just the stereocenter showing the location of the four priorities:

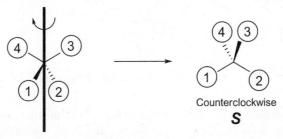

Now we need to rotate the molecule so that the fourth priority is on a dash. To do this, imagine spearing the molecule with a pencil and then rotating the pencil 90°:

Now the 4 is on a dash, so we can look at 1, 2, and 3, and we see that they go counterclockwise. Therefore, the configuration is *S*.

Let's see one more example:

We redraw just the stereocenter showing the location of the four priorities, and then we spear the molecule with a pencil and rotate 180° to put the 4 on a dash:

Now, the 4 is on a dash, so we can look at 1, 2, and 3, and we see that they go clockwise. Therefore, the configuration is *R*.

And now, for the trick. If you were able to see all of that, great! But if you had trouble seeing the molecules in 3D, there is a simple trick that will help you get the answer every time. To understand how the trick works, you need to realize that if you redraw the molecule so that any two of the four groups are switched, then you have switched the configuration (*R* turns into *S*, and *S* turns into *R*):

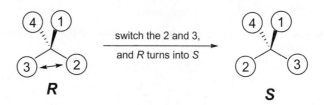

You can use this idea to your advantage. Here is the trick: Switch the number 4 with whatever group is on the dash—then your answer is the opposite of what you see. Let's do an example:

This looks tough because the 4 is on a wedge. But let's do the trick: switch the 4 with whichever group is on the dash; in this case, we switch the 4 with the 1:

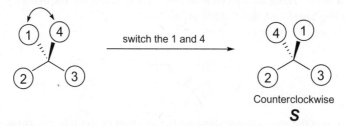

After doing the switch, the 4 is on a dash, and it becomes easy to figure out. It is counterclockwise, which means *S*. We had to do one switch to make it easy to figure out, which means that we changed the configuration. So if it became *S* after the switch, then it must have been *R* before the switch. That's the trick. *But be careful.* This trick will work every time, but you must not forget that the answer you immediately get is the opposite of the real answer, because you did one switch.

Now, let's practice determining *R* or *S* when you are given the numbers, so that we can make sure you know how to do this step. You can either visualize the molecule in 3D, or you can use the trick—whatever works best for you.

PROBLEMS In each case, assign the correct configuration (*R* or *S*).

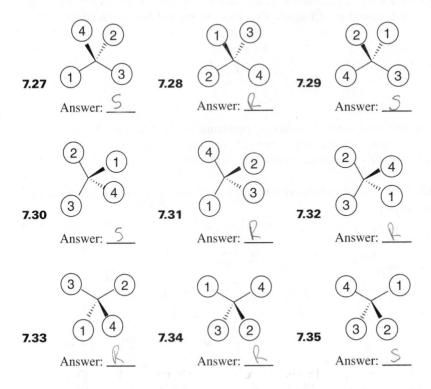

7.27 Answer: _S_

7.28 Answer: _R_

7.29 Answer: _S_

7.30 Answer: _S_

7.31 Answer: _R_

7.32 Answer: _R_

7.33 Answer: _R_

7.34 Answer: _R_

7.35 Answer: _S_

So we know how to assign priorities (1–4), and we know how to use those priorities to determine configuration. Now, let's do some real problems:

EXERCISE 7.36 The compound below has one stereocenter. Find it, and determine whether it is *R* or *S*:

Answer The carbon atom bearing the two chlorine atoms cannot be the stereocenter because there are two Cl atoms (two of the same group). The stereocenter is the carbon atom with the OH group attached. It has four different groups attached to it. Now that we found it, we need to assign priorities.

The O on the dash gets priority number 1, and the hydrogen atom (not shown, but it is on a wedge) gets number 4. Between the two carbon atoms, the one on the right is connected to two Cl atoms. This wins. So the numbers go like this:

The 4 is on a wedge, which makes the problem a little bit difficult. So let's use the trick. If we switch the 4 and the 1, we get something that is *R*. So, it must be that it was *S* before we switched the groups and our answer is *S*.

PROBLEMS For each compound below, find all stereocenters, and determine their configuration.

7.37   7.38   7.39

7.40   7.41   7.42

7.3 NOMENCLATURE

When we learned how to name compounds, we said that we would skip over the naming of stereocenters until we learned how to determine configuration. Now that we know how to determine whether a stereocenter is *R* or *S*, we can now see how to include this in the name of a compound. It is actually quite simple. If there is only one stereocenter, then you simply place either (*R*) or (*S*) at the beginning of the name. For example, 2-butanol has one stereocenter, and can either be (*R*)-2-butanol or (*S*)-2-butanol, depending on the configuration of the stereocenter. If more than one stereocenter is present, then you must also use numbers to identify the location of each stereocenter. Consider the following example:

Based on everything we learned in the second chapter (nomenclature), we would name this compound 3,4-dimethylhexan-2-one. Now we must also add the configurations to the name. The stereocenter on the left is R, and the stereocenter on the right is S. We use the numbering system of the parent chain to determine where the stereocenters are. Since the parent chain was numbered left to right, we add (3R,4S) to the name in the section on stereoisomerism:

Stereoisomerism	Substituents	Parent	Unsaturation	Functional group

so the name is: (3R,4S)-3,4-dimethylhexan-2-one. As we saw earlier, we italicize stereoisomerism when it is part of a name.

Now let's turn to a different type of stereoisomerism, one that we already discussed in the chapter on nomenclature. Let's look at double bonds. Recall we indicate the presence of a double bond using the term –en–:

Stereoisomerism	Substituents	Parent	Unsaturation	Functional group

And we indicated the position of the double bond with the numbering system. But then we saw that there are often two ways for the atoms of a double bond to connect to each other in 3D space. We saw a system for distinguishing these possibilities, using the terms *cis* and *trans*:

cis   *trans*

This was indicated in the first part of the name (stereoisomerism):

Stereoisomerism	Substituents	Parent	Unsaturation	Functional group

This system of identifying double-bond stereoisomers is very limited, because you need two groups that are the same to use the *cis/trans* system of naming. So what do you do if you have four different groups on a double bond? There are still two possible stereoisomers:

but we cannot use the *cis/trans* system here. So, we have another system that allows us to differentiate between these two compounds. This other system uses the same numbering based on priorities that we used for stereocenters (based on atomic numbers). We look at both sides of the double bond; each side has two groups:

Two groups on this side   Two groups on this side

We begin with one side (let's start with the left), and we ask which of the two groups on the left has priority:

Which of these gets the priority?

The oxygen atom gets priority over the carbon atom, based on atomic number. When comparing the two groups on the right side, the fluorine atom gets priority over the hydrogen atom, again based on atomic numbers. So now we know which group gets the priority on each side:

In the example above, it was easy to assign priorities, but what about when it gets a little more complicated:

In this example, we have to compare carbon atoms to each other. The groups are all different, so we need to find a way to assign priorities. To do so, we follow the same rules that we did when assigning priorities with R and S:

1. If the atoms are the same on one side, then just move farther out and analyze again.
2. One oxygen beats three carbon atoms (remember to look for the first point of difference).
3. A double bond counts as two individual bonds.

Let's go back to our first example. We look at the priority group on one side and the priority group on the other side, and we ask: are these groups on the same side (like *cis*) or on opposite sides (like *trans*)?

E   **Z**

The same side is called Z (for the German word "zusammen" meaning together), and opposite sides are called E (for the German word "entgegen" meaning opposite).

This way of naming double bonds is far superior to *cis/trans* nomenclature because you can use this *E/Z* system for any double bond, even if all four groups are different. *Cis/trans* nomenclature requires two groups to be the same.

We include this information in the name of a molecule, very much like we did for R and S configurations. For example, if the double bond is between carbons numbered 5 and 6 on a parent chain, then we would add the term (5E) or (5Z) at the beginning of the name.

EXERCISE 7.43 Name the following compound, including stereochemistry in the beginning of the name.

[handwritten: 4,fluoro- S (3Z)(5S)- 3,5dimethylheptene]

Answer Remember that we go through all five parts of the name, starting at the end and working our way backwards:

Stereoisomerism	Substituents	Parent	Unsaturation	Functional group

We begin with the functional group. There is no functional group (so the suffix is –e). Moving backward in the name, we look for any unsaturation, and there is one double bond (so the unsaturation is –en-). Then, we choose the longest parent that includes the double bond, which is seven carbon atoms long, so the parent is –hept-. There are three substituents (two methyl groups and a fluorine), so we add fluoro and dimethyl before the parent. Then we put in the numbers. We give the double bond the lowest number, so we number from left to right. This gives us

    4-fluoro-3,5-dimethylhept-3-ene

If you do not remember how to do that, then you should review Chapter 5 on nomenclature. Now we are ready to put in the stereochemistry. The double bond is Z:

and there is a stereocenter that is S:

When we number the parent chain, we see that the double bond is at the third carbon in the parent chain, and the stereocenter is at the fifth carbon in the parent chain:

So the name is

(3Z,5S)-4-fluoro-3,5-dimethylhept-3-ene

PROBLEMS Name each of the following compounds. Be sure to include stereoisomerism at the beginning of every name:

7.44   Name: (2Z)- 2-Flouro-pent-2-ene

7.45   Name: (1R,3R)- 3-methyl-cyclohexanol

7.46   Name: (3S)- 3-methyl-pent-1-ene

7.47   Name: (3E)- 2,3-dimethyl-4,ethyl- hept-3-ene

7.48   Name: (2Z,4E)- hept-2,4-diene

7.49   Name: (2E,4Z,6Z,8E)- deca-2,4,6,8-tetraene

7.4 DRAWING ENANTIOMERS

We mentioned before that enantiomers are two compounds that are nonsuperimposable mirror images. Let's first clear up the term "enantiomers," since students will often use this word incorrectly in a sentence. Let's compare it to people again. If two boys are born to the same parents, those boys are called brothers. Each one is the brother of the other. If you had to describe both of them, you say that they are brothers. Similarly, when you have two compounds that are nonsuperimposable mirror images, they are called enantiomers. Each one is the enantiomer of the other. Together, they are a pair of enantiomers. But what do we mean by "nonsuperimposable mirror images"? Let's go back to the brother analogy to explain it.

Imagine that the two brothers are twins. They are identical in every way except one. One of them has a mole on his right cheek, and the other has a mole on his left cheek. This allows you to distinguish them from each other. They are mirror images of each other, but they don't look exactly the same (one cannot be superimposed on top of the other). It is very important to be able to see the relationship between different compounds. It is important to be able to draw enantiomers. Later in the course, you will see reactions where a stereocenter is created and both enantiomers are formed. To predict the products, you must be able to draw both enantiomers. In this section, we will see how to draw enantiomers.

The first thing you need to realize is that enantiomers always come in pairs. Remember that they are mirror images of each other. There are only two sides to a mirror, so there can be only two different compounds that have this relationship (nonsuperimposable mirror images). This is very much like the twin brothers. Each brother only has one twin brother, not more.

So we must learn how to draw one enantiomer when we are given the other. When we see the different ways of doing this, we will begin to recognize when compounds are enantiomers and when they are not.

The simplest way to draw an enantiomer is to redraw the carbon skeleton, but invert all stereocenters. In other words, change all dashes into wedges and change all wedges into dashes. For example,

The compound above has a stereocenter (what is the configuration?). If we wanted to draw the enantiomer, we would redraw the compound, but we would turn the wedge into a dash:

This is a pretty simple procedure for drawing enantiomers. It works for compounds with many stereocenters just as easily. For example,

The enantiomer of 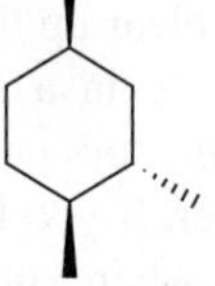   is

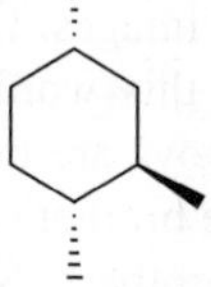

We simply invert all stereocenters. This is actually what we would see if we placed a mirror directly behind the first compound and then looked into the mirror. The carbon skeleton would look the same, but the stereocenters would all be inverted:

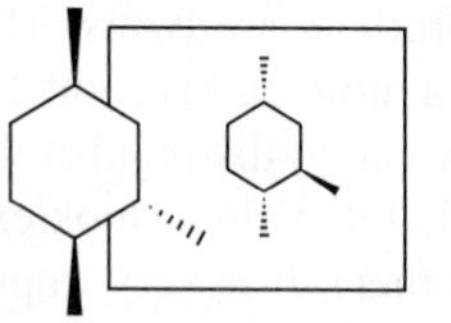

EXERCISE 7.50 Draw the enantiomer of the following compound:

OH   OH   OH   OH

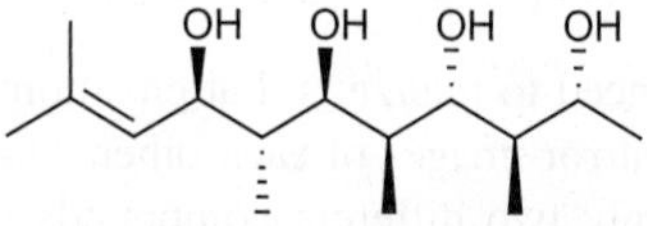

Answer Redraw the molecule, but invert every stereocenter. Convert all wedges into dashes, and convert all dashes into wedges:

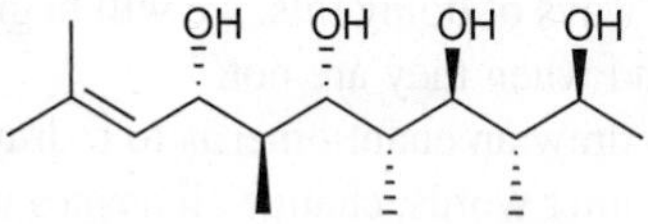

OH   OH   OH   OH

PROBLEMS Draw the enantiomer of each of the following compounds.

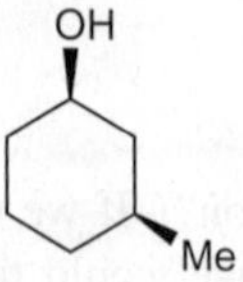

7.51

OH

Me

Answer:

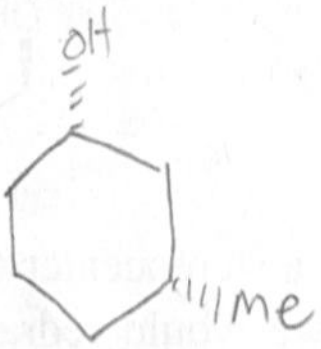

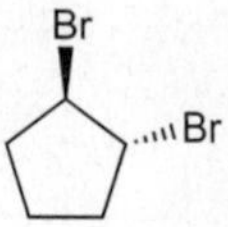

7.52

Br

Br

Answer:

7.53 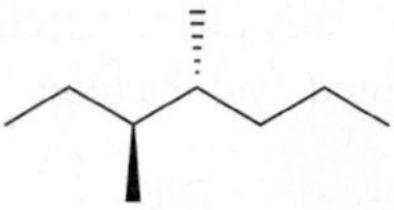   Answer:

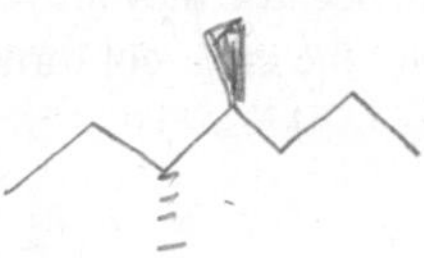

7.54 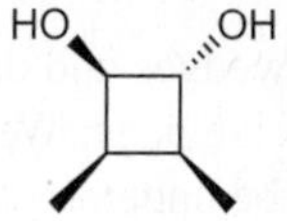   Answer:

7.55 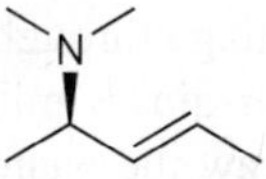   Answer:

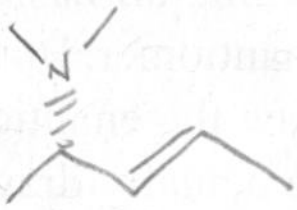

7.56 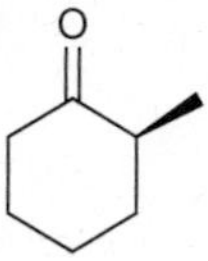   Answer:

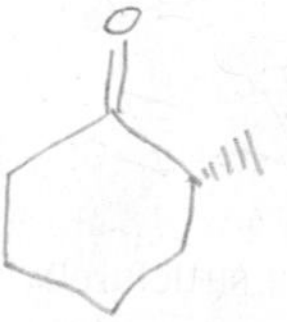

There is another way to draw enantiomers. In the previous method, we placed an imaginary mirror *behind* the compound, and we looked into that mirror to see the reflection. In the second method for drawing enantiomers, we place the imaginary mirror *on the side of* the compound, and we look into the mirror to see the reflection. Let's see an example:

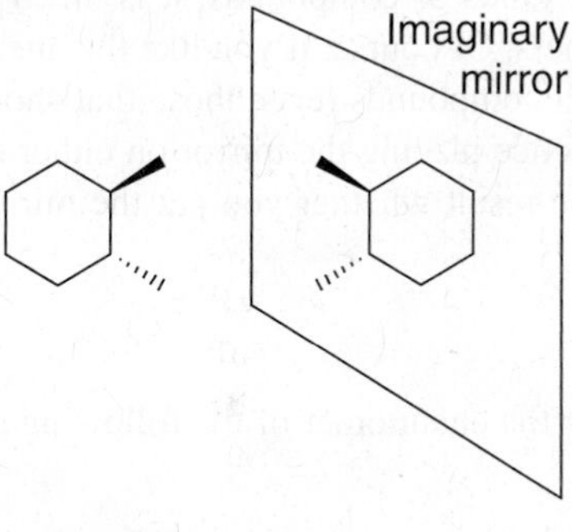

But why do we need a second way of drawing enantiomers? Didn't the first method seem good enough? The first method (switching all dashes with wedges) was pretty simple to do. But there are times when the first method will not work so well. There are a few examples of cyclic and bicyclic carbon skeletons where the wedges and dashes are not drawn, because they are implied. We have actually already seen an example of one of these: the chair conformation of a substituted cyclohexane.

In this drawing, all of the lines are drawn as straight lines (no wedges and dashes) even though we know that the bonds are not all in the plane of the page. We don't need to draw the wedges and dashes because the geometry can be understood from the drawing. We could try to draw the enantiomer by converting the drawing into a hexagon-style drawing (with wedges and dashes), then drawing the enantiomer using the first method (switching all dashes for wedges), and then redrawing the chair conformation of that enantiomer. But that is a lot of steps to go through when there is a simpler way to draw the enantiomer—just put the imaginary mirror on the side (there is no need to actually draw the mirror), and draw the enantiomer like this:

Whenever we have a structure where the wedges and dashes are implied but not drawn, it is much easier to use this method. There are other examples of carbon skeletons that, by convention, do not show the wedges and dashes. Most of these examples are rigid bicyclic systems. For example,

When dealing with these kinds of compounds, it is much easier to use the second method to draw enantiomers. Of course, if you like this method, you can always use this second method for all compounds (even those that show wedges and dashes).

You should get practice placing the mirror on either side (and you should notice that you get the same result whether you put the mirror on the left side or the right side).

EXERCISE 7.57 Draw the enantiomer of the following compound:

Answer This is a rigid bicyclic system, and the dashes and wedges are not shown. Therefore, we will use the second method for drawing enantiomers. We will place the mirror on the side of the compound, and draw what would appear in the mirror:

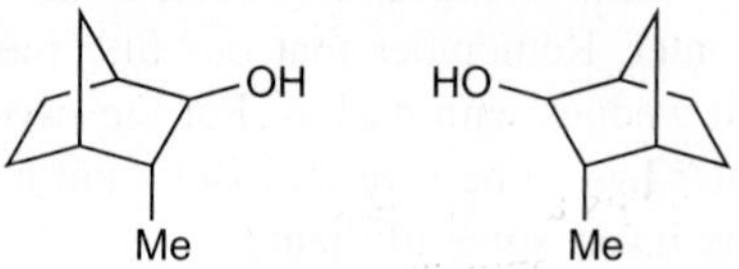

PROBLEMS Draw the enantiomer of each of the following compounds.

7.58 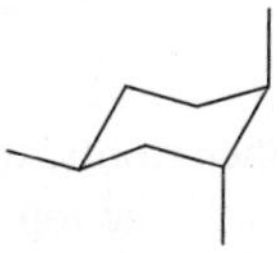   Answer: _____

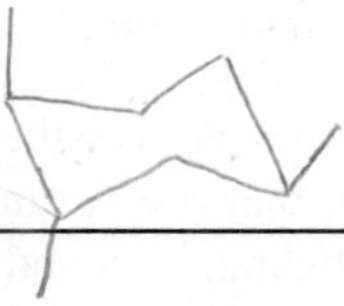

7.59 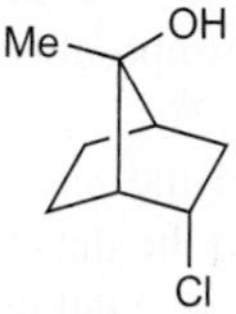   Answer: _____

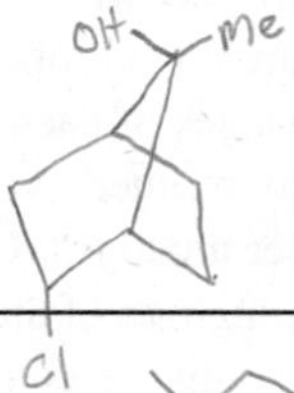

7.60 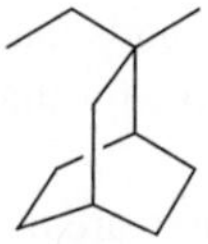   Answer: _____

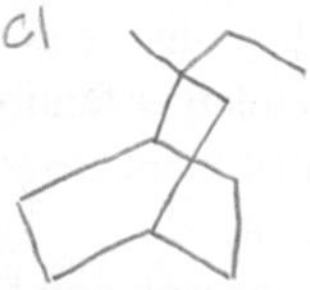

7.61 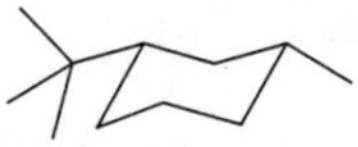   Answer: _____

7.62 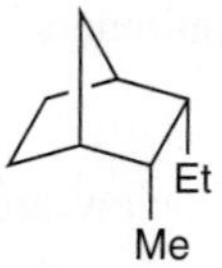   Answer: _____

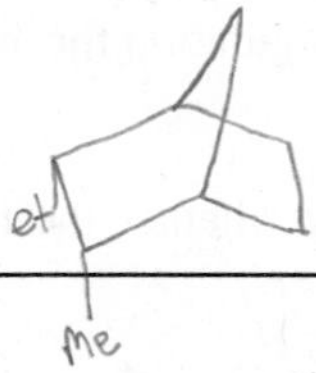

7.63 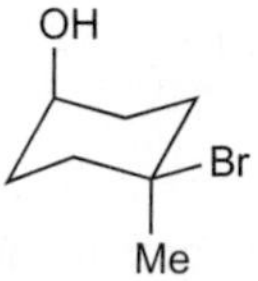   Answer: _____

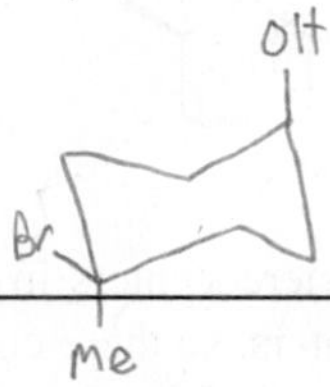

DIASTEREOMERS

In all of our examples so far, we have been comparing two compounds that are mirror images. For them to be mirror images, they need to have different configurations for every single stereocenter. Remember that our first method for drawing enantiomers was to switch all wedges with dashes. For the two compounds to be enantiomers, every stereocenter had to be inverted. But what happens if we have many stereocenters and we only invert some of them?

Let's start off with a simple case where we only have two stereocenters. Consider the two compounds below:

We can clearly see that they are not the same compound. In other words, they are nonsuperimposable. But, they are *not* mirror images of each other. The top stereocenter has the same configuration in both compounds. If they are not mirror images, then they are not enantiomers. So what is their relationship? They are called diastereomers. Diastereomers are any compounds that are nonsuperimposable stereoisomers that are not mirror images of each other.

We use the term "diastereomer" very much like we used the term "enantiomer" (remember the brother analogy). One compound is called the diastereomer of the other, and you can have a group of diastereomers. When we were talking about enantiomers, we saw that they always come in pairs, never more than two. But diastereomers can form a much larger family. We can have 100 compounds that are all diastereomers of each other (if there are enough stereocenters to allow for that many permutations of the stereocenters).

E/Z isomers (or *cis/trans* isomers) fall under this category. They are called diastereomers, because they are stereoisomers that are not mirror images of each other:

If you are given two stereoisomers, you should be able to determine whether they are enantiomers or diastereomers. All you need to look at are the stereocenters. They must all be of different configuration for the compounds to be enantiomers.

EXERCISE 7.64 Identify whether the two compounds shown below are enantiomers or diastereomers:

Answer There are two stereocenters in each compound. The configurations are different for both stereocenters, so these compounds are enantiomers. In fact, if you were given the first compound only, you could have drawn the enantiomer by using the first method—switching all wedges and dashes.

PROBLEMS   For each pair compounds below, determine whether the pair are enantiomers or diastereomers.

7.65

Answer: ___enantiomers___

7.66

Answer: ___diastereomers___

7.67

Answer: ___enantiomers___

7.68

Answer: ___enantiomers___

7.69

Answer: ___diastereomers___

7.70

Answer: ___diastereomers___

7.6 MESO COMPOUNDS

This is a topic that notoriously confuses students, so let's start off with an analogy. Let's use the analogy of the twin brothers who look identical except for one feature: one of them has a mole on the left side of his face, and the other has a mole on the right side of his face. You can tell them apart based on the mole, and the brothers are mirror images of each other. Imagine that their parents had other sets of twins, lots of sets of twins. So, all in all, there are a lot of siblings (who are all brothers and sisters of each other), but they are paired up, 2 in a group. Each child has *only one* twin sibling, who is his or her mirror image. Now imagine that the parents, out of nowhere, have one more child who is born without a twin—just a regular one-baby birth. When you look at this family, you would see a lot of sets of twins, and then one child who has no twin (and has two moles—one on each side of his face). You might ask that child, where is your twin? Where is your mirror image? He would answer: I don't have a twin. I am the mirror image of myself. That's why the family has an odd number of children, instead of an even number.

The analogy goes like this: when you have a lot of stereocenters in a compound, there will be many stereoisomers (brothers and sisters). But, they will be paired up into sets of enantiomers (twins). Any one molecule will have many, many diastereomers (brothers and sisters), but it will have only one enantiomer (its mirror image twin). For example, consider the following compound:

This compound has five stereocenters, so it will have many diastereomers (compounds where only some of the wedges have been inverted). There are many, many possible compounds that fit that description, so this compound will have many brothers and sisters. But this compound will only have one twin—only one enantiomer (there is *only one* mirror image of the compound above):

It is possible for a compound to be its own mirror image. In such a case, the compound will not have a twin. It will be all by itself, and the total number of stereoisomers will be an odd number, rather than an even number. That one lonely compound is called a *meso* compound. If you try to draw the enantiomer (using either of the methods we have seen), you will find that your drawing will be the same compound as what you started with.

So how do you know if you have a *meso* compound?

A *meso* compound has stereocenters, but the compound also has symmetry that allows it to be the mirror image of itself. Consider *cis*-1,2-dimethylcyclohexane as an example. This molecule has a plane of symmetry cutting the molecule in half. Everything on the left side of the plane is mirrored by everything on the right side:

If a molecule has an internal plane of symmetry, then it is a *meso* compound. If you try to draw the enantiomer (using either one of the two methods we saw), you will find that you are drawing the same thing again. This molecule does not have a twin. It is its own mirror image:

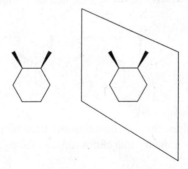

So, to be *meso*, the compound needs to be the same as its mirror image. We have seen that this can happen when we have an internal plane of symmetry. It can also happen when the compound has a center of inversion. For example,

This compound does not possess a plane of symmetry, but it does have a center of inversion. If we invert everything around the center of the molecule, we regenerate the same thing. Therefore, this compound will be superimposable on its mirror image, and the compound is *meso*. You will rarely see an example like this one, but it is not correct to say that the plane of symmetry is the only symmetry element that makes a compound *meso*. In fact, there is a whole class of symmetry elements (to which the plane of symmetry and center of inversion belong) called S_n axes, but we will not get into this, because it is beyond the scope of the course. For our purposes, it is enough to look for planes of symmetry.

There is one fail-safe way to tell if a compound is a *meso* compound: simply draw what you think should be the enantiomer and see if you can rotate the new drawing in any way to superimpose on the original drawing. If you can, then the compound will be *meso*. If not, then your second drawing is the enantiomer of the original compound.

EXERCISE 7.71 Is the following a *meso* compound?

Answer We need to try to draw the mirror image and see if it is just the same compound redrawn. If we use the second method for drawing enantiomers (placing the mirror on the side), then we will be able to see that the compound we would draw is the same thing:

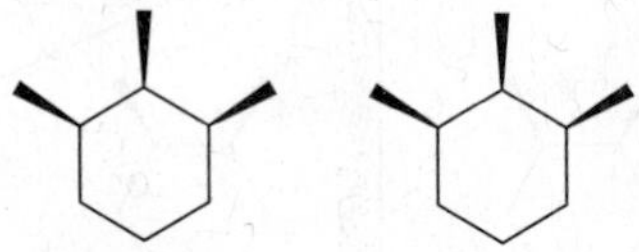

Therefore, it is a *meso* compound.

A simpler way to draw the conclusion would be to recognize that the molecule has an internal plane of symmetry that chops through the center of one of the methyl groups:

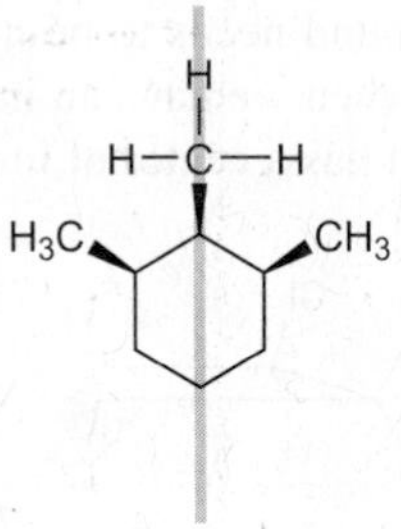

PROBLEMS Identify which of the following compounds is a *meso* compound.

7.72   *meso*

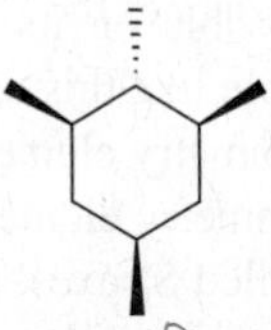

7.73   *not meso*

7.74   *meso*

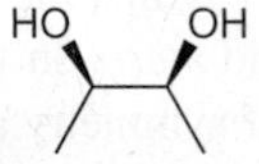

7.7 DRAWING FISCHER PROJECTIONS

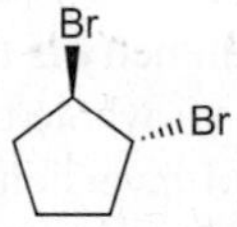

There is an entirely different way to draw stereocenters (instead of using regular bond-line drawings with dashes and wedges). Fischer projections are helpful for drawing molecules that have many stereocenters, one after another. These drawings look like this:

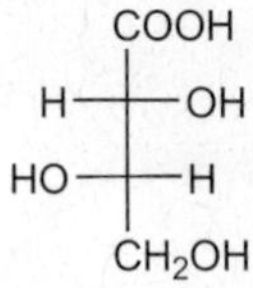

	COOH	
H—	—OH	
HO—	—H	
	CH₂OH	

2 stereocenters

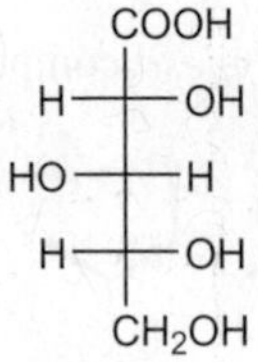

	COOH	
H—	—OH	
HO—	—H	
H—	—OH	
	CH₂OH	

3 stereocenters

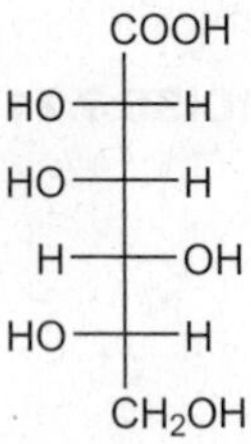

	COOH	
HO—	—H	
HO—	—H	
H—	—OH	
HO—	—H	
	CH₂OH	

4 stereocenters

First we need to understand exactly what these drawings mean, and then we will learn a step-by-step method for drawing them properly.

Using Fischer projections saves time because we don't have to draw all of the dashes and wedges for each stereocenter. Instead, we draw only straight lines, with the idea that all horizontal lines are coming out at us and all vertical lines are going away from us. Let's see exactly how this works. Consider the following molecule, which is drawn in a zigzag format (R_1 and R_2 represent groups whose identities are not being defined yet, because it does not matter for now):

Remember that all of the single bonds are all freely rotating, so there are a large number of conformations that the molecule can assume. When we rotate a single bond, the dashes and wedges change, but this is *not* because the configuration has changed. Remember that configurations do not change when a molecule twists and bends. Watch what happens when we rotate one of the single bonds:

Notice that R_1 is now pointing straight down, and the OH is now on a dash. The configuration has *not* changed. If you need to convince yourself of this, determine the configuration of that stereocenter in each of two drawings. You will see that it has not changed.

Now let's draw another of the possible conformations for this molecule. If we rotate a couple more single bonds until the carbon skeleton is looping around like a bracelet, we get the following conformation:

The molecule is twisting and bending around all of the time, and the conformation with the bracelet-shaped skeleton is just one of the possible conformations. The molecule probably spends very little of its time like this (it is a relatively high energy conformation), but this is the conformation that we will use to draw our Fischer projection.

Now imagine piercing a pencil through R_1 and R_2 (the pencil is represented by the dotted line below). If you grab the ends of the pencil and rotate, you will find that

R_1 and R_2 will stay in the page, but the rest of the molecule will pop out in front of the page:

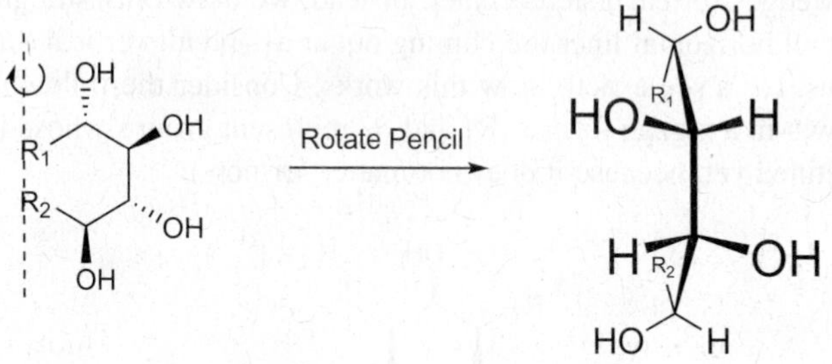

Now we imagine flattening the skeleton into a straight vertical line, and we redraw the molecule using only straight lines for the groups:

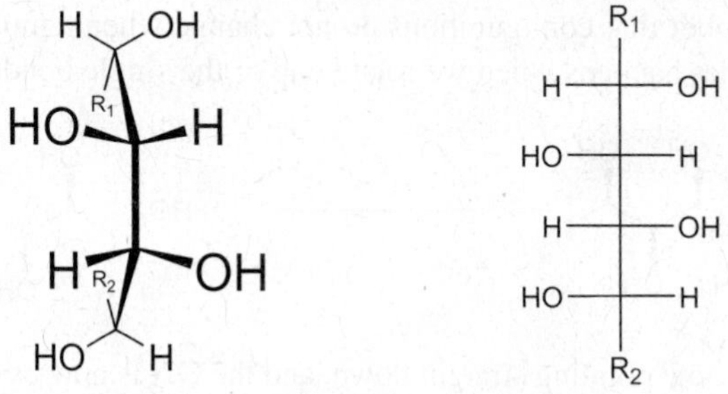

This is our Fischer projection. All of the configurations can be seen on this drawing, because we are able to picture in our minds what the 3D shape is. So the rule is that all horizontal lines are coming out at us, and all vertical lines are going away from us:

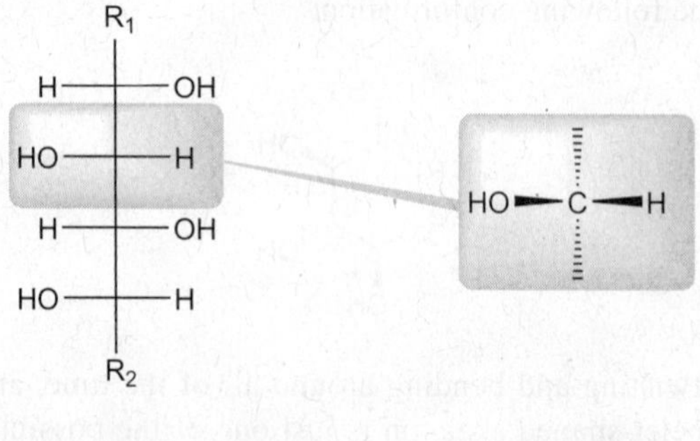

You might be wondering how you would determine the configuration of a stereocenter when you are given a Fischer projection. If each stereocenter is drawn as two wedges and two dashes, how do you figure out how to look at the stereocenter? The answer is simple. Just choose one dash and one wedge, and draw them as straight lines. It doesn't matter which ones you choose—you will get the answer right regardless:

$$
\begin{array}{ccccccc}
& CH_3 & & CH_3 & & CH_3 & & CH_3 \\
HO\!-\!\!\!\!\raisebox{0pt}{$\vert$}\!\!\!\!-H & = & HO\!\!\blacktriangleright\!C\!\blacktriangleleft\!H & = & HO\!-\!C\!\blacktriangleleft\!H & \text{or} & HO\!\!\blacktriangleright\!C\!-\!H & \text{etc.} \\
& CH_2CH_3 & & CH_2CH_3 & & CH_2CH_3 & & CH_2CH_3
\end{array}
$$

Once you have a drawing with two straight lines, one dash and one wedge, then you should be able to determine whether the stereocenter is *R* or *S*. If you cannot, then you should go back and review the section on assigning configuration.

Fischer projections can also be used for compounds with just one stereocenter, as above, but they are usually used to show compounds with multiple stereocenters. You will utilize Fischer projection heavily when you learn about carbohydrates at the end of the course.

Now we can understand why we cannot draw a Fischer projection sideways. If we did, we would be inverting the stereocenter. To draw the enantiomer of a Fischer projection, do not turn the drawing sideways. Instead, you should use the second method we saw for drawing enantiomers (place the mirror on the side of the compound and draw the reflection). Recall that this was the method that we used for drawings where wedges and dashes were implied but not shown. Fischer projections are another example of drawings that fit this criterion:

$$
\begin{array}{cc}
\begin{array}{c}
COOH \\
HO\!-\!\!\!\raisebox{0pt}{$\vert$}\!\!\!-H \\
HO\!-\!\!\!\raisebox{0pt}{$\vert$}\!\!\!-H \\
H\!-\!\!\!\raisebox{0pt}{$\vert$}\!\!\!-OH \\
HO\!-\!\!\!\raisebox{0pt}{$\vert$}\!\!\!-H \\
CH_2OH
\end{array}
&
\begin{array}{c}
COOH \\
H\!-\!\!\!\raisebox{0pt}{$\vert$}\!\!\!-OH \\
H\!-\!\!\!\raisebox{0pt}{$\vert$}\!\!\!-OH \\
HO\!-\!\!\!\raisebox{0pt}{$\vert$}\!\!\!-H \\
H\!-\!\!\!\raisebox{0pt}{$\vert$}\!\!\!-OH \\
CH_2OH
\end{array}
\end{array}
$$

Enantiomers

EXERCISE 7.75 Determine the configuration of the stereocenter below. Then draw the enantiomer.

$$
\begin{array}{c}
CH_2OH \\
H\!-\!\!\!\raisebox{0pt}{$\vert$}\!\!\!-Cl \\
Me
\end{array}
$$

Answer We begin by drawing the stereocenter as it is implied by the Fischer projection:

$$
\begin{array}{c}
CH_2OH \\
H\!\!\blacktriangleright\!C\!\blacktriangleleft\!Cl \\
Me
\end{array}
$$

Next, we choose one dash and one wedge, and we turn them into straight lines (it doesn't matter which dash or which wedge we choose):

$$\begin{array}{c} CH_2OH \\ | \\ H-C\blacktriangleleft Cl \\ \vdots \\ Me \end{array}$$

Then we assign priorities based on atomic number:

$$\begin{array}{c} ② \\ CH_2OH \\ | \\ ④H-C\blacktriangleleft Cl\ ① \\ \vdots \\ Me \\ ③ \end{array}$$

The 4 is not on a dash, so we switch it with the 3 so it can be on a dash, and we see that the configuration is *S*. Since we had to do a switch to get this, the configuration of the original stereocenter (before the switch) was *R*:

therefore:

Now, we need to draw the enantiomer. For Fischer projections, we use the method where we place a mirror on the side, and then we draw the reflection:

$$\begin{array}{c} CH_2OH \\ H-\!\!\!\!+\!\!\!\!-Cl \\ Me \end{array} \qquad \begin{array}{c} CH_2OH \\ Cl-\!\!\!\!+\!\!\!\!-H \\ Me \end{array}$$

Enantiomer

PROBLEMS For each compound below, determine the configuration of the stereo-center, and then draw the enantiomer.

7.76
$$\begin{array}{c} Et \\ H-\!\!\!\!+\!\!\!\!-Br \\ Me \end{array}$$

7.77
$$\begin{array}{c} CH_2OH \\ Me-\!\!\!\!+\!\!\!\!-Br \\ Et \end{array}$$

7.78

$$
\begin{array}{c}
O\diagdown^H \\
H{-}{-}OH \\
CH_2OH
\end{array}
$$

PROBLEMS For each compound below, determine the configuration of every stereocenter. Then draw the enantiomer of each compound below (the COOH group is a carboxylic acid group).

7.79

$$
\begin{array}{c}
COOH \\
H{-}{-}OH \\
HO{-}{-}H \\
CH_2OH
\end{array}
$$

7.80

$$
\begin{array}{c}
COOH \\
H{-}{-}OH \\
HO{-}{-}H \\
H{-}{-}OH \\
CH_2OH
\end{array}
$$

7.81

$$
\begin{array}{c}
COOH \\
H{-}{-}Cl \\
Br{-}{-}H \\
H{-}{-}OH \\
HO{-}{-}H \\
CH_2OH
\end{array}
$$

7.8 OPTICAL ACTIVITY

Students confuse *R/S* with +/− all of the time, so let's conclude our chapter by clearing up the difference. Compounds are chiral if they have stereocenters and they are not *meso* compounds. A chiral compound will have an enantiomer (a nonsuperimposable mirror image). An interesting thing happens when you take a chiral compound and subject it to plane-polarized light. The plane of the polarized light rotates as it passes through the sample. If this rotation is clockwise, then we say the rotation is +. If the rotation is counterclockwise, then we say the rotation is −. If we want to refer to a racemic mixture (an equal mixture of both enantiomers), we will often say (+/−) in the beginning of the name to mean that both enantiomers are present in solution (and the rotations cancel each other).

Do not confuse clockwise rotation of plane-polarized light with clockwise ordering of atomic numbers when determining configurations. They are not related. When determining configuration, we impose a set of man-made rules to help us distinguish between the two possible configurations. By using these rules, we will always be able to communicate which configuration we are referring to, and we only need one letter to communicate this (*R* or *S*) if we use the rules properly. However, +/− is totally different.

The rotation of plane polarized light (either + or −) is not a man-made convention. It is a physical effect that is measured in the lab. It is impossible to predict whether a compound will be + or − without actually going into the lab and trying. If a stereocenter is *R*, this does not mean that the compound will be +. It could just as easily be −. In fact, whether a compound is + or − will depend on temperature. So a compound can be + at one temperature and − at another temperature. But clearly, temperature has nothing to do with *R* and *S*. So, don't confuse *R/S* with +/−. They are totally independent and unrelated concepts.

You will never be expected to look at a compound that you have never seen and then predict in which direction it will rotate plane-polarized light (unless you know how the enantiomer rotates plane-polarized light, because enantiomers have opposite effects). But you will be expected to assign configurations (*R* and *S*) for stereocenters in compounds that you have never seen.

CHAPTER 8

MECHANISMS

Mechanisms are your key to success in this course. If you can master the mechanisms, you will do very well in this class. If you don't master mechanisms, you will do poorly in this class. What are mechanisms and why are they so important?

When two compounds react with each other to form new and different products, we try to understand *how* the reaction occurred. Every reaction involves the flow of electron density—electrons move to break bonds and form new bonds. Mechanisms illustrate how the electrons move during a reaction. The flow of electrons is shown with curved arrows; for example,

These arrows show us how the reaction took place. For most of the reactions that you will see this semester, the mechanisms are well understood (although there are some reactions whose mechanisms are still being debated today). You should think of a mechanism as "bookkeeping of electrons." Just as an accountant will do the bookkeeping of a company's cash flow (money coming in and money going out), the mechanism of a reaction is the bookkeeping of the flow of electrons.

When you understand a mechanism, you will understand why the reaction took place, why the stereocenters turned out the way they did, and so on. If you do not understand the mechanism, then you will find yourself memorizing the exact details of every single reaction. Unless you have a photographic memory, that will be a very difficult challenge. By understanding mechanisms, you will be able to make more sense of the course content, and you will be able to better organize all of the reactions in your mind.

The mechanisms that you will learn in the first half of your course are the most critical ones. This is the time when you will either master arrow pushing and mechanisms or you will not master them. If you don't, you will struggle with all mechanisms in the rest of the course, which will turn your organic chemistry experience into a nightmare. It is absolutely critical that you master the mechanisms for the early reactions that you cover. That way, you will have the tools that you need to understand all of the other mechanisms in your course.

In this chapter, we will *not* learn every mechanism that you need to know. Rather, we will focus on the tools that you need to properly read a mechanism and abstract the important information. You will learn some of the basic ideas behind arrow pushing in mechanisms, and these ideas will help you conquer the early mechanisms

165

that you will learn. The second half of this chapter provides a place for you to keep a list of mechanisms as you progress through the course. This list (which you will fill out as you go along) is arranged so that you will have the key information at your fingertips, and you will be able to use the list as a study guide for your exams.

8.1 CURVED ARROWS

We have already gotten quite a bit of experience with curved arrows in Chapter 2 (Resonance). There is one very major difference between curved arrows for drawing resonance structures and curved arrows for drawing mechanisms. With resonance structures, we saw that the electrons were not really moving at all. We were pretending that they were moving so that we could draw all of the resonance structures. By contrast, the curved arrows that we use in mechanisms refer to the actual *movement of electrons*. Electrons are moving to break and form bonds (hence the term *chemical reaction*). Why are we stressing this difference? We first need to understand what arrows represent before we can move on to the rules of pushing arrows.

When we learned how to draw resonance structures, we saw two commandments that we must not violate: (1) never break a single bond, and (2) never exceed an octet for second-row elements. When drawing mechanisms, we are trying to understand where the electrons actually moved to break and form bonds. Therefore, it is OK to break single bonds. In fact, it happens in almost every reaction. So when drawing mechanisms there is only one commandment to follow: never exceed an octet for second-row elements.

Now that we have some of the ground rules down, let's just have a quick review of curved arrows, and the different types of arrows that you can draw. Every curved arrow has a *head* and a *tail*. It is essential that the head and tail of every arrow be drawn in precisely the proper place. *The tail shows where the electrons are coming from, and the head shows where the electrons are going:*

Tail · · · · · · · · · Head

Therefore, there are only two things that you have to get right when drawing each arrow. The tail needs to be in the right place and the head needs to be in the right place. Remember that electrons exist in orbitals, either as lone pairs or as bonds. So the *tail* of an arrow can only come *from a bond* or *from a lone pair:* The *head* of an arrow can only be drawn to *make a bond* or to *make a lone pair:* In total, this gives us four possibilities:

1. Lone pair → bond
2. Bond → lone pair
3. Bond → bond
4. Lone pair → lone pair

The last possibility does not work, because we cannot push electrons from one lone pair to another (at least not in one step). So we only have to consider the first three possibilities. Every arrow you will see will belong to one of these three categories, so let's see examples of each of the three categories.

From a Lone Pair to a Bond

Consider the following step, in which a single bond is formed:

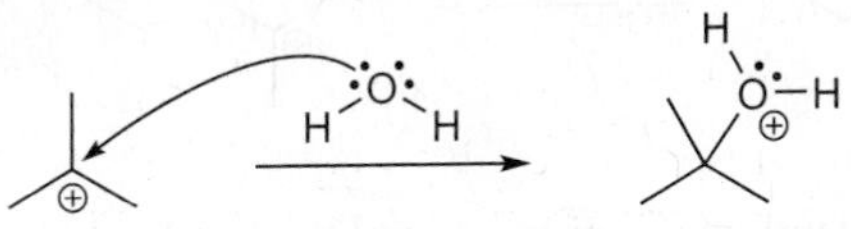

The tail of the arrow is coming from a lone pair on the oxygen atom, and the head of the arrow is going to form a bond between oxygen and carbon. Since the head of the arrow is placed on an atom, it might *seem like* the electrons are going from a lone pair to a lone pair, but they are not. The electrons are going from the oxygen lone pair to form a bond to the carbon atom. If this makes you unhappy, there is an alternative way of drawing the arrow that shows it more clearly:

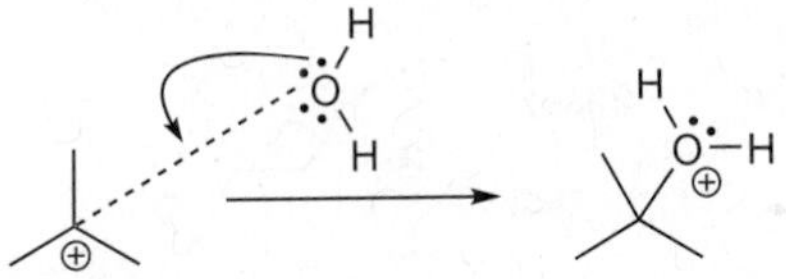

The dotted line shows the bond that is about to form, and we draw the arrow to that dotted line. In this drawing it is very clear that the head of the arrow is going to form a bond. When you see an arrow drawn the first way (where it looks like it is going to an atom rather than to form a bond), don't be confused by this—it really is just going to form a bond.

From a Bond to a Lone Pair

Now consider the following step, in which a single bond is broken:

The tail of the arrow is on a bond, and the head of the arrow is forming a lone pair on the chlorine atom. The two electrons of the bond used to be shared between the carbon atom and the chlorine atoms. But now, both electrons are going on the chlorine atom. So carbon has lost an electron, and chlorine has gained one. This is why carbon ends up with a positive charge, and chlorine gets a negative charge.

By the way, a chlorine atom with a negative charge is called a chloride ion (-ide- implies the negative charge). So, in this reaction, chloride leaves to generate a carbocation (a carbon with a positive charge).

From a Bond to a Bond

Consider the first arrow in the example below, where we are using the electrons of the pi bond to attack a proton (H^+), and expelling Cl^- in the process:

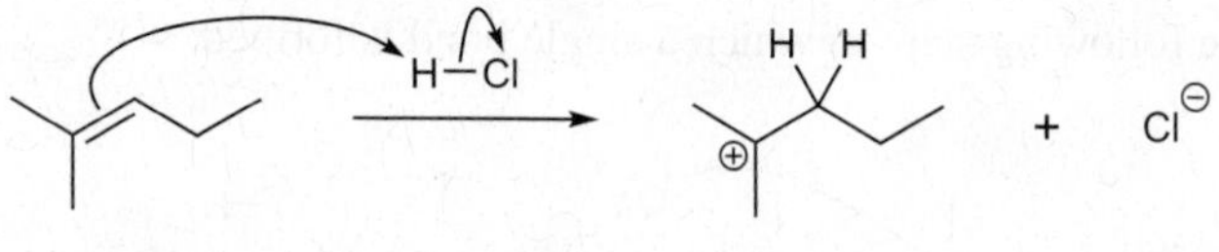

The first arrow has its tail on the pi bond, and the head is being used to form a bond between a carbon atom and the proton.

You will notice in the example above that there are two arrows. The first arrow is going from a bond to a bond. But the second arrow is going from a bond to form a lone pair. So we see that you can have more than one type of arrow together in one step of a mechanism.

In fact, it is possible to have all three types of arrows in one step of a mechanism. Consider the example below:

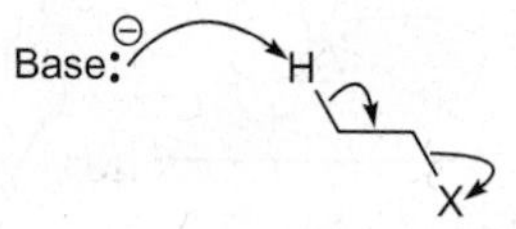

Notice that there is one long flow of electron density, illustrated with three arrows. We begin at the tail of the arrow on the base, because that is where the flow starts. This arrow is going from a lone pair to form a bond. The second arrow goes from a bond to form a bond, and the third arrow goes from a bond to form a lone pair on X. This type of reaction is called an elimination reaction, because we are *eliminating* H^+ and X^- to form a double bond:

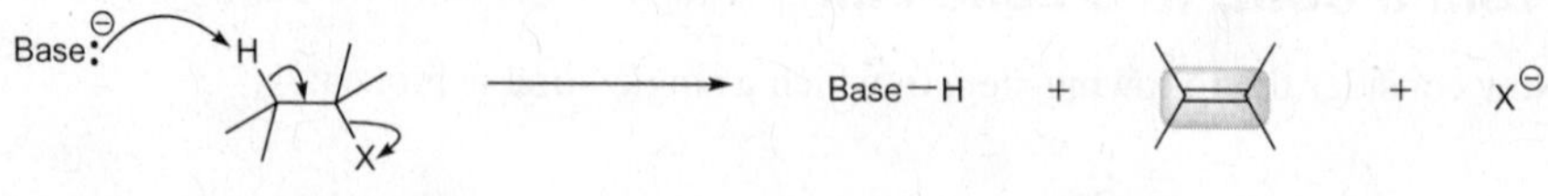

Notice that the arrows are all going from one end of the molecule to the other. *Never* draw arrows going in opposite directions. That would not make any sense! To see what we mean by this, consider the example below:

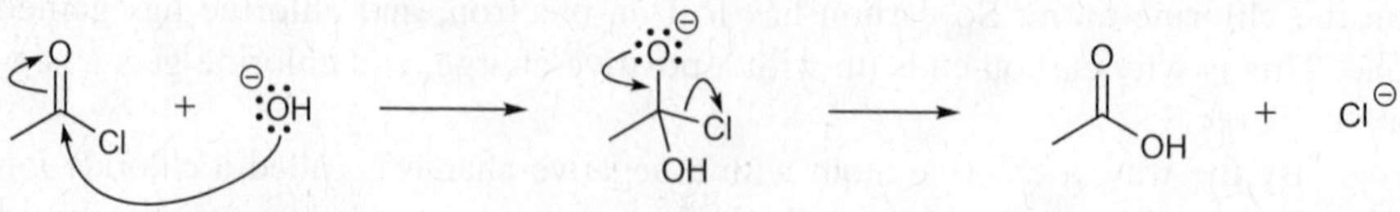

This type of reaction will be covered much later on in your course, but let's use it now as an example. In the first step, we have two arrows: from a lone pair to form a bond, and then from a bond to form a lone pair:

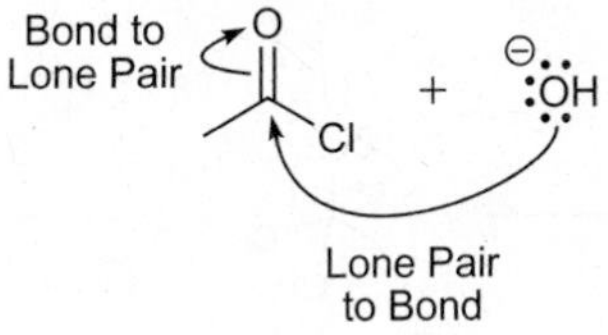

In the second step of the mechanism, we also have two arrows: from a lone pair to form a bond, and then from a bond to form a lone pair:

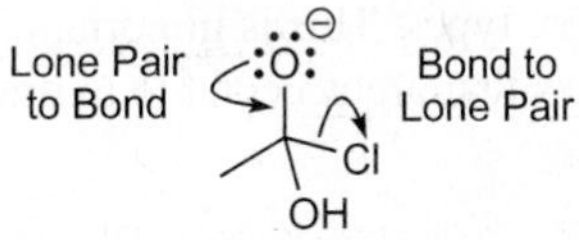

If we consider the overall reaction, we notice that the HO⁻ is replacing the Cl. If we look at how the electrons flowed, we see that it all started at the negative charge of the attacking HO⁻. This charge flowed up temporarily on to the oxygen atom of the C=O in step 1 of the mechanism, and then the charge flowed back down to expel Cl⁻:

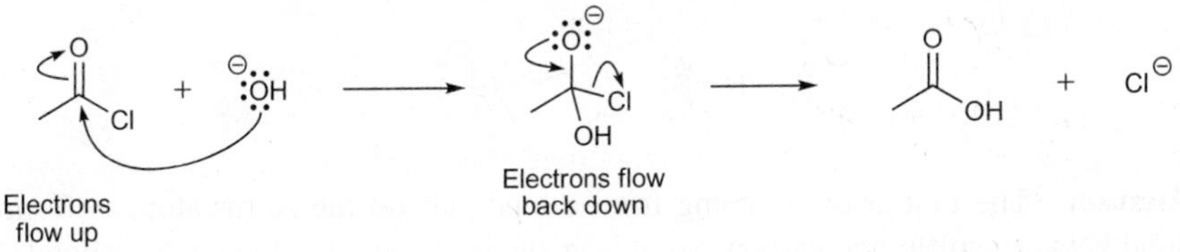

When we consider how the charge flowed throughout the whole reaction, it might be tempting to draw it all in one step, like this:

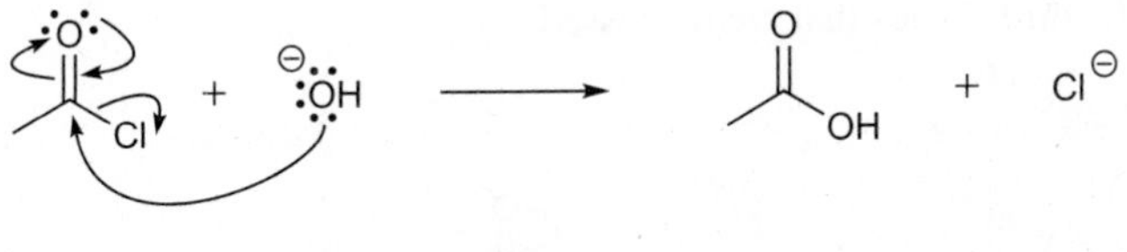

However, this is no good, because we have two arrows going in opposite directions:

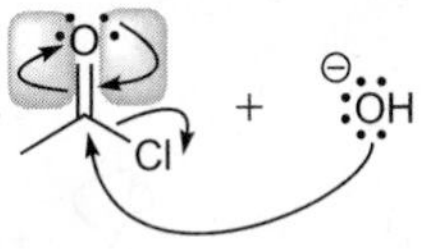

Never draw arrows in opposite directions. That would imply that the electrons were flowing in opposite directions *at the same time.* That is not possible. In this reaction, the electrons first flowed up, and then they flowed back down. So we have to draw it as two steps:

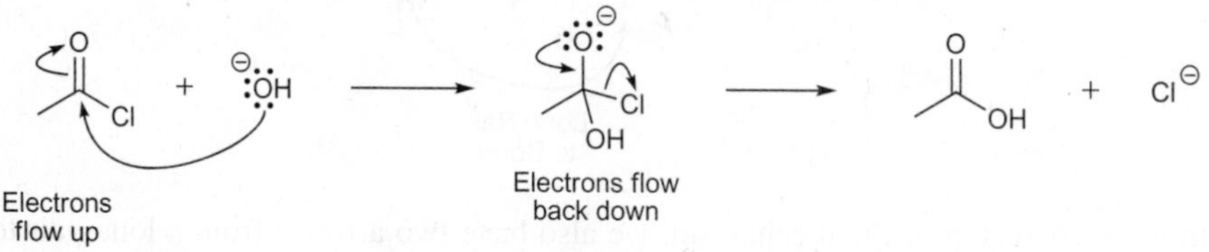

Electrons flow up

Electrons flow back down

Before we can practice drawing arrows, we first need to make sure that we can identify the three different arrow types. This is important, because it will get you accustomed to the types of arrows that are acceptable to draw.

EXERCISE 8.1 In the example below, classify each arrow that you see into one of the following three types:

1. Bond → bond
2. Bond → lone pair
3. Or lone pair → bond

Answer The first arrow is going from a lone pair on the sulfur atom to form a bond between sulfur and carbon. So, this arrow is of the type: lone pair → bond.

The second arrow is going from a bond to form a lone pair, so the second arrow is of the type: bond → lone pair.

PROBLEMS For each of the following examples, classify each arrow that you see into one of the three types that we discussed.

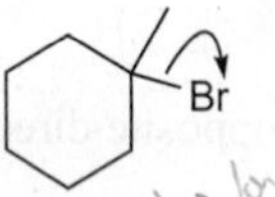

8.2

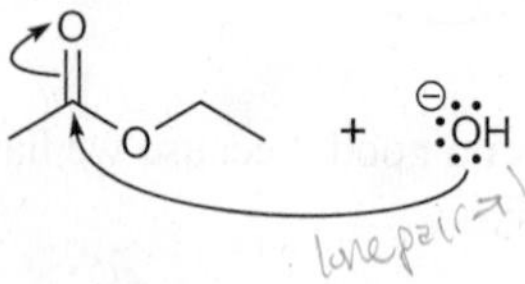

8.3

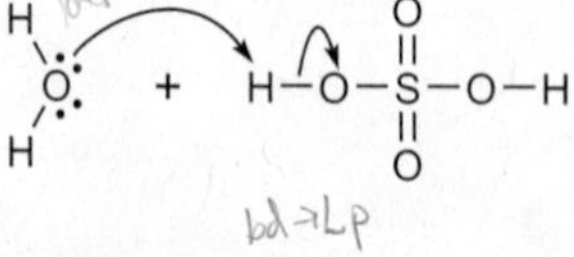

8.4

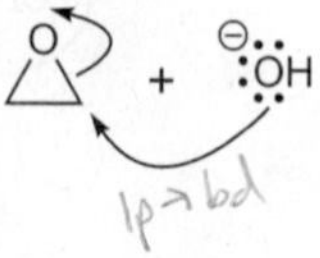

8.5

8.6   **8.7**

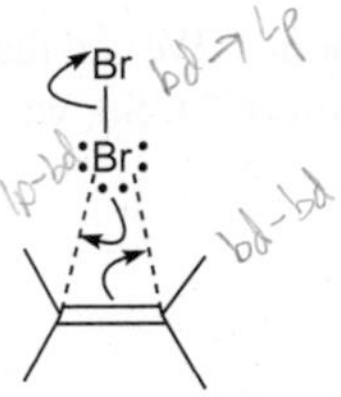

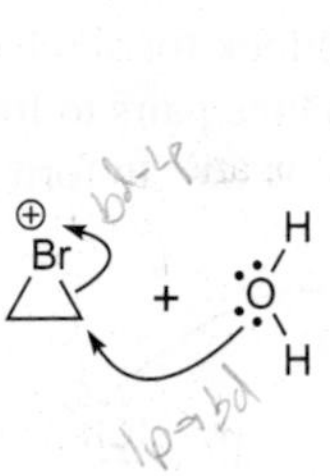

8.2 ARROW PUSHING

Now that we know what kinds of arrows are acceptable, we can begin to practice drawing them (or "pushing" them, as its called). To do this, we need to learn how to analyze a step in a mechanism, and train our eyes to look for all of the lone pairs and all of the bonds. We have said that all arrows are coming from or going to either lone pairs or bonds. So it makes sense that we need to be able to look at a step in a mechanism and determine which bonds have changed and which lone pairs have changed. Let's see this in an example.

EXERCISE 8.8 Complete the mechanism of the following reaction by drawing the proper arrows in each step:

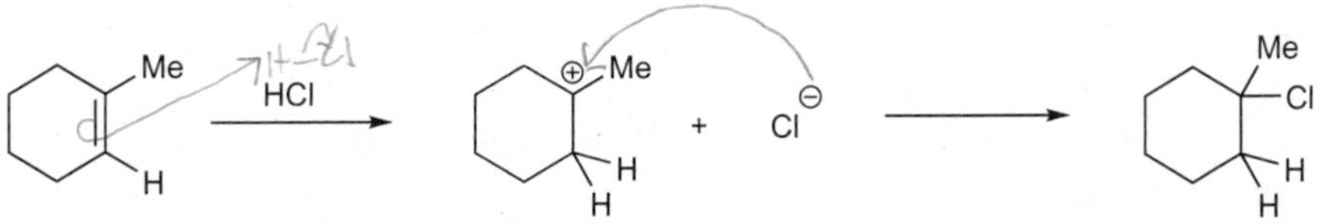

Answer We need to look for all changes for bonds or lone pairs. In the first step, the double bond is disappearing, one of the carbon atoms of the double bond is forming a new bond to a proton (H$^+$), and we are breaking the H—Cl bond to expel Cl$^-$. So we have broken two bonds (C=C, and H—Cl) and we have formed one bond (C—H) and one extra lone pair (on Cl). Therefore, we will need two arrows to make this happen. Where do we start?

Keep in mind that electron density always flows in one direction. In this example we can see which direction the flow went, because a positive charge was formed on the carbon atom, and a negative charge developed on chlorine. We can use that information to figure out the direction of the flow. The first arrow needs to show the double bond going to form a bond to the proton (from a bond to a bond) and then we need a second arrow to show the bond from the H—Cl going to form a negative charge on the chlorine atom (from a bond to a lone pair):

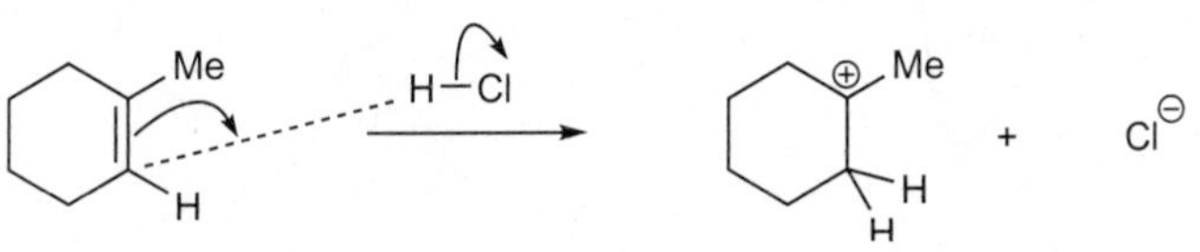

In the next step, again we look for all changes to lone pairs or bonds. We see that the Cl is giving up one of its lone pairs to form a bond with a carbon (C$^+$). So, we need only one arrow, from a lone pair to form a bond:

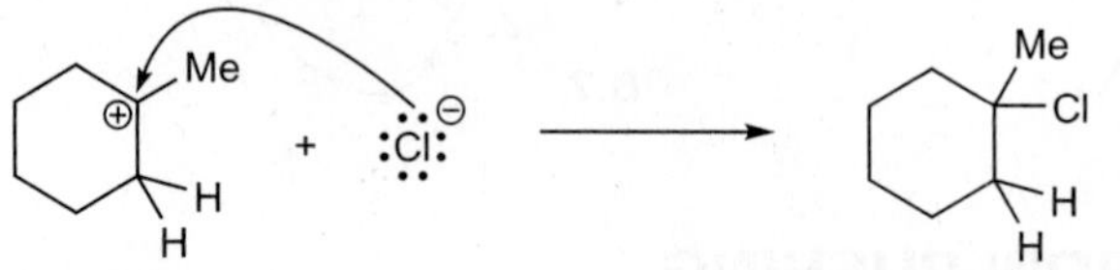

PROBLEMS   For each transformation below, complete the mechanism by drawing the proper arrows.

8.9

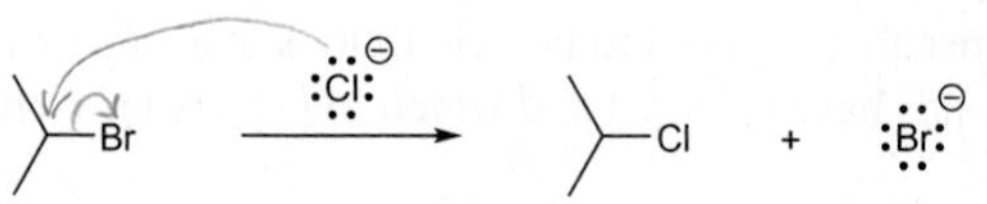

8.10

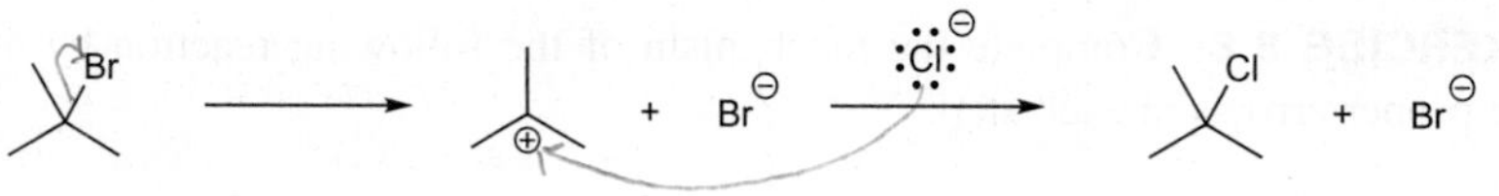

8.11

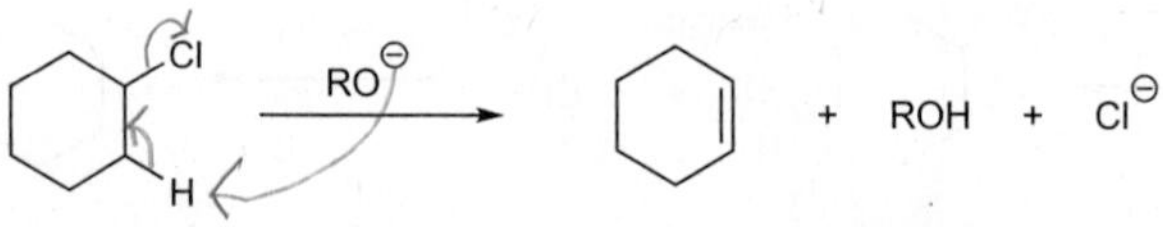

8.12

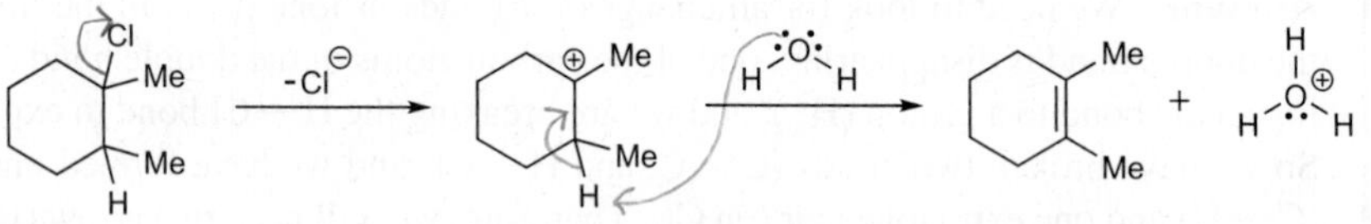

Consider the second step of problem 8.12. A lone pair from the oxygen atom is removing a proton to form a double bond:

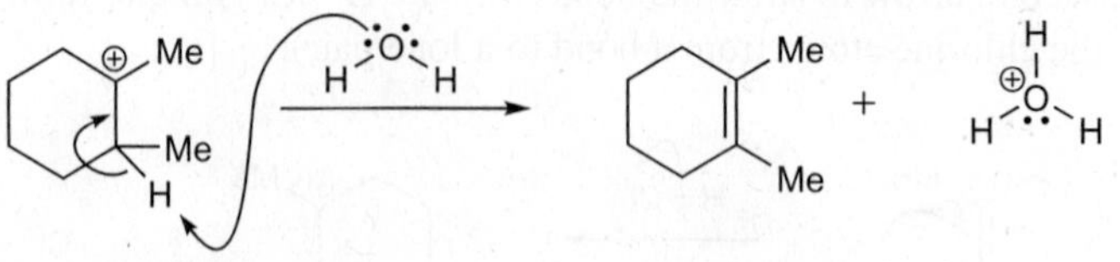

Remember that arrows indicate the flow of electrons. Arrows do not show where atoms went. Many students will accidentally draw it like this:

Students often make this mistake because they want to show where the H is going. But this is wrong. Remember that arrows show the *movement of electrons, not atoms.* The H was able to move only because the electrons came from the oxygen atom and grabbed the H.

8.3 DRAWING INTERMEDIATES

We have seen the different types of arrows and how to draw them. Now we need to get practice drawing intermediates when we are given the arrows. Intermediates are structures that exist for a very short time before reacting further. Let's consider an analogy. Imagine that you are trying to climb a mountain and it is very cold (below freezing). You are wearing a hat that keeps your ears warm, but it is loose and keeps slipping off. Your friend offers you a spare hat that he brought, and you borrow it. Now you need to take your old hat off to replace it with the new hat. If someone were to take a picture of you while you have nothing on your head, the picture would look very strange. There you are, in the freezing cold, with no hat on. You were only like that for 3 seconds, but it was long enough for someone to take a picture. Intermediates of reactions are similar.

Intermediates are intermediate structures in going from the starting material to the product. They do not live for very long, and it is rare that you can isolate one and store it in a bottle, but they do exist for very short periods of time. Their structures are often critical in understanding the next step of the reaction. Going back to the analogy, if I saw the picture of you without your hat on, and I knew how cold it was on that mountain, then I would have been able to predict that you put on a hat right after the picture was taken. I would have known this because I would have been able to immediately identify an uncomfortable situation, and I could have predicted what resolution must have taken place to alleviate the problem. The same is true of intermediates. If we can look at an intermediate and determine which part of the intermediate is unstable, and we also know what options are available to alleviate the instability, then we can predict the products of the reaction based on an analysis of the intermediate. That's why they are so important.

So let's get practice drawing intermediates. If you look closely at any step of a mechanism, you will see that the arrows tell you exactly how to draw the intermediate. Since you know how to classify every arrow into one of three categories (previous section of this chapter), now you will be able to read each arrow as a road map of how to draw the intermediate. Here's an example:

Let's read the arrows. The first arrow is from a lone pair to form a bond. The arrow shows electrons in a lone pair on a nucleophile (a structure that is electron rich) forming a bond with a carbon atom. The second arrow is from a bond to a bond. The third arrow goes from a bond to form a lone pair. All in all, these arrows serve as a road map for drawing the intermediate:

The trickiest part is getting the formal charges correct. If you have trouble assigning formal charges, then you will need to go back and review the sections on formal charges in Chapter 1 and Chapter 2 of this book. Assigning formal charges is a very important part of drawing the intermediates. Drawing the structure without the charges would be like taking the photograph in the analogy above, but digitally removing all of the snow. Without the snow, I wouldn't know that it was cold, so I would not be able to predict that you put a hat on shortly after the picture was taken. If you don't draw the source of instability on the intermediate, then what good is it?

One trick will help you in some situations when you have a flow of electrons represented by a few arrows (as in the example above). Notice that the only change in formal charges comes on the first and last atom of the system where the electrons are flowing. In our example above, the nucleophile loses its negative charge by using its lone pair to form a bond with a carbon atom. At the other end of the system, oxygen is gaining a negative charge as a bond is converted into a lone pair on oxygen. Notice the conservation of charge. If the overall charge is negative at the beginning of the reaction, then it must also be negative at the end of the reaction. If something starts off with no charge, then it can split up into a positive charge and a negative charge, because the *total* charge is still conserved.

EXERCISE 8.13   Look at the arrows below, and draw the intermediate that you get after pushing the arrows:

Answer   We need to read the arrows like a road map: the first arrow is going from a lone pair on HO⁻ to form a bond with the carbon atom of the C=O. The second arrow goes from the C=O bond to form a lone pair on oxygen. We use this info to draw the resulting intermediate:

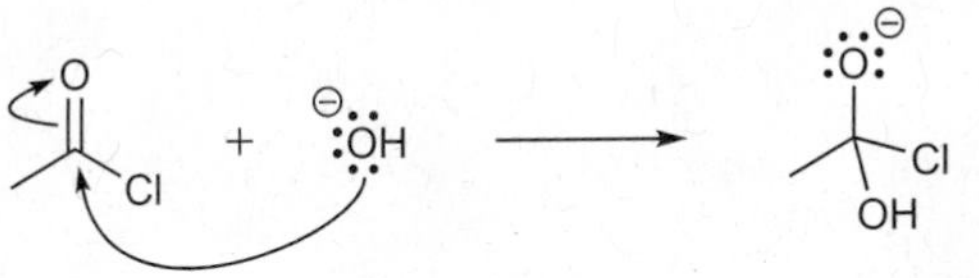

The hard part was assigning formal charges. Notice that we had two arrows moving in a flow. We had a negative charge in the beginning, so we must have a negative charge in the end. It started off on the first atom in the flow of arrows, and it ended on the last atom of the flow (oxygen).

PROBLEMS   For each problem below, draw the intermediate that would result from pushing the curved arrows as shown.

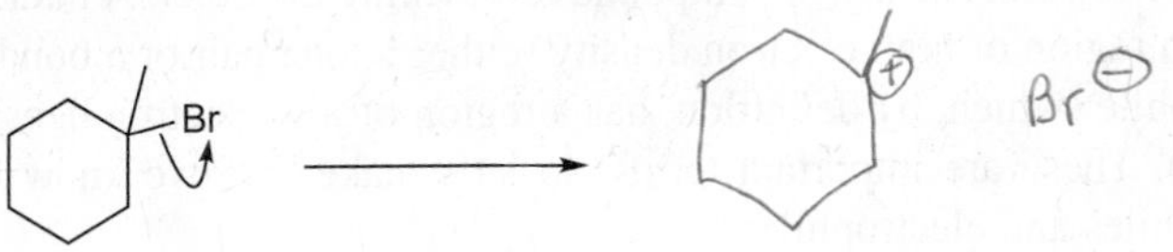

8.14

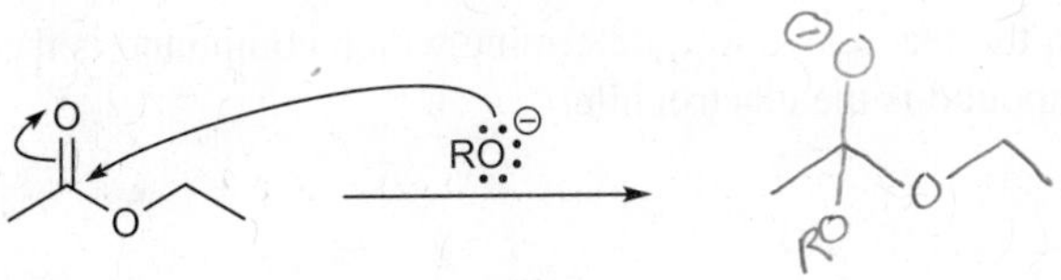

8.15

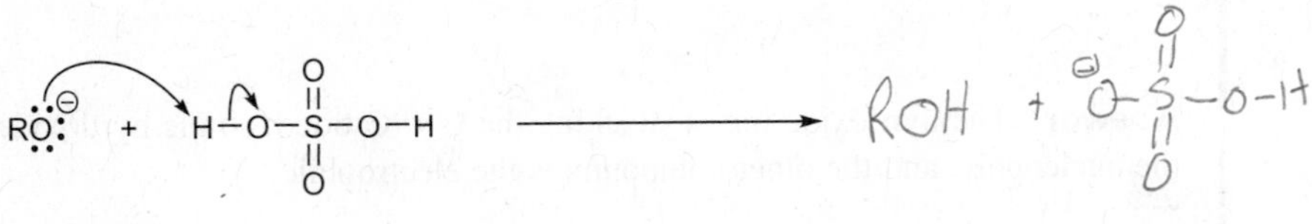

8.16

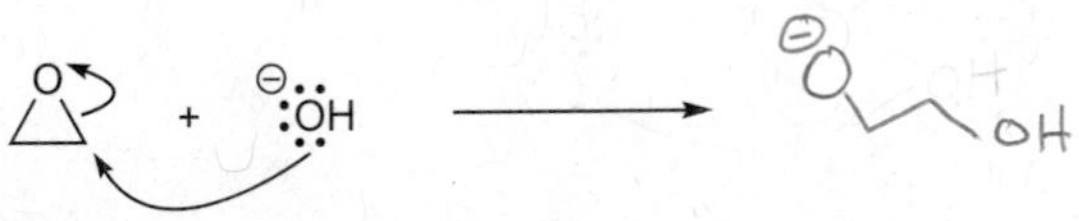

8.17

8.18

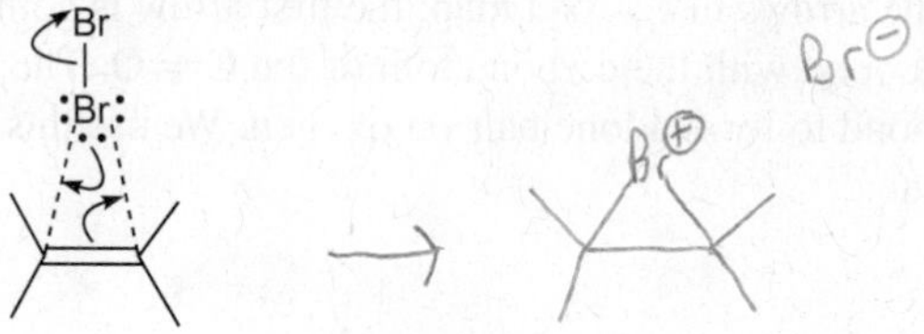

8.19

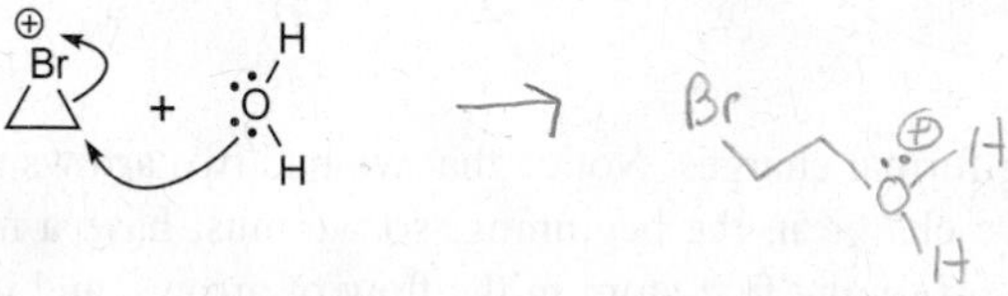

8.4 NUCLEOPHILES AND ELECTROPHILES

Whenever one compound uses its electrons to attack another compound, we call the attacker a *nucleophile,* and we call the compound being attacked an *electrophile.* It is very simple to tell the difference between an electrophile and a nucleophile. You just look at the arrows and see which compound is attacking the other. A nucleophile will always use a region of high electron density (either a lone pair or a bond) to attack the electrophile (which, by definition, has a region of low electron density that can be attacked). These are important terms, so let's make sure we know how to identify nucleophiles and electrophiles.

EXERCISE 8.20 In the reaction below, determine which compound is the nucleophile and which compound is the electrophile:

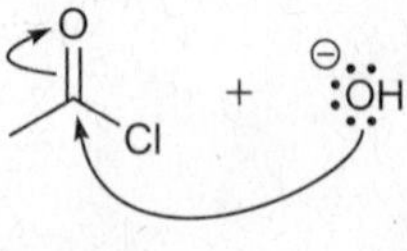

Answer The hyrdoxide ion is attacking the C=O bond, so the hyrdoxide ion is the nucleophile and the other compound is the electrophile:

Electrophile   Nucleophile

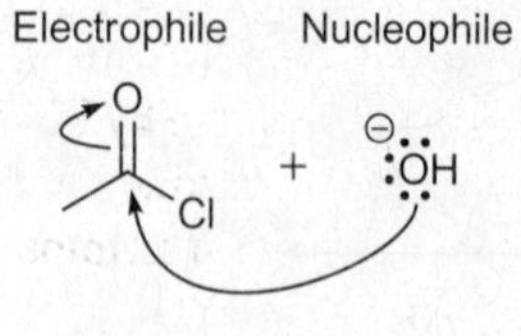

PROBLEMS   In each of the following steps, identify the nucleophile and electrophile.

8.21

electro

:ÖH
nucleo

OH

8.22

electro
Br

H
:Ö
H
nucleo

8.23

nucleo
:Ö
H   H
elect

8.24

nucleo
:Ö:

Cl electro

8.5   BASES VERSUS NUCLEOPHILES

Students are often unclear about the difference between nucleophiles and bases. Since most mechanisms involve the use of nucleophiles and bases, it will be worth our time to clear up the difference.

Consider the hydroxide ion (HO⁻). Sometimes it acts like a base and removes a proton from another compound:

HÖ: H

X

H_2O +   + $X^{\ominus}$

At other times it acts like a nucleophile and attacks another compound (forming a new bond to an atom in that compound):

HÖ: Cl

OH   + $Cl^{\ominus}$

The difference between basicity and nucleophilicity is a difference of *function*. In other words, the hydroxide ion can function in two ways: as a base (which means it is pulling off a proton and then running away with that proton) or as a nucleophile (latching onto a compound). In some cases, the hydroxide ion might function mostly as a base; while in other situations, the hydroxide ion might function mostly as a nucleophile. To understand mechanisms well, it is important to be able to distinguish between the two roles. Let's see an example.

EXERCISE 8.25 Below you will find the first two steps of a mechanism. In each step, determine whether the hydroxide ion is functioning as a nucleophile or as a base:

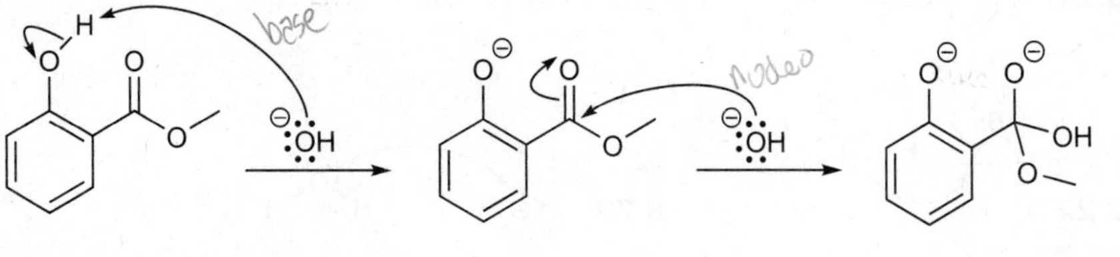

Answer In the first step, the hydroxide ion is removing a proton, so it is functioning as a base. In the second step, it is attacking the $C=O$ and latching on to the compound, so it is functioning as a nucleophile.

PROBLEMS In each step below, determine whether the hydroxide ion is functioning as a nucleophile or as a base.

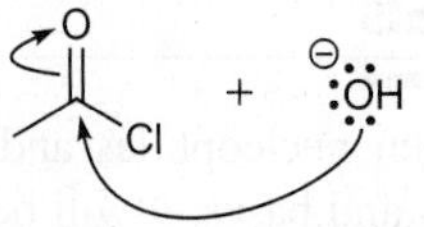

8.26

Answer: _nucleo_

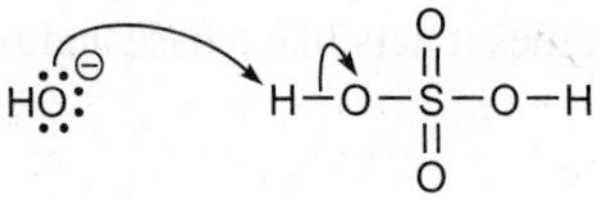

8.27

Answer: _base_

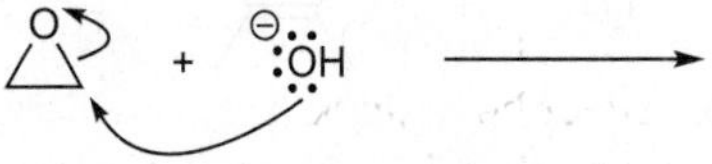

8.28

Answer: _nucleo_

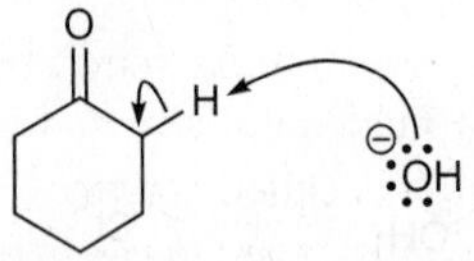

8.29

Answer: _base_

PROBLEMS In each step below, determine whether the methoxide ion (MeO⁻) is functioning as a nucleophile or as a base.

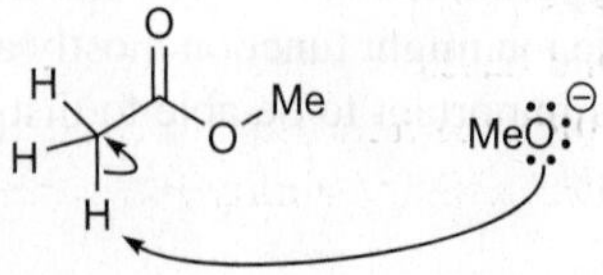

8.30

Answer: _base_

8.31   Answer: ___nucleo___

PROBLEMS   In each step below, determine whether water is functioning as a nucleophile or as a base.

8.32   Answer: ___nucleo___

8.33   Answer: ___base___

There is another subtle difference between nucleophiles and bases that is worth mentioning, because it illustrates a common theme in organic chemistry. We can see the difference by defining the terms nucleophilicity and basicity.

Once we determine that a reagent is acting as a nucleophile, we measure how fast it functions that way with the term *nucleophilicity*. Nucleophilicity measures *how quickly* a reagent will attack another compound. For example, we saw above that water can function as a nucleophile because it has lone pairs that can attack a compound. But the hydroxide ion will clearly be more nucleophilic—the hydroxide ion has a negative charge, so it will attack compounds *faster.*

Basicity measures base strength (or how unstable the base is) by *the position of equilibrium.* The term *basicity* does not reflect how quickly the equilibrium was reached. The equilibrium might have been established in a fraction of a second or it could have taken several hours. It doesn't matter, because we are not measuring speed of reaction. We are measuring stability and the position of the equilibrium.

Now we can understand the difference between nucleophilicity and basicity. Nucleophilicity measures how fast things happen, which is called *kinetics.* Basicity measures stability and the position of equilibrium, which is called *thermodynamics.* Throughout your course, you will see many reactions where the product is determined by kinetic concepts, and you will also see many reactions where the product is determined by thermodynamic concepts. In fact, there will even be times where these two factors are competing with each other and you will need to make a choice of which factor wins: kinetics or thermodynamics.

So the difference between nucleophiles and bases is a difference of function. And now we can also appreciate that nucleophilicity is a measure of a kinetic phenomenon (rate of reaction), while basicity is a measure of stability (thermodynamic phenomenon).

8.6 THE REGIOCHEMISTRY IS CONTAINED WITHIN THE MECHANISM

Regiochemistry refers to *where* the reaction takes place. In other words, in what region of the molecule is the reaction taking place? Let's see examples of this for different types of reactions. In the process, we will uncover some new terminology as we learn about different reactions.

Let's consider elimination reactions. When we eliminate H and X (where X is some leaving group that can leave with a negative charge, like Cl or Br), it is possible to form the double bond in different locations. Consider the following compound:

This compound can undergo two possible elimination reactions (to make it easier to see, we are drawing the H that gets eliminated in each case, even though we usually do not draw hydrogen atoms on bond-line drawings):

Where does the double bond form? This is a question of regiochemistry. We distinguish between these two possibilities by considering how many groups are attached to each double bond. Double bonds can have anywhere from 1 to 4 groups attached to them:

*Mono*substituted   *Di*substituted   *Tri*substituted   *Tetra*substituted

So if we look back at the reaction above, we find that the two possible products are monosubstituted and disubstituted double bonds. Whenever we have an elimination reaction where more than one possible double bond can be formed, we have names for the different products based on which one is more substituted and which one is less substituted. The more substituted product is called the Zaitsev product, and the less substituted product is called the Hoffmann product. Usually, we get the Zaitsev product, but under special circumstances we get the Hoffman product. You will learn about this in detail in your textbook when you cover elimination reactions. For now, you just need to realize that this is an issue of regiochemistry. The difference between the Zaitsev product and the Hoffman has to do with where the double bond formed. This is regiochemistry.

Let's consider another example of regiochemistry, in a completely different type of reaction. Consider the addition reaction of HCl across a double bond:

There are two possible ways to add the H and the Cl. Which product do we get?

One possibility would be to put the Cl on the less substituted carbon (carbon connected to two other carbon atoms), and the other possibility would be to put the Cl on the more substituted carbon (carbon connected to three other carbon atoms). If we put the Cl on the more substituted carbon, we call this a Markovnikov addition. If we put the Cl on the less substituted carbon, we call it an anti-Markovnikov addition. How do we know whether we get Markovnikov addition or anti-Markovnikov addition? This is an issue of regiochemistry.

For the reaction above, let's analyze the two possible outcomes. In each case, the first step involves the electrons of the double bond attacking the proton of HCl to form a carbocation (a carbon with a positive charge). The difference between the two possibilities is where the carbocation is formed:

Recall that alkyl groups are electron donating, so the carbocation on the bottom (called a tertiary carbocation because it has three alkyl groups) will be more stable than the carbocation on the top (called a secondary carbocation because it has only two alkyl groups).

Tertiary
(More stable)

Secondary
(Less stable)

Therefore, possibility 2 is a better mechanism (because it involves a more stable intermediate. If we follow the last step of the mechanism for possibility 2, we see that the Cl will attach where the carbocation is, which will be at the more substituted carbon:

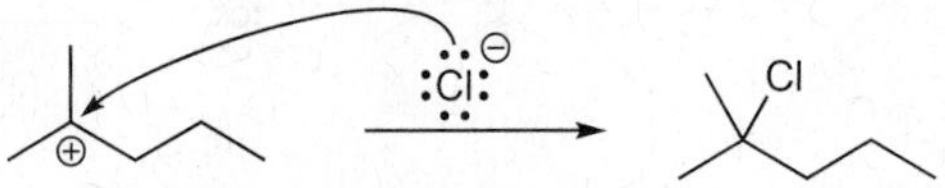

We see that the final position of chlorine is determined by the stability of the intermediate carbocation, which becomes evident as we work through the mechanism. Since chlorine ends up at the more substituted carbon, we call this a Markovnikov addition. The mechanism for this reaction helped explain the regiochemistry of the reaction.

Sometimes regiochemistry is not an issue. For example, if we are adding H and H across a double bond, then it does not matter which carbon gets the first H and which carbon gets the second H. Either way, they both end up with an H. Similarly, if we add two OH groups across a double bond, regiochemistry also does not matter. Any time we add two of the same group across a double bond, we do not have to worry about the regiochemistry.

Here is where we get back to mechanisms. Whether we are talking about Zaitsev vs. Hoffman elimination reactions or about Markovnikov vs. anti-Markovnikov addition reactions, the explanation of the regiochemistry for every reaction is contained within the mechanism. If we completely understand the mechanism, then we will understand why the regiochemistry had to be the way it turned out. By understanding the mechanism, we eliminate the need to memorize the regiochemistry for every reaction. With every reaction you encounter, you should consider the regiochemistry of the reaction and look at the mechanism for an explanation of the regiochemistry.

PROBLEMS   You will, over the course of your studies, learn mechanisms for the following reactions. In the meantime, you will be given the regiochemical information that you need to answer each of the problems below. These problems are intended to ensure that you understand what regiochemistry means.

8.34 Consider the reaction shown. If you were to add HBr across the double bond, what would the product be? Assume a Markovnikov addition.

8.35 When you do the same reaction (as above) in the presence of peroxides (R-O-O-R), you get an anti-Markovnikov addition of HBr across the double bond. Draw the product of an anti-Markovnikov addition.

8.36 Consider the elimination reaction below, which uses a strong base. The product will be a double bond. This reaction will produce two Zaitsev products. One will be *cis* and one will be *trans*. Draw these products, and identify which is *cis* and which is *trans*.

8.37 Consider the elimination reaction below, which uses a strong, sterically hindered base (LDA). The product will be a double bond. This reaction will produce the Hoffmann product. Draw this product.

8.7 THE STEREOCHEMISTRY IS CONTAINED WITHIN THE MECHANISM

Stereochemistry is all about configurations of stereocenters (R vs. S) and double bonds (E vs. Z). Whenever we have a reaction where we are forming a stereocenter, we need to ask whether we get a racemic mixture (equal amounts of R and S) or only one configuration. And, if so, why? Also, whenever we form a double bond, we need to ask whether we get both E and Z isomers or only one of them? And, if so, why?

Whenever a mechanism is proposed to explain a reaction, the proposed mechanism must explain the observed stereochemical outcome of the reaction. Therefore, if you *understand* the accepted mechanism, then you will not have to memorize the stereochemical outcome. It will simply make sense. Let's see an example. Consider the addition of Br and Br across a double bond. We already saw that we don't need to worry about the regiochemistry of this reaction, because we are adding two of the same group. But what about the stereochemistry? We are creating two new stereocenters:

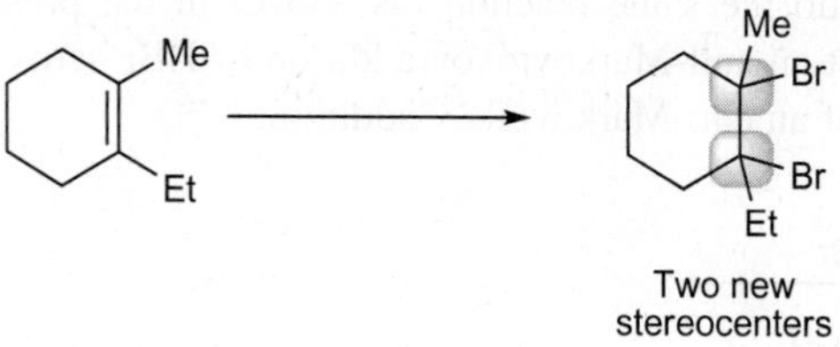

Two new
stereocenters

Each stereocenter has two possibilities (*R* or *S*). Since there are two stereocenters, we will have four total possibilitites: *SR*, *RS*, *RR*, and *SS*. These four compounds represent two sets of enantiomers:

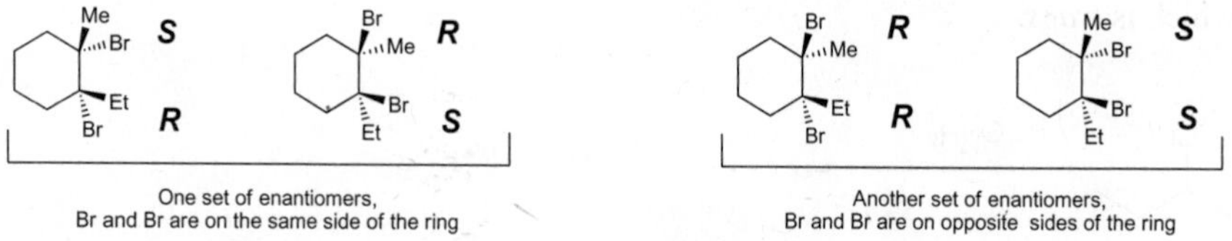

One set of enantiomers,	Another set of enantiomers,
Br and Br are on the same side of the ring	Br and Br are on opposite sides of the ring

How many of them do we get? Do we get both sets of enantiomers as our products (meaning all four products), or do we only get one set (meaning two out of the four possible products)? This depends on how the reaction took place.

If an addition reaction can take place only through a mechanism that allows a *syn* addition, then the two groups must be added across the same side of the double bond. So we will get only that one set of enantiomers:

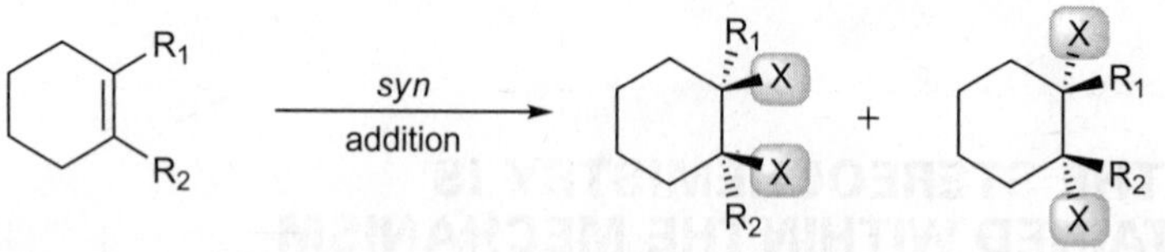

If a reaction can go through only an *anti* addition, then the two groups must be added across opposite sides of the double bond. So we will get only the set of enantiomers where the groups are on opposite sides:

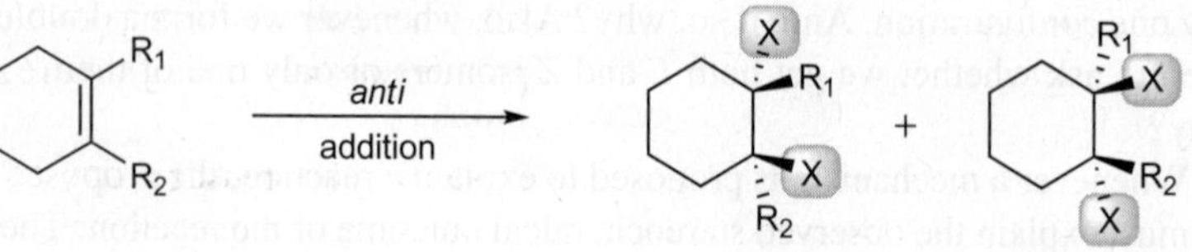

Sometimes, the reaction is not stereoselective. In other words, we get both *syn* and *anti* addition. So we get all four products (both sets of enantiomers).

Each reaction will be different. Some will give only *syn* addition, some will give only *anti* addition, and others will not be stereospecific. For every addition reaction, we need to know the stereochemistry of the addition, and that information is contained within the mechanism.

So let's go back to our example above with the addition of Br and Br across a double bond. This reaction is an *anti* addition, so we get only the set of enantiomers that has the two Br groups on opposite sides of the ring:

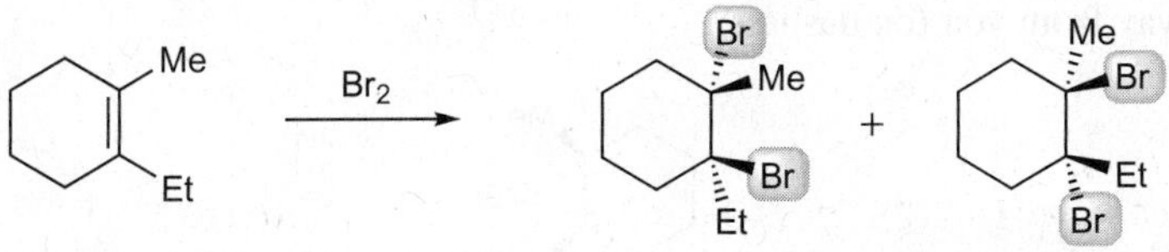

Let's look at the mechanism to understand why. In the first step, we form a bridged intermediate, called a bromonium ion:

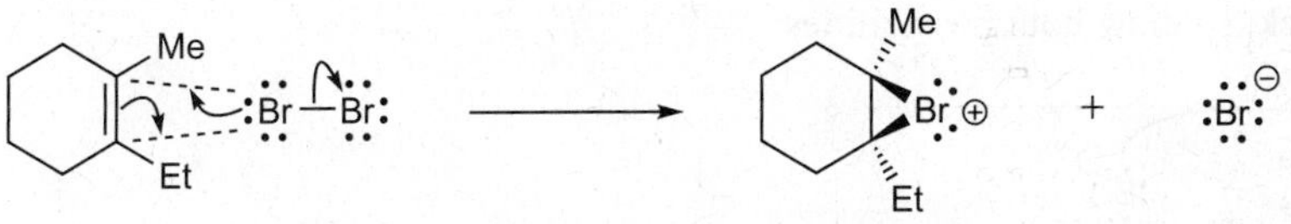

In this step, the double bond is acting as a nucleophile that attacks Br_2 (the electrophile in this reaction). The arrows are not going in opposite directions—they are actually moving in a small circle to form a ring.

Then, in the next step, the bromide (formed in the first step) comes back and attacks the bromonium ion, opening up the bridge. The bromide can attack either carbon (both possibilities shown below):

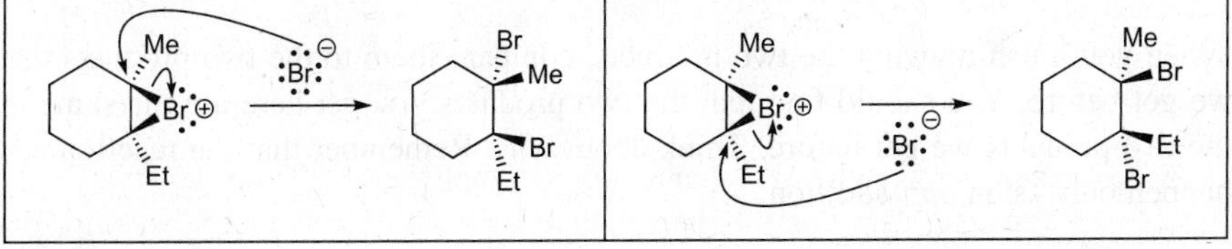

When the bromide attacks, it must attack on the other side of the ring (not the side of the bromonium bridge, but rather on the opposite side of the ring) to break open the bridge. So the addition must be an *anti addition.*

We see that the mechanism explains why the addition must be *anti.* For every reaction, the stereochemistry will always be explained by the mechanism.

PROBLEM 8.38 In the reaction above, we saw that the first step involved formation of a bromonium ion.

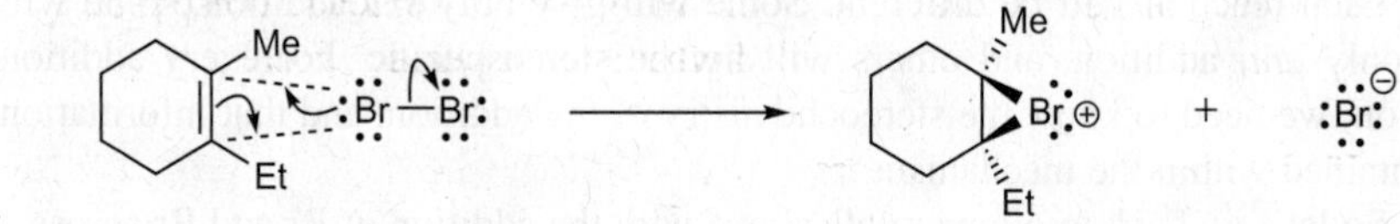

You will notice that the bromonium ion has the bridge coming out toward you (on wedges), but we did not say at the time that it could also have formed with the bridge going away from you (on dashes):

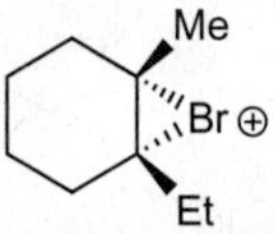

We did not talk about this at the time, because the end products would still have been the same as the way we did it before. Draw what happens if bromide (Br⁻) attacks this other bromonium ion. Remember that there are two carbon atoms that bromide could attack, so draw both possibilities:

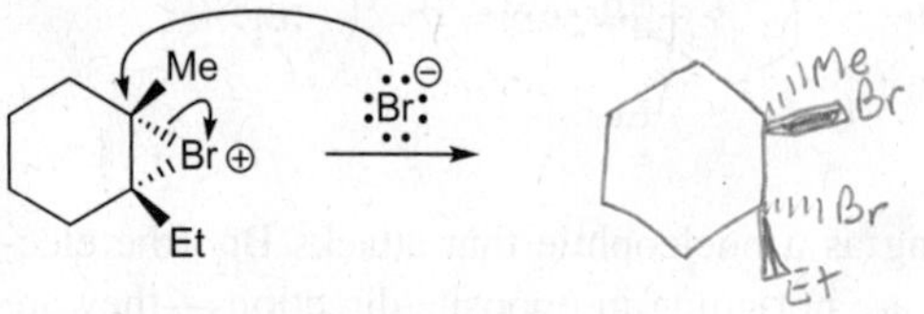

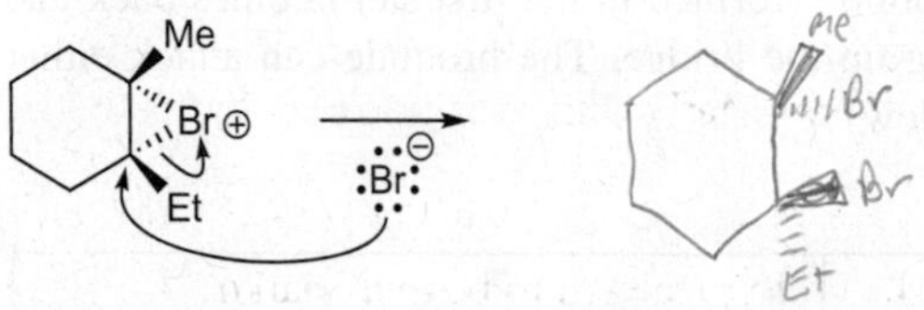

When you finish drawing the two products, compare them to the two products that we got before. You should find that the two products you get here are the same as the two products we got before. Think about why. Remember that the reaction can happen only as an *anti* addition.

Every new class of reactions (additions, eliminations, substitutions, etc.) has its own terminology for stereochemistry. As you learn each of these classes of reactions, keep a watchful eye on what terminology is used to describe the stereochemistry. Then, look at the mechanism of each reaction within each class, and try to understand how the mechanism explains the stereochemistry.

PROBLEMS   Over the course of your studies, you will learn mechanisms for each of the following reactions. In the meantime, you will be given the stereochemical information that you need to answer each of the following problems. These problems are intended to ensure that you understand what stereochemistry means.

8.39   If you add OH and OH across the following double bond in a *syn* addition, what will the products be?

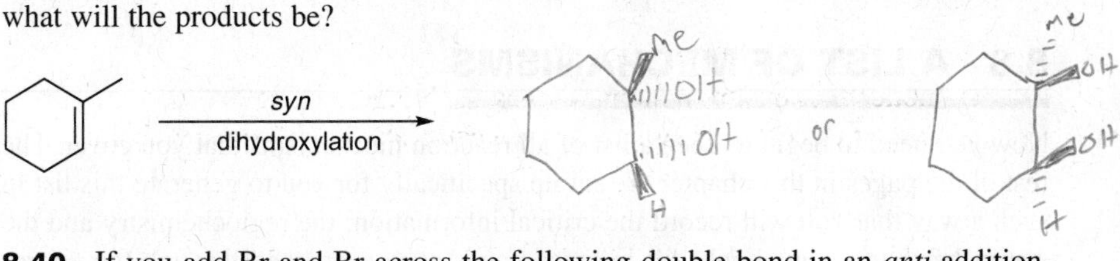

8.40   If you add Br and Br across the following double bond in an *anti* addition, what will the products be?

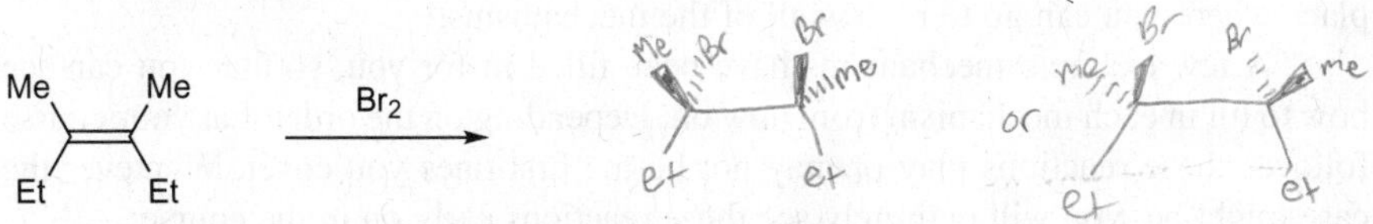

8.41   If you add Br and Br across the following double bond in an *anti* addition, you get only one product. If you draw the two products that you would expect, you will find that they are the same compound (a *meso* compound). Draw this product.

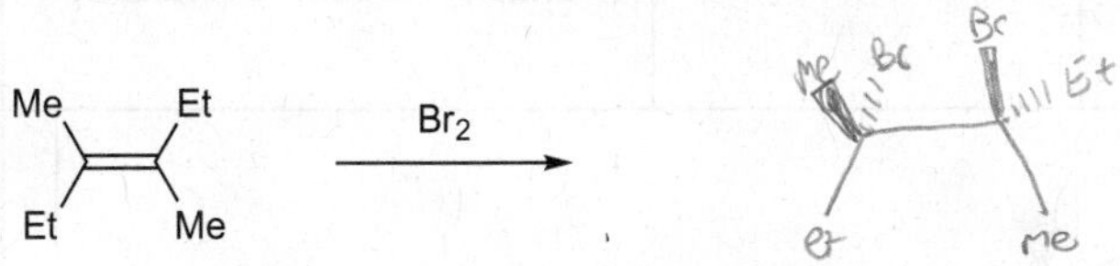

Do not confuse the concepts of regiochemistry and stereochemistry. For instance, in addition reactions, the term "*anti*-Markovnikov addition" refers to the *regiochemistry* of the addition, but the term "*anti*" refers to the *stereochemistry* of the addition. Students often confuse these concepts (probably because both terms have the word "*anti*"). It is possible for an addition reaction to be *anti*-Markovnikov and a *syn* addition (hydroboration is an example that you will learn about at some point in time). You must realize that regiochemistry and stereochemistry are two totally different concepts.

8.42   In the following reaction, we will add H and OH across a double bond. The regiochemistry is *anti*-Markovnikov, and the stereochemistry is a *syn* addition. Draw the products you would expect now that you know all of the information.

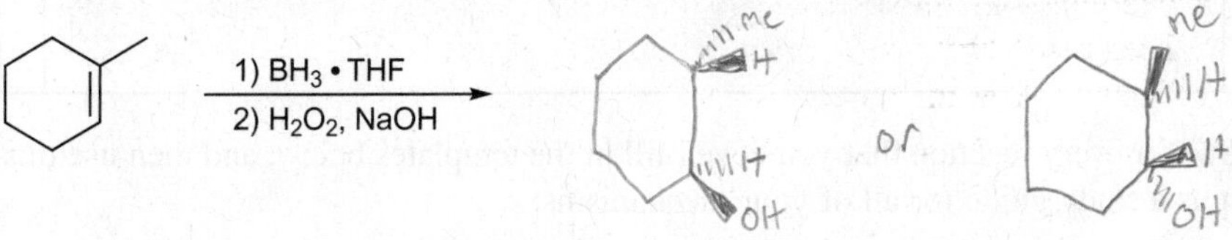

You must know the stereochemistry and regiochemistry for every reaction, and each of them is contained within the mechanism. In the problems above, you were told what to expect for the stereochemistry and the regiochemistry. When you are doing problems in your textbook and on your exams, you will be expected to know what these pieces of information are simply from looking at the reagents. A solid understanding of every mechanism will be an invaluable asset to you in this course.

8.8 A LIST OF MECHANISMS

Now you need to begin to keep a list of all reaction mechanisms that you cover. The rest of the pages in this chapter are set up specifically for you to generate this list in such a way that you will record the critical information: the regiochemistry and the stereochemistry. You should fill in these pages as you proceed through the course and you learn more mechanisms. As your list gets larger, you will have one central place where you can go to review all of the mechanisms.

A few example mechanisms have been filled in for you, so that you can see how to fill in each mechanism from now on. Depending on the order that your course follows, these reactions may or may not be the first ones you cover. Whatever the case might be, you will definitely see these reactions early on in the course:

Reaction type	Stereochemistry	Regiochemistry
Substitution (S_N2)	Inversion	Not applicable (nucleophile attacks carbon next to LG)

Reaction type	Stereochemistry	Regiochemistry
Substitution (S_N1)	Racemization	Not applicable (nucleophile attacks carbon next to LG)

Now, for every reaction that you cover, fill in the templates below, and then use this list as a study guide for all of your mechanisms:

Reaction type	Stereochemistry	Regiochemistry
Hydroboration-oxidation (BH3)	Syn	OH = anti-Markovnikov

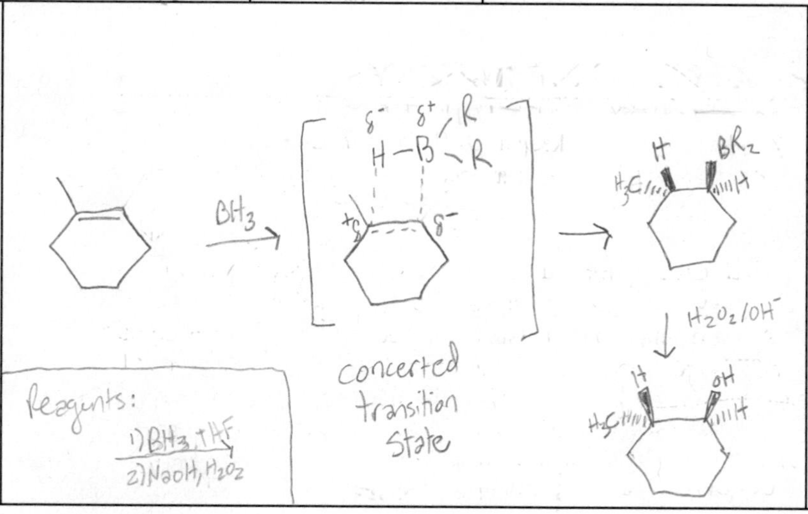
BH3
δ- δ+ R
H—B—R
+δ δ-
H BR2
H3C H
H2O2/OH-
H OH
H3C H
concerted transition state
Reagents:
1) BH3, THF
2) NaOH, H2O2

Reaction type	Stereochemistry	Regiochemistry
Hydrogenation (catalytic)	Syn	n/a

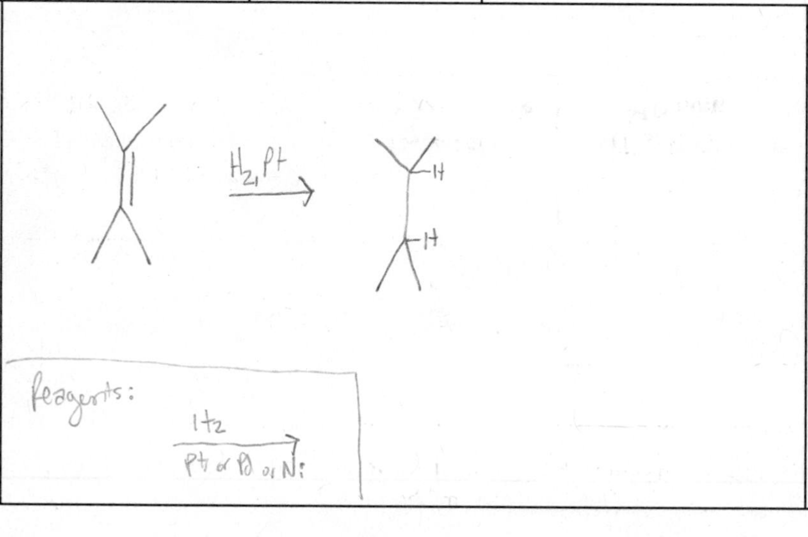
H2, Pt
H
H
Reagents:
H2
Pt or Pd or Ni

Reaction type	Stereochemistry	Regiochemistry
Hydration (acid catalyzed)	Both syn and anti	OH = Markovnikov

Reagents: $\xrightarrow[H_2SO_4]{H_2O}$

Reaction type	Stereochemistry	Regiochemistry
Bromine addition (Br_2) to alkene	anti	First halogen = anti markovnikov

Reagent: $\xrightarrow{X_2}$
X= halogen

Reaction type	Stereochemistry	Regiochemistry
Oxymercuration-reduction	anti	Hg = antimarkounikov

R-CH=CH₂   :Ö-OAc   :Hg-OAc → R-ĊH-CH₂   :ÖAc

⤷ Hg-OAC   R-CH-CH₂   :ÖH₂

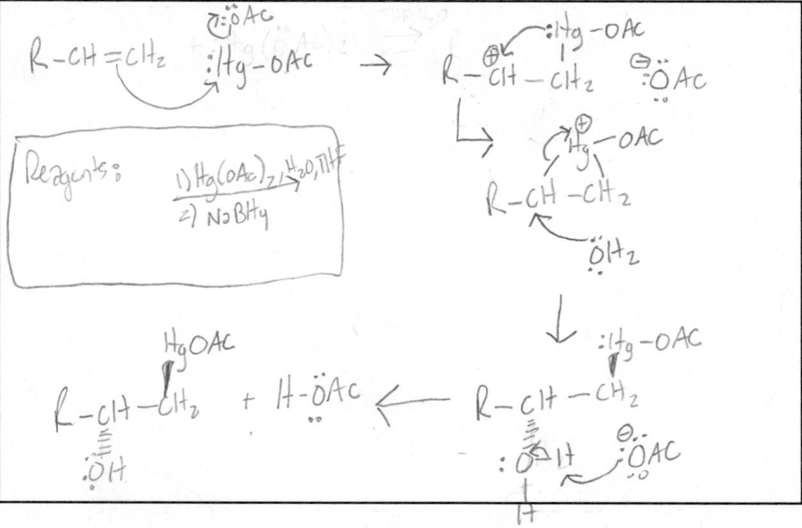

Reagents: 1) Hg(OAc)₂, H₂O, THF   2) NaBH₄

↓ :Hg-OAc

HgOAC   R-CH-CH₂ + H-ÖAC ← R-CH-CH₂   :ÖH :ÖH :ÖAC

Reaction type	Stereochemistry	Regiochemistry
HBr addition (Hydrohalogenation)	Syn and anti in equal amts	Br = Markounikov

[structure] + H-Br → [structure] Br⁻

↓

Br [structure]

Reagents: HX

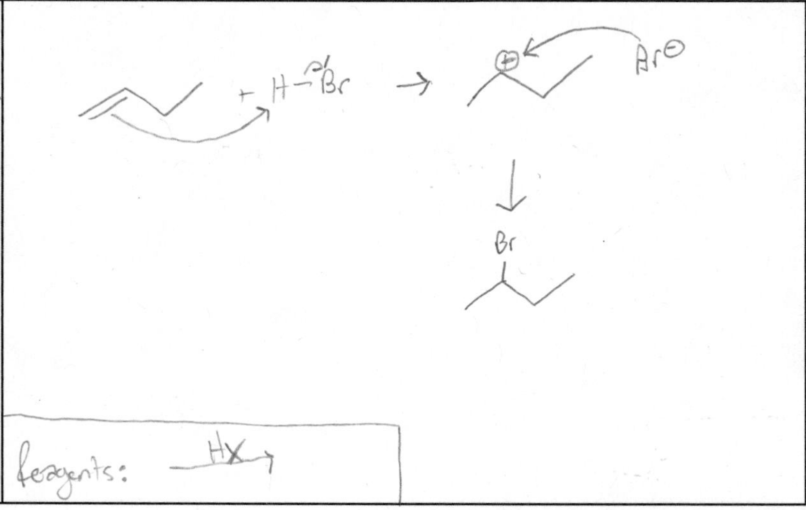

Reaction type	Stereochemistry	Regiochemistry
Ozonolysis		

Reaction type	Stereochemistry	Regiochemistry
halohydrin formation	anti	OH = Markovnikov

Reagents:

$$\frac{X_2}{H_2O} \longrightarrow$$

Reaction type	Stereochemistry	Regiochemistry
Formation of Grignard reagents		N/a

$$\text{cyclohexyl—Cl} + Mg \xrightarrow[\text{or any ether}]{\text{THF}} \text{cyclohexyl—Mg—Cl}$$

Reaction type	Stereochemistry	Regiochemistry
Formation of organolithium reagents		n/a

$$CH_3CH_2-Cl + 2Li \xrightarrow[\text{(hydrocarbons)}]{\text{hexane}} CH_3CH_2-Li + LiCl$$

Reaction type	Stereochemistry	Regiochemistry
Protonolysis of Grignard and organolithium reagents		N/a

$$CH_3\overset{\ominus}{\ddot{C}} \quad \overset{+}{M_g}X \qquad H-\overset{\cdot\cdot}{\underset{\cdot\cdot}{O}}R \rightarrow CH_3CH_2-H + R\overset{\cdot\cdot}{\underset{\cdot\cdot}{O}}\overset{\ominus}{:} + \overset{\oplus}{M_g}X$$

Reaction type	Stereochemistry	Regiochemistry
Free-Radical Halogenation of Alkanes		N/a

1) $:\overset{\cdot\cdot}{\underset{\cdot\cdot}{C}l}-\overset{\cdot\cdot}{\underset{\cdot\cdot}{C}l}: \overset{light}{\longrightarrow} :\overset{\cdot\cdot}{\underset{\cdot\cdot}{C}l}\cdot + \cdot\overset{\cdot\cdot}{\underset{\cdot\cdot}{C}l}:$

2) $:\overset{\cdot\cdot}{\underset{\cdot\cdot}{C}l}\cdot \quad H-CH_3 \longrightarrow :\overset{\cdot\cdot}{\underset{\cdot\cdot}{C}l}-H + \cdot CH_3$

$:\overset{\cdot\cdot}{\underset{\cdot\cdot}{C}l}-\overset{\cdot\cdot}{\underset{\cdot\cdot}{C}l}: \quad \cdot CH_3 \longrightarrow :\overset{\cdot\cdot}{\underset{\cdot\cdot}{C}l}\cdot + :\overset{\cdot\cdot}{\underset{\cdot\cdot}{C}l}-CH_3$

3) Cycle ends when the chlorine radical pair with another one of itself made from another rxn like this

Reaction type	Stereochemistry	Regiochemistry
H+		

Reaction type	Stereochemistry	Regiochemistry
Free radical addition of HBr	Both syn and anti	Br = anti-markovnikov

R-O-O-R → R-O• R-O• + H-Br → ROH Br•

Br• →   Br H-Br →   H Br

Reagents: HBr / ROOR, light

Reaction type	Stereochemistry	Regiochemistry
Dehydration of alcohols		

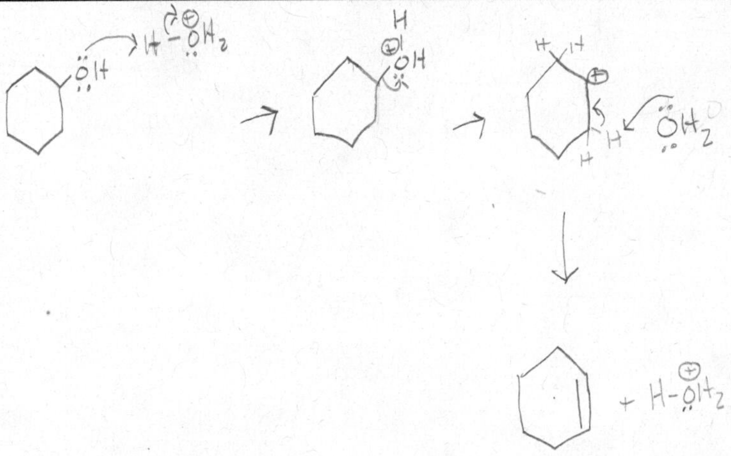

Reaction type	Stereochemistry	Regiochemistry
Cleavage of Ethers (primary and secondary) SN₂	inversion	N/a

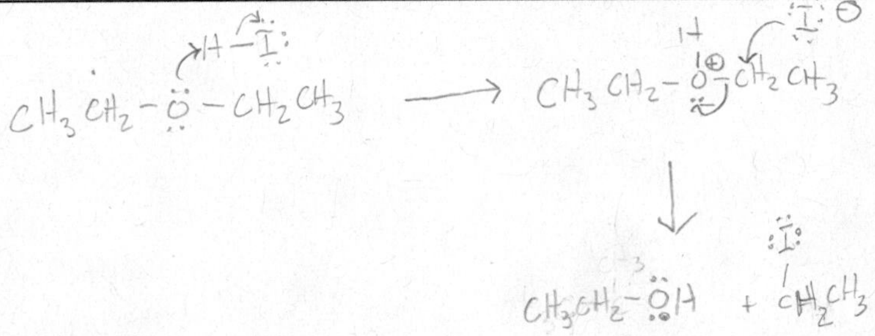

Reagents
Heat
strong acid

Reaction type	Stereochemistry	Regiochemistry
Cleavage of Ethers (tertiary) SN_1	racemization	n/a

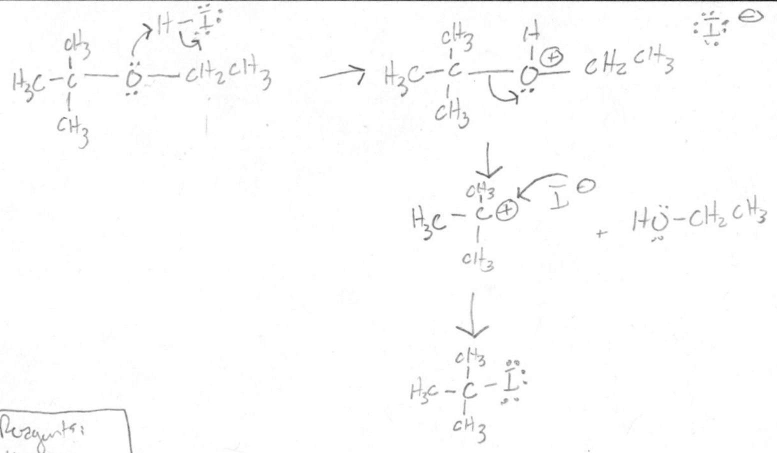

Reagents:
Heat
acid

Reaction type	Stereochemistry	Regiochemistry
Nucleophilic substitution of epoxides ($SN2$)	inversion	Nucleophile attacks less subs. carbon (basic conditions) opp for acidic conditions

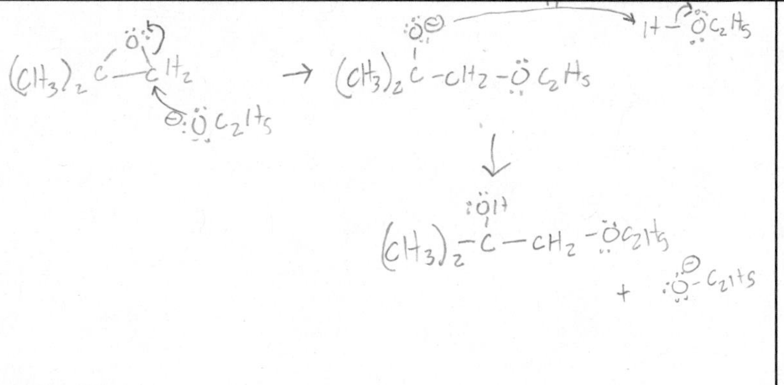

Nucleophile in alcohol solvent

Reaction type	Stereochemistry	Regiochemistry
Acid-catalyzed epoxide hydrolysis	trans	n/a

Reaction type	Stereochemistry	Regiochemistry

Reaction type	Stereochemistry	Regiochemistry

Reaction type	Stereochemistry	Regiochemistry

Reaction type	Stereochemistry	Regiochemistry

Reaction type	Stereochemistry	Regiochemistry

Reaction type	Stereochemistry	Regiochemistry

Reaction type	Stereochemistry	Regiochemistry

Reaction type	Stereochemistry	Regiochemistry

Reaction type	Stereochemistry	Regiochemistry

Reaction type	Stereochemistry	Regiochemistry

Reaction type	Stereochemistry	Regiochemistry

Reaction type	Stereochemistry	Regiochemistry

Reaction type	Stereochemistry	Regiochemistry

Reaction type	Stereochemistry	Regiochemistry

Reaction type	Stereochemistry	Regiochemistry

Reaction type	Stereochemistry	Regiochemistry

Reaction type	Stereochemistry	Regiochemistry

Reaction type	Stereochemistry	Regiochemistry

Reaction type	Stereochemistry	Regiochemistry

Reaction type	Stereochemistry	Regiochemistry

Reaction type	Stereochemistry	Regiochemistry

CHAPTER *9*

SUBSTITUTION REACTIONS

In the last chapter we saw the importance of understanding mechanisms. We said that mechanisms are the keys to understanding everything else. In this chapter, we will see a very special case of this. Students often have difficulty with substitution reactions—specifically, being able to predict whether a reaction is an S_N2 or an S_N1. These are different types of substitution reactions and their mechanisms are very different from each other. By focusing on the differences in their mechanisms, we can understand why we get S_N2 in some cases and S_N1 in other cases.

Four factors are used to determine which reaction takes place. These four factors make perfect sense when we understand the mechanisms. So, it makes sense to start off with the mechanisms.

9.1 THE MECHANISMS

Ninety-five percent of the reactions that we see in organic chemistry occur between a nucleophile and an electrophile. A nucleophile is a compound that either is negatively charged or has a region of high electron density (like a lone pair or a double bond). An electrophile is a compound that either is positively charged or has a region of low electron density. When a nucleophile encounters an electrophile, a reaction can occur.

In both S_N2 and S_N1 reactions, a *nucleophile* is attacking an electrophile, giving a *substitution* reaction. That explains the S_N part of the name. But what do the "1" and "2" stand for? To see this, we need to look at the mechanisms. Let's start with S_N2:

On the left, we see a nucleophile. It is attacking a compound that has an electrophilic carbon atom that is attached to a leaving group (LG). A *leaving group* is any group that can be expelled (we will see examples of this very soon). The leaving group serves two important functions: 1) it withdraws electron density from the carbon atom to which it is attached, rendering the carbon atom electrophilic, and 2) it is capable of stabilizing the negative charge after being expelled.

An S_N2 mechanism employs two curved arrows: one going from a lone pair on the nucleophile to form a bond between the nucleophile and carbon, and the other going from the bond between the carbon atom and the LG to form a lone pair on the LG. Notice that the configuration at the carbon atom gets inverted in this reaction. So the stereochemistry of this reaction is inversion of configuration. Why does this happen? It is kind of like an umbrella flipping in a strong wind. It takes a good force to do it, but it is possible to flip the umbrella. The same is true here. If the nucleophile is good enough, and if all of the other conditions are just right, a reaction can take place in which the configuration of the stereocenter is inverted (by bringing the nucleophile in on one side, and kicking off the LG on the other side).

Now we get to the meaning of "2" in S_N2. Remember from the last chapter that nucleophilicity is a measure of kinetics (how fast something happens). Since this is a *nucleophilic* substitution reaction, then we care about how fast the reaction is happening. In other words, what is the rate of the reaction? This mechanism has only one step, and in that step, two things need to find each other: the nucleophile and the electrophile. So it makes sense that the rate of the reaction will be dependent on how much electrophile is around *and* how much nucleophile is around. In other words, the rate of the reaction is dependent on the concentrations of two entities. The reaction is said to be "second order," and we signify this by placing a "2" in the name of the reaction.

Now let's look at the mechanism for an S_N1 reaction:

Racemic

In this reaction, there are two steps. The first step has the LG leaving all by itself, without any help from an attacking nucleophile. This generates a carbocation, which then gets attacked by the nucleophile in step 2. This is the major difference between S_N2 and S_N1 reactions. In S_N2 reactions, everything happens in one step. In S_N1 reactions, it happens in two steps, and we are forming a carbocation in the process. The existence of the carbocation as an intermediate in *only* the S_N1 mechanism is the key. By understanding this, we can understand everything else.

For example, let's look at the stereochemistry of S_N1 reactions. We already saw that S_N2 reactions proceed via inversion of configuration. But S_N1 reactions are very different. Recall that a carbocation is sp^2 hybridized, so its geometry is trigonal planar. When the nucleophile attacks, there is no preference as to which side it can attack, and we get both possible configurations in equal amounts. Half of the molecules would have one configuration and the other half would have the other configuration. We learned before that this is called a racemic mixture. Notice that we can explain the stereochemical outcome of this reaction by understanding the nature of the carbocation intermediate that is formed.

This also allows us to understand why we have the "1" in S_N1. There are two steps in this reaction. The first step is very slow (the LG just leaves on its own to form C^+ and LG^-), and the second step is very fast. Therefore, the rate of the second step is irrelevant. Let's use an analogy to understand this.

Imagine that you have an hourglass with two openings that the sand had to pass by:

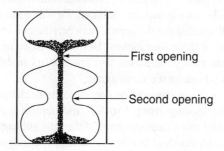

The first opening is much smaller, and the sand can travel through this opening only at a certain speed. The size of the second opening doesn't really matter. If you made the second opening a little bit wider, it would not help the sand get to the bottom any faster. As long as the top opening is smaller, the rate of the falling sand will depend only on the size of the top opening.

The same is true in a two-step reaction. If the first step is slow and the second step is fast, then the speed of the second step is irrelevant. The rate of product formation will depend *only* on the rate of the first step (the slow step). So in our S_N1 reaction, the first step is the slow step (loss of the LG to form the carbocation) and the second step is fast (nucleophile attacking the carbocation). Just as we saw in the hourglass, the second step of our mechanism will not affect the rate of the reaction. Notice that the nucleophile does not appear in the mechanism until the second step. If we added more nucleophile, it would not affect the rate of the first step. Adding more nucleophile would only speed up the second step. But we already saw that the rate of the second step does not matter for the overall reaction rate. Speeding up the second step will not change anything. So the concentration of nucleophile does not affect the rate of the reaction.

Of course, it is important that we have a nucleophile present, but how much we have doesn't matter. So now we can understand the "1" in S_N1. The *rate* of the reaction is dependent only on the concentration of the electrophile, and not that of the nucleophile. The rate is dependent on the concentration of only one entity, and the reaction is said to be "first order." We signify this by placing a "1" in the name. Of course, this does not mean that you *only* need the electrophile. You still need the nucleophile for the reaction to happen. You still need two different things (nucleophile and electrophile). The "1" simply means that the rate is not dependent on the concentration of both of them. The rate is dependent on the concentration of only one of them.

The mechanisms of S_N1 and S_N2 reactions helped us understand the stereochemistry of each reaction, and we were also able to see why we call them S_N1 and S_N2 reactions (based on reaction rates that are justified by the mechanisms). So, the mechanisms really do explain a lot. This should make sense, because a proposed mechanism must successfully explain the experimental observations. So, of course the mechanism explains the reason for racemization in an S_N1 process. That is what makes the mechanism plausible.

We mentioned before that we need to consider four factors when choosing whether a reaction will go by an S_N1 or S_N2 mechanism. These four factors are: electrophile, nucleophile, leaving group, and solvent. We will go through each factor one at a time, and we will see that the difference between the two mechanisms is the key to understanding each of these four factors. Before we move on, it is very important that you understand the two mechanisms. For practice, try to draw them in the space below without looking back to see them again.

Remember, an S_N2 mechanism has one step: the nucleophile attacks the electrophile, expelling the leaving group. An S_N1 mechanism has two steps: first the leaving group leaves to form a carbocation, and then the nucleophile attacks that carbocation. Also remember that S_N2 involves inversion of configuration, while S_N1 involves racemization. Now, try to draw them.

S_N2:

S_N1:

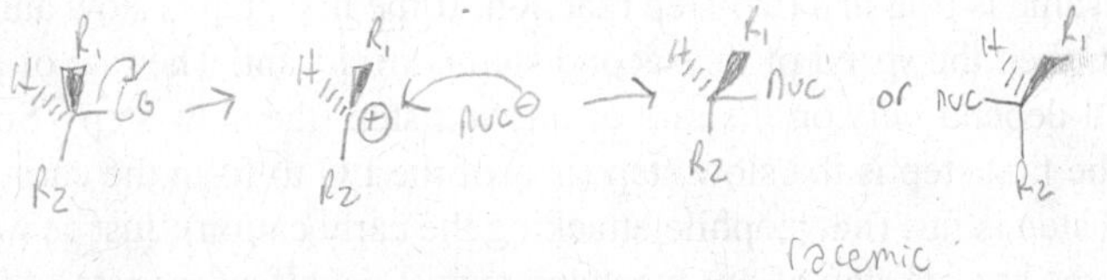

9.2 FACTOR 1—THE ELECTROPHILE (SUBSTRATE)

The electrophile is the compound being attacked by the nucleophile. In substitution and elimination reactions (which we will see in the next chapter), we generally refer to the electrophile as the *substrate*.

Remember that carbon has four bonds. So, other than the bond to the leaving group, the carbon atom that we are attacking has three other bonds:

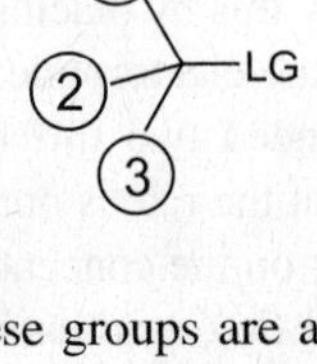

The question is, how many of these groups are alkyl groups (methyl, ethyl, propyl, etc.)? We represent alkyl groups with the letter "R." If there is one alkyl group, we call the substrate "primary" ($1°$). If there are two alkyl groups, we call the substrate "secondary" ($2°$). And if there are three alkyl groups, we call the substrate "tertiary" ($3°$):

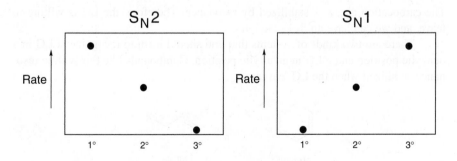

Primary   Secondary   Tertiary

In an S_N2 reaction, alkyl groups make it very crowded at the electrophilic center where the nucleophile needs to attack. If there are three alkyl groups, then it is virtually impossible for the nucleophile to get in and attack (this is an argument based on sterics):

So, for S_N2 reactions, 1° is better, 2° is OK, and 3° rarely happens.

But S_N1 reactions are totally different. The first step is not attack of the nucleophile. The first step is loss of the leaving group to form the carbocation. Then the nucleophile attacks the carbocation. Remember that carbocations are trigonal planar, so it doesn't matter how big the groups are. The groups go out into the plane, so it is easy for the nucleophile to attack. Sterics is not a problem.

In S_N1 reactions, the stability of the carbocation is the paramount issue. Recall that alkyl groups are electron donating. Therefore, 3° is best because the three alkyl groups stabilize the carbocation. 1° is the worst because there is only one alkyl group to stabilize the carbocation. This has nothing to do with sterics; this is an argument of electronics (stability of charge). So we have two opposite trends, for completely different reasons:

These charts show the rate of reaction. If you have a 1° substrate, then the reaction will proceed via an S_N2 mechanism, with inversion of configuration. If you have a 3° substrate, then the reaction will proceed via an S_N1 mechanism, with racemization. What do you do if the substrate is 2°? You move on to factor 2.

EXERCISE 9.1 Identify whether the following substrate is more likely to participate in an S_N2 or S_N1 reaction.

(structure: cyclohexane with ethyl chloride chain, labeled S_N2)

ANSWER The substrate is primary, so we predict an S_N2 reaction.

PROBLEMS Identify whether each of the following substrates is more likely to participate in an S_N2 or S_N1 reaction.

9.2 *(cyclohexane with Br)* ___Both___

9.3 Cl *(chain)* ___S_N2___

9.4 *(structure with Br)* ___Both___

9.5 *(structure with I)* ___S_N1___

There is one other way to stabilize a carbocation (other than alkyl groups)—resonance. If a carbocation is resonance stabilized, then it will be easier to form that carbocation:

(reaction scheme showing resonance stabilized carbocation)

The carbocation above is stabilized by resonance. Therefore, the LG is willing to leave, and we can have an S_N1 reaction.

There are two kinds of systems that you should learn to recognize: a LG in a benzylic position and a LG in an allylic position. Compounds like this will be resonance stabilized when the LG leaves:

(structures showing benzylic and allylic compounds with X)

Benzylic   Allylic

If you see a double bond near the LG and you are not sure if it is a benzylic or allylic system, just draw the carbocation you would get and see if there are any resonance structures.

EXERCISE 9.6 In the compound below, circle the LGs that are benzylic or allylic:

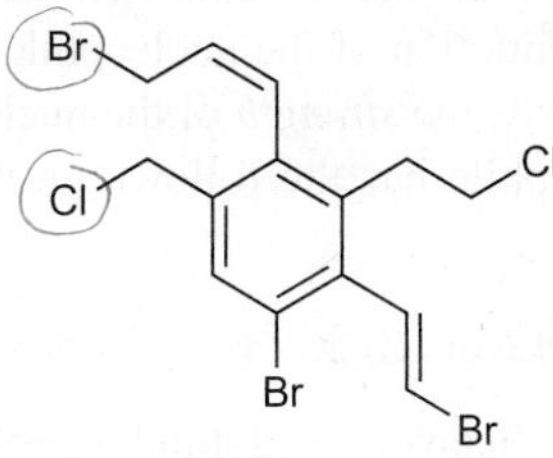

Answer

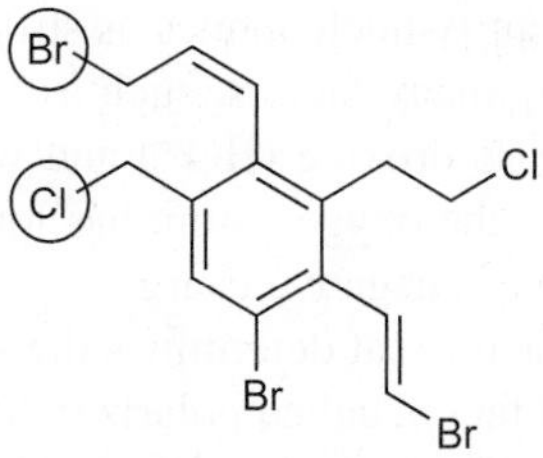

PROBLEMS For each compound below, determine whether the LG leaving would form a resonance-stabilized carbocation. If you are not sure, try to draw resonance structures of the carbocation you would get if the leaving group is expelled.

9.7

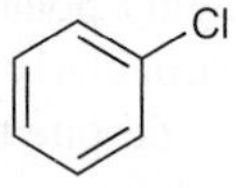

no

9.8

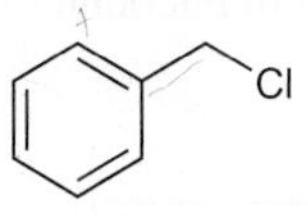

yes

9.9

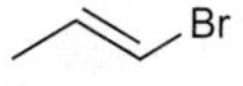

no

9.10

yes

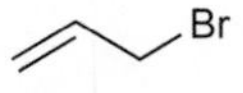

9.3 FACTOR 2 – THE NUCLEOPHILE

The rate of an S_N2 process is dependent on the strength of the nucleophile. A strong nucleophile will speed up the rate of an S_N2 reaction, while a weak nucleophile will slow down the rate of an S_N2 reaction. In contrast, an S_N1 process is not affected by the strength of the nucleophile. Why not? Recall that the "1" in S_N1 means that the rate of reaction is dependent only on the substrate, *not* on the nucleophile (remember the hourglass analogy). The concentration of the nucleophile is not relevant in determining the rate of reaction. Similarly, the *strength* of the nucleophile is also not relevant.

In summary, the nucleophile has the following effect on the competition between S_N2 and S_N1:

- A strong nucleophile favors S_N2.

- A weak nucleophile disfavors S_N2 (and thereby allows S_N1 to compete successfully).

We must therefore learn to identify nucleophiles as strong or weak. The strength of a nucleophile is determined by many factors, such as the presence or absence of a negative charge. For example, hydroxide (HO^-) and water (H_2O) are both nucleophiles, because in both cases, the oxygen atom has lone pairs. But hydroxide is a stronger nucleophile since it has a negative charge.

Charge is not the only factor that determines the strength of a nucleophile. In fact, there is a more important factor, called polarizability, which describes the ability of an atom or molecule to distribute its electron density unevenly in response to external influences. For example, sulfur is highly polarizable, because its electron density can be unevenly distributed when it comes near an electrophile. Polarizability is directly related to the size of the atom (and more specifically, the number of electrons that are distant from the nucleus). Sulfur is very large and has many electrons that are distant from the nucleus, and it is therefore highly polarizable. Iodine shares the same feature. As a result, I^- and HS^- are particularly strong nucleophiles. For similar reasons, H_2S is also a strong nucleophile, despite the fact that it lacks a negative charge.

Below are some strong and weak nucleophiles that we will encounter often:

Common Nucleophiles

Strong			Weak
:Ï: ⊖	HṢ: ⊖	HÖ: ⊖	:F̈: ⊖
:B̈r: ⊖	H₂S̈:	RÖ: ⊖	H₂Ö:
:C̈l: ⊖	RṠH	N≡C: ⊖	RÖH

EXERCISE 9.11 Identify whether the following nucleophile will favor S_N2 or S_N1:

⌒⌒⌒SH

S_N2

ANSWER   This compound has a sulfur atom with lone pairs. A lone pair on a sulfur atom will be strongly nucleophilic, even without a negative charge, because sulfur is large and highly polarizable. Strong nucleophiles favor S_N2 reactions.

PROBLEMS   Identify whether each of the following nucleophiles will favor S_N2 or S_N1.

9.12   Answer: _____SN2_____   **9.13**   Answer: _____SN1_____

9.14   Answer: _____SN1_____   **9.15**   Answer: _____SN2_____

9.16   Answer: _____SN2_____   **9.17**   :C≡N   Answer: _____SN2_____

9.4   FACTOR 3 – THE LEAVING GROUP

Both S_N1 and S_N2 mechanisms are sensitive to the identity of the leaving group. If the leaving group is bad, then neither mechanism can operate, but S_N1 reactions are generally more sensitive to the leaving group than S_N2 reactions. Why? Recall that the rate-determining step of an S_N1 process is loss of a leaving group to form a carbocation and a leaving group:

We have already seen that the rate of this step is very sensitive to the stability of the carbocation, so it should make sense that it is also sensitive to the stability of the leaving group. The leaving group must be highly stabilized in order for an S_N1 process to be effective.

What determines the stability of a leaving group? As a general rule, good leaving groups are the conjugate bases of strong acids. For example, iodide (I^-) is the conjugate base of a very strong acid (HI):

Strong Acid   **Conjugate Base**
                                    (Weak)

Iodide is a very weak base because it is highly stabilized. As a result, iodide can function as a good leaving group. In fact, iodide is one of the best leaving groups. The following figure shows a list of good leaving groups, all of which are the conjugate bases of strong acids:

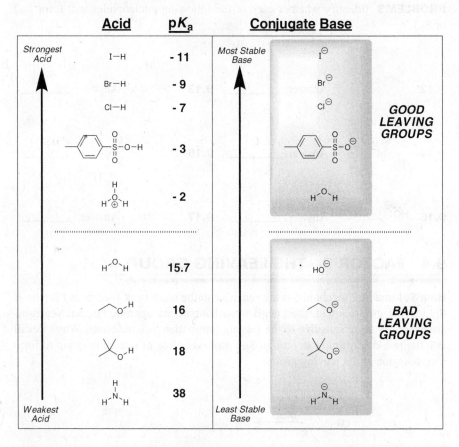

In contrast, hydroxide is a bad leaving group, because it is not a stabilized base. In fact, hydroxide is a relatively strong base, and, therefore, it rarely functions as a leaving group. It is a bad leaving group. But under certain circumstances, it is possible to convert a bad leaving group into a good leaving group. For example, when treated with a strong acid, an OH group is protonated, converting it into a good leaving group:

Bad leaving group   Good leaving group

The most commonly used leaving groups are halides and sulfonate ions:

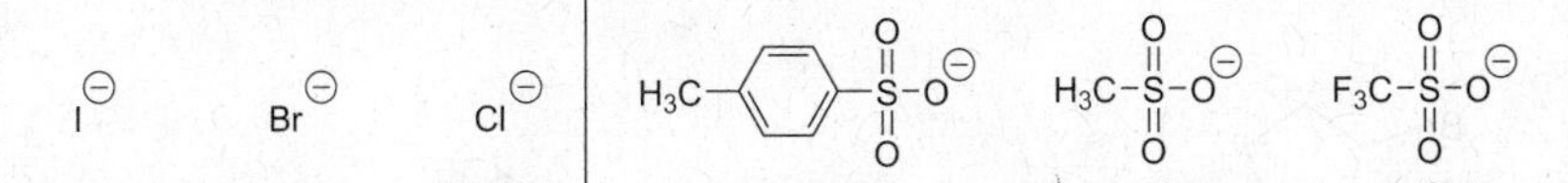

halides			sulfonate ions		
iodide	bromide	chloride	tosylate	mesylate	triflate

Among the halides, iodide is the best leaving group because it is a weaker base (more stable) than bromide or chloride. Among the sulfonate ions, the best leaving group is the triflate group, but the most commonly used is the tosylate group. It is abbreviated as OTs. When you see OTs connected to a compound, you should recognize the presence of a good leaving group.

EXERCISE 9.18 Identify the leaving group in the following compound:

ANSWER We have seen that hydroxide is not a good leaving group, because its conjugate acid (H$_2$O) is not a strong acid. As a result, hydroxide is not a weak base, so it does not function as a leaving group. In contrast, chloride is a good leaving group because its conjugate acid (HCl) is a strong acid. Therefore, chloride is a weak base, so it can serve as a leaving group.

PROBLEMS Identify the best leaving group in each of the following compounds:

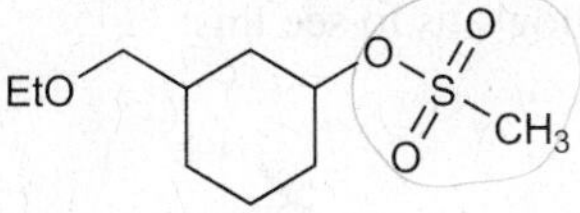

9.19   Answer: _mesylate_

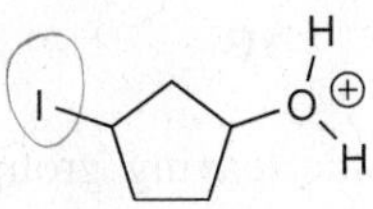

9.20   Answer: _Iodide_

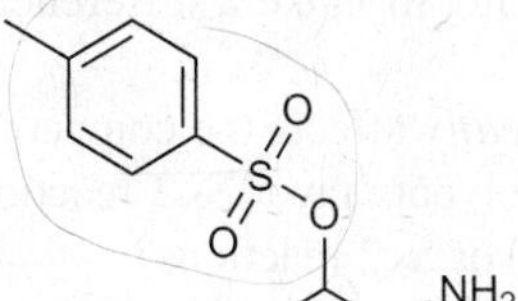

9.21   Answer: _tosylate_

9.22

Answer: _cl_

9.23

Answer: _Br_

9.24

Answer: _Br_

9.25 Compare the structures of 3-methoxy-3-methylpentane and 3-iodo-3-methylpentane, and identify which compound is more likely to undergo an S_N1 reaction.

9.26 When 3-ethyl-3-pentanol is treated with excess chloride, no substitution reaction is observed, because hydroxide is a bad leaving group. If you wanted to force an S_N1 reaction, using 3-ethyl-3-pentanol as the substrate, what reagent would you use to change the leaving group into a better leaving group and provide chloride ions at the same time?

HCl

9.5 FACTOR 4—THE SOLVENT

So far, we have explored the substrate, the nucleophile, and the leaving group. This takes care of all of the parts of the compounds that are reacting with each other. Let's summarize substitution reactions in a way that allows us to see this:

Nuc   Nuc

Substrate   Substrate

LG   LG

So, by talking about the substrate, the nucleophile, and the leaving group, we have covered almost everything. But there is one more thing to take into account. What solvent are these compounds dissolved in? It can make a difference. Let's see how.

There is a really strong solvent effect that *greatly* affects the competition between S_N1 and S_N2, and here it is: polar aprotic solvents favor S_N2 reactions. So, what are polar aprotic solvents, and why do they favor S_N2 reactions?

Let's break it down into two parts: *polar* and *aprotic*. Hopefully, you remember from general chemistry what the term "polar" means, and you should also remember that "like dissolves like" (so polar solvents dissolve polar compounds, and nonpolar solvents dissolve nonpolar compounds). Therefore, we really need a polar solvent to run substitution reactions. S_N1 desperately needs the polar solvent to stabilize the carbocation, and S_N2 needs a polar solvent to dissolve the nucleophile. S_N1 certainly needs the polar solvent more than S_N2 does, but you will rarely see a substitution reaction in a nonpolar solvent. So, let's focus on the term aprotic.

Let's begin by defining a protic solvent. We will need to jog our memories about acid–base chemistry. Recall that in Chapter 3 we talked about the acidity of protons (these are hydrogen atoms without the electrons, symbolized by H^+), and we saw that protons can be removed from a compound if the compound can stabilize the negative charge that develops when H^+ is removed. A protic solvent is a solvent that has a proton connected to an electronegative atom (for example, H_2O or EtOH). It is called protic because the solvent can serve as a source of protons. In other words, the solvent can give a proton because the solvent can stabilize a negative charge (at least a little bit). So what is an aprotic solvent?

Aprotic means that the solvent does *not* have a proton on an electronegative atom. The solvent can still have hydrogen atoms, but none of them are connected to electronegative atoms. The most common examples of polar aprotic solvents are acetone, DMSO, DME, and DMF:

Acetone

Dimethylsulphoxide (DMSO)

Dimethoxyethane (DME)

Dimethylformamide (DMF)

There are, of course, other polar aprotic solvents. You should look through your textbook and your class notes to determine if there are any other polar aprotic solvents that you will be expected to know. If there are any more, you can add them to the drawing above. You should learn to recognize these solvents when you see them.

So why do these solvents speed up the rate of S_N2 reactions? To answer this question, we need to talk about a solvent effect that is usually present when we dissolve a nucleophile in a solvent. A nucleophile with a negative charge, when dissolved in a polar solvent, will get surrounded by solvent molecules in what is called a *solvent shell:*

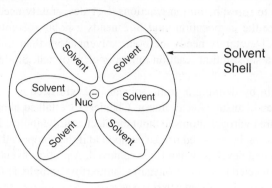

This solvent shell is in the way, holding back the nucleophile from doing what it is supposed to do (go attack something). For the nucleophile to do its job, the nucleophile must first shed this solvent shell. This is always the case when you dissolve a nucleophile in a polar solvent, *except* when you use a polar aprotic solvent.

Polar aprotic solvents are not very good at forming solvent shells around negative charges. So if you dissolve a nucleophile in a polar aprotic solvent, the nucleophile is said to be a "naked" nucleophile, because it does not have a solvent shell. Therefore, it does not need to first shed a solvent shell before it can react with something. It never had a solvent shell to begin with. This effect is drastic. As you can imagine, a nucleophile with a solvent shell is going to spend most of its existence with the solvent shell, and there will be only brief moments every now and then when it is free to react. By allowing the nucleophile to react all of the time, we are greatly speeding up the reaction. S_N2 reactions performed with nucleophiles in polar aprotic solvents occur about 1000 times faster than those in regular protic solvents.

Bottom line: Whenever a solvent is indicated, you should look to see if it is one of the polar aprotic solvents listed above. If it is, it is a safe bet that the reaction is going to be S_N2.

EXERCISE 9.27 Predict whether the reaction below will occur via an S_N2 or an S_N1 mechanism:

Answer This reaction utilizes DMSO, which is a polar aprotic solvent, so we expect an S_N2 reaction even though the substrate is secondary.

PROBLEM 9.28 Go back to the list of polar aprotic solvents, study the list, and then try to copy the list here without looking back.

9.6 USING ALL FOUR FACTORS

Now that we have seen all four factors individually, we need to see how to put them all together. When analyzing a reaction, we need to look at all four factors and make a determination of which mechanism, S_N1 or S_N2, is predominating. It may not be just one mechanism in every case. Sometimes both mechanisms occur and it is difficult to predict which one predominates. Nevertheless, it is a lot more common to see situations that are obviously leaning toward one mechanism over the other. For example, it is clear that a reaction will be S_N2 if we have a primary substrate with a strong nucleophile in a polar aprotic solvent. On the flipside, a reaction will clearly be S_N1 if we have a tertiary substrate with a weak nucleophile and an excellent leaving group.

Your job is to look at all of the factors and make an informed decision. Let's put everything we saw into one chart. Review the chart. If there are any parts that do not make sense, you should return to the section on that factor and review the concepts.

Substrate	Nucleophile	Leaving group	Solvent
1°—Only S_N2, No S_N1	Strong—S_N2	Bad—Neither	Polar aprotic—S_N2
2°—Both	Both	Good—Both (but more S_N2)	
3°—Only S_N1 No S_N2	Weak—S_N1	Excellent—S_N1	

EXERCISE 9.29 For the reaction below, look at all of the reagents and conditions, and determine if the reaction will proceed via an S_N2 or an S_N1, or both or neither.

—strong nucleophile
- 1°
-cl⊖ S_N2

Answer The substrate is primary, which immediately tells us that it needs to be S_N2. On top of that, we see that we have a strong nucleophile, which also favors S_N2. The LG is good, which doesn't tell us much. The solvent is not indicated. So, taking everything into account, we predict that the reaction follows an S_N2 mechanism.

PROBLEMS For each reaction below, look at all of the reagents and conditions, and determine if the reaction will proceed via an S_N2 or an S_N1, or both or neither.

9.30
$\underset{\text{DME}}{\overset{\text{HO}^{\ominus}}{\longrightarrow}}$ SN_2

9.31
$\overset{\text{H}_2\text{O}}{\longrightarrow}$ SN_1

9.32
$\overset{\text{H}-\text{Br}}{\longrightarrow}$ SN_1

9.33
$\overset{\text{H}_2\text{O}}{\longrightarrow}$ neither

9.34
$\underset{\text{DMSO}}{\overset{\text{Cl}^{\ominus}}{\longrightarrow}}$ SN_2

9.35
$\overset{\text{ROH}}{\longrightarrow}$ SN_1

9.7 SUBSTITUTION REACTIONS TEACH US SOME IMPORTANT LESSONS

S_N1 and S_N2 reactions produce almost the same products. In both reactions, a leaving group is replaced by a nucleophile. The difference in products between S_N1 and S_N2 reactions arises when the leaving group is attached to a stereocenter. In this situation, the S_N2 mechanism will invert the stereocenter, while the S_N1 mechanism will produce a racemic mixture. That's the main difference—the configuration of one stereocenter. It seems like a lot of work to go through to determine the configuration of one stereocenter (which matters only some of the time).

So the obvious question is, why did we go through all of that trouble to learn how to determine whether a reaction is S_N1 or S_N2? There are many answers to this question, and it is important to spend some time on this, because it will help frame the rest of the course for you. Let's go through some answers one at a time.

First we learned the important concept that everything is located in the mechanisms. By understanding the mechanisms completely, everything else can be justified based on the mechanisms. All of the factors that influence the reaction can be understood by carefully examining the mechanism. This is true for every reaction you will see from now on. Now you have had some practice thinking this way.

Next we learned that there are multiple factors at play when analyzing a reaction. Sometimes the factors can all be pointing in the same direction, while at other times the factors can be in conflict. When they are in conflict, we need to weigh them and decide which factors win out in determining the path of the reaction. This concept of competing factors is a theme in organic chemistry. The experience of going through S_N1 and S_N2 mechanisms has prepared you for thinking this way for all reactions from now on.

Finally we learned that if we analyze the first factor (substrate), we will find two effects at play: electronics and sterics. We saw that S_N2 reactions require primary or secondary substrates because of sterics—it is too crowded for the nucleophile to attack a tertiary substrate. On the other hand, S_N1 reactions did not have a problem with sterics, but electronics was a bigger issue. Tertiary was the best, because the alkyl groups were needed to stabilize the carbocation.

These two effects (sterics and electronics) are major themes in organic chemistry. Much of what you learn in the rest of the course can be explained with either an electronic or a steric argument. The sooner you learn to consider these two effects in every problem you encounter, the better off you will be. Electronics is usually the more complicated effect. In fact, the other three factors that we saw (nucleophile, leaving group, and solvent effects) were all electronic arguments. Once you get the hang of the kinds of electronic arguments that are generally made, you will begin to see common threads in all of the reactions that you will encounter in this course.

Don't get me wrong—it is very important to be able to predict whether a stereocenter gets inverted or not when a substitution reaction takes place. That alone would have been enough of a reason to learn all of the factors in this chapter. But I also want you to keep your eye on some of the "bigger picture" issues. They will help you as you move through the course.

CHAPTER 10

ELIMINATION REACTIONS

In the previous chapter, we saw that a substitution reaction can occur when a compound possesses a leaving group. In this chapter, we will explore another type of reaction, called *elimination*, which can also occur for compounds with leaving groups. In fact, substitution and elimination reactions frequently compete with each other, giving a mixture of products. At the end of this chapter, we will learn how to predict the products of these competing reactions. For now, let's consider the different outcomes for substitution and elimination reactions:

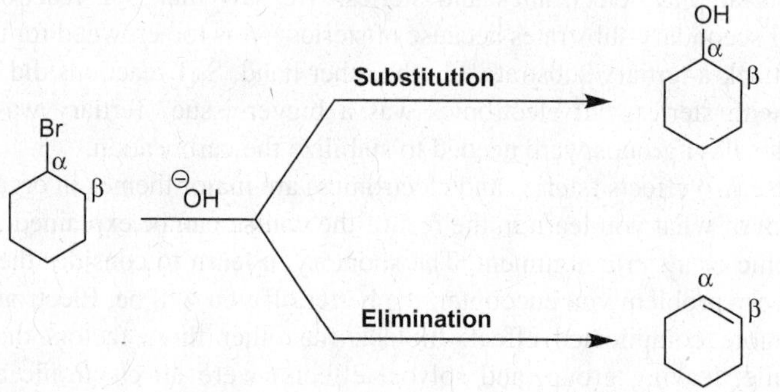

In a substitution reaction, the leaving group is replaced with a nucleophile. In an elimination reaction, a beta (β) proton is removed together with the leaving group, forming a double bond. In the previous chapter, we saw two mechanisms for substitution reactions (S_N1 and S_N2). In a similar way, we will now explore two mechanisms for elimination reactions, called E1 and E2. Let's begin with the E2 mechanism.

10.1 THE E2 MECHANISM

In an E2 process, a base removes a proton, causing the simultaneous expulsion of a leaving group:

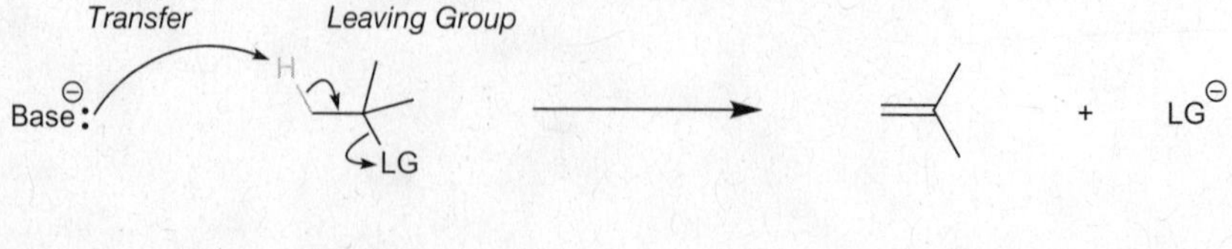

Notice that there is only one mechanistic step (no intermediates are formed), and that step involves both the substrate and the base. Because that step involves two chemical entities, it is said to be bimolecular. Bimolecular elimination reactions are called E2 reactions, where the "2" stands for "bimolecular."

Now let's consider the effect of the substrate on the rate of an E2 process. Recall from the previous chapter that S_N2 reactions generally do not occur with tertiary substrates, because of steric considerations. But E2 reactions are different than S_N2 reactions, and in fact, tertiary substrates often undergo E2 reactions quite rapidly. To explain why tertiary substrates will undergo E2 but not S_N2 reactions, we must recognize that the key difference between substitution and elimination is the role played by the reagent. In a substitution reaction, the reagent functions as a nucleophile and attacks an electrophilic position. In an elimination reaction, the reagent functions as a base and removes a proton, which is easily achieved even with a tertiary substrate. In fact, tertiary substrates react even more rapidly than primary substrates.

10.2 THE REGIOCHEMICAL OUTCOME OF AN E2 REACTION

Recall from Chapter 8 that the term "regiochemistry" refers to *where* the reaction takes place. In other words, in what *region* of the molecule is the reaction taking place? When H and X are eliminated (where X is some leaving group), it is sometimes possible for more than one alkene to form. Consider the following example, in which two possible alkenes can be formed:

Where does the double bond form? This is a question of regiochemistry. The way we distinguish between these two possibilities is by considering how many groups are attached to each double bond. Double bonds can have anywhere from 1 to 4 groups attached to them:

Monosubstituted   **Di**substituted   **Tri**substituted   **Tetra**substituted

So, if we look back at the reaction above, we find that the two possible products are trisubstituted and disubstituted:

Trisubstituted   **Di**substituted

For an elimination reaction where there is more than one possible alkene that can be formed, we have names for the different products based on which alkene is more substituted and which is less substituted. The more substituted alkene is called the Zaitsev product, and the less substituted alkene is called the Hofmann product. Usually, the Zaitsev product is the major product:

Major   +   *Minor*

However, there are many exceptions in which the Zaitsev product (the more-substituted alkene) is not the major product. For example, if the reaction above is performed with a sterically hindered base (rather than using ethoxide as the base), then the major product will be the less-substituted alkene:

Minor   +   *Major*

In this case, the Hofmann product is the major product, because a sterically hindered base was used. This case illustrates an important concept: *The regiochemical outcome of an E2 reaction can often be controlled by carefully choosing the base.* Below are two examples of sterically hindered bases that will be encountered frequently throughout your organic chemistry course:

Potassium *tert*-butoxide
(*t*-BuOK)

Lithium diisopropylamide
(LDA)

PROBLEMS Draw the Zaitsev and Hofmann products that are expected when each of the following compounds is treated with a strong base to give an E2 reaction. For the following problems, don't worry about identifying which product is major and which is minor, since the identity of the base has not been indicated. Just draw both possible products:

10.1

Zaitsev   Hofmann

10.2

Zaitsev

Hofmann

10.3

Zaitsev

Hofmann

10.3 THE STEREOCHEMICAL OUTCOME OF AN E2 REACTION

The examples in the previous section focused on regiochemistry. We will now focus our attention on stereochemistry. For example, consider performing an E2 reaction with the following substrate:

This substrate has two identical β positions so regiochemistry is not an issue in this case. Deprotonation of either β position produces the same result. But in this case, stereochemistry is relevant, because two stereoisomeric alkenes are possible:

EtO$^{\ominus}$

Major   +   **Minor**

Both stereoisomers (*cis* and *trans*) are produced, but the *trans* product predominates. This specific example is said to be *stereoselective*, because the substrate produces two stereoisomers in unequal amounts.

In the previous example, the β position had two different protons:

In such a case, we saw that both the *cis* and the *trans* isomers were produced, with the *trans* isomer being favored. Now let's consider a case where the β position contains only one proton. In such a case, only one product is formed. The reaction is said to be *stereospecific* (rather than stereoselective), because the proton and the leaving group must be *antiperiplanar* during the reaction. This is best illustrated using Newman projections, which allow us to draw the compound in a conformation in which the proton and the leaving group are antiperiplanar. This conformation then shows you which stereoisomer you get. The following example will illustrate how this is done.

EXERCISE 10.4 Draw the expected product(s) when the following substrate is treated with a strong base to give an E2 reaction:

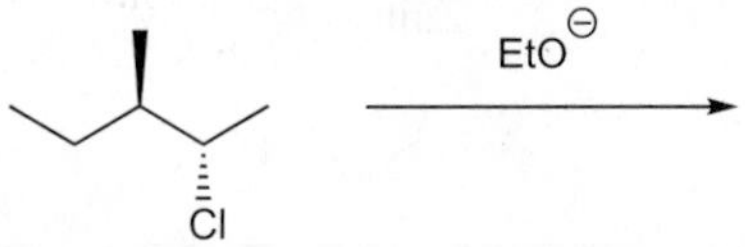

ANSWER Let's first consider the expected regiochemical outcome of the reaction. The reaction does not employ a sterically hindered base, so we expect formation of the more substituted alkene (the Zaitsev product):

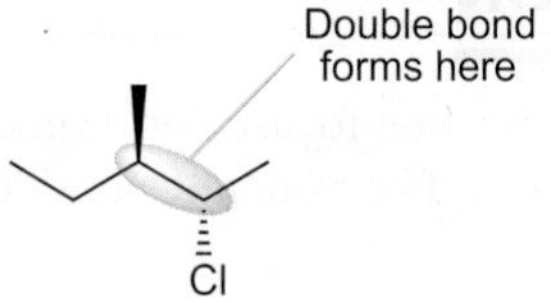

Double bond forms here

Now let's consider the stereochemical outcome. In this case, the beta position (where the reaction is taking place) has only one proton:

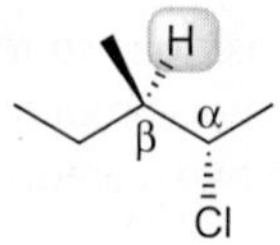

So, we expect this reaction to be stereospecific, rather than stereoselective. That is, we expect only one alkene, rather than a mixture of stereoisomeric alkenes. In order to determine which alkene is obtained, we begin by drawing the Newman projection:

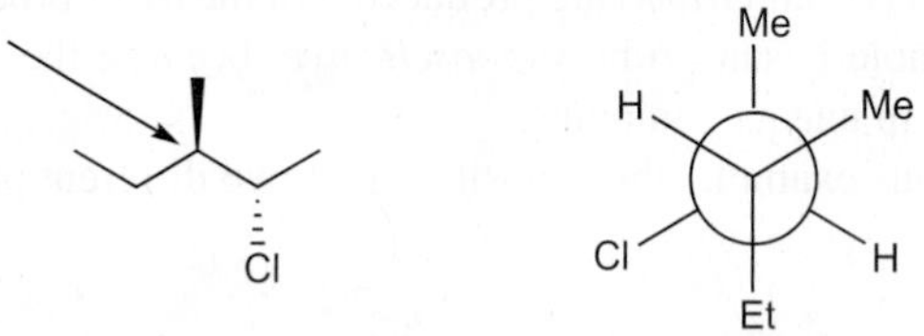

Next, we need to draw the conformation in which the H (on the front carbon) and the leaving group (Cl) are antiperiplanar:

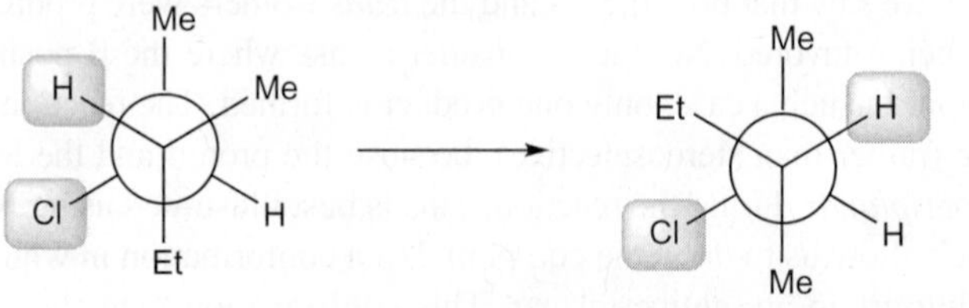

Antiperiplanar conformation

This is the conformation from which the reaction can take place. The double bond is being formed between the front carbon and the back carbon, and this Newman projection shows us the stereochemical outcome (look carefully at the dotted ovals, which are drawn to help you see the outcome more clearly):

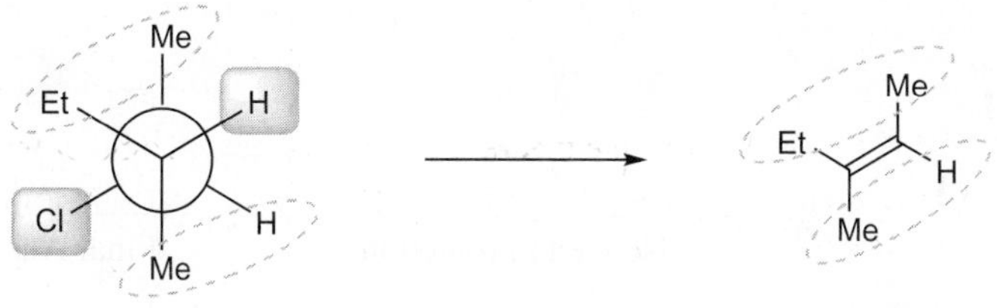

This is the Zaitsev product that we expect. The stereoisomer of this alkene is not produced, because the E2 process is stereospecific:

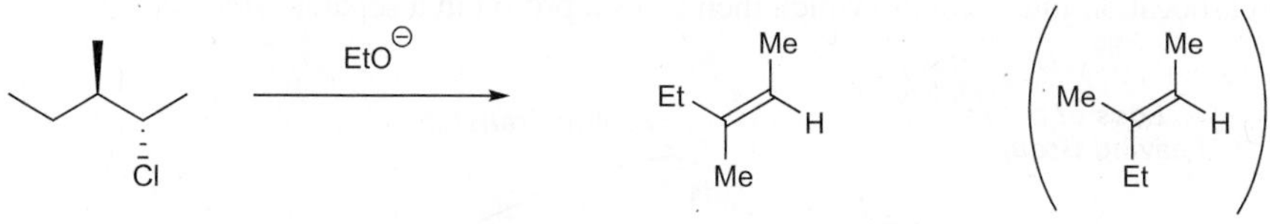

not
expected

You need to get into the habit of drawing Newman projections so that you can determine the stereoisomer that is expected from an E2 reaction. If you are rusty on Newman Projections, you should go back and review the first two sections in Chapter 6 in this book. Then come back to here, and try to use Newman projections to determine the stereochemical outcome of the following reactions.

PROBLEMS   Draw the major product that is expected when each of the following substrates is treated with ethoxide (a strong base that is not sterically hindered) to give an E2 reaction:

10.5

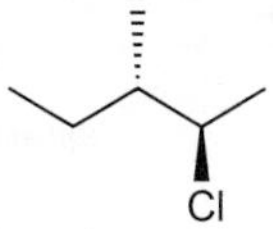

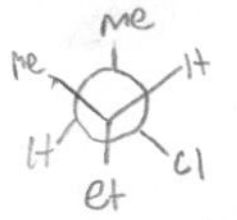

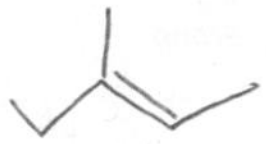

_____   _____
Newman projection   Final Answer

10.6

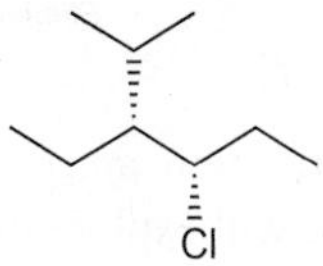

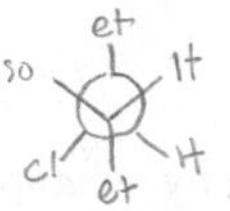

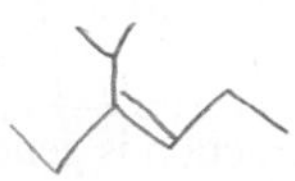

_____   _____
Newman projection   Final Answer

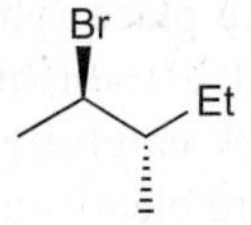

10.7

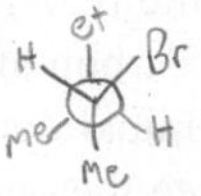

Br [structure with Et]

[Newman projection with H, et, Br, me, Me, H]

[Final Answer: alkene with et]

_____ _____
Newman projection 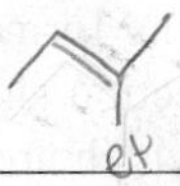   Final Answer

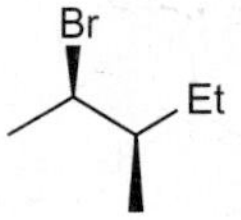

10.8

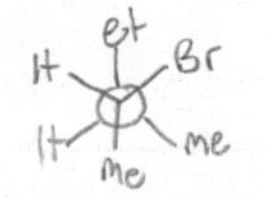

Br [structure with Et]

[Newman projection with H, et, Br, H, me, me, Me]

[Final Answer: alkene with et]

_____ _____
Newman projection 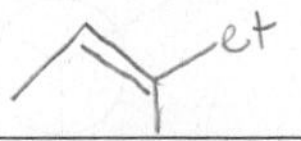   Final Answer

10.4 THE E1 MECHANISM

In an E1 process, there are two separate steps: the leaving group first leaves, generating a carbocation intermediate, which then loses a proton in a separate step:

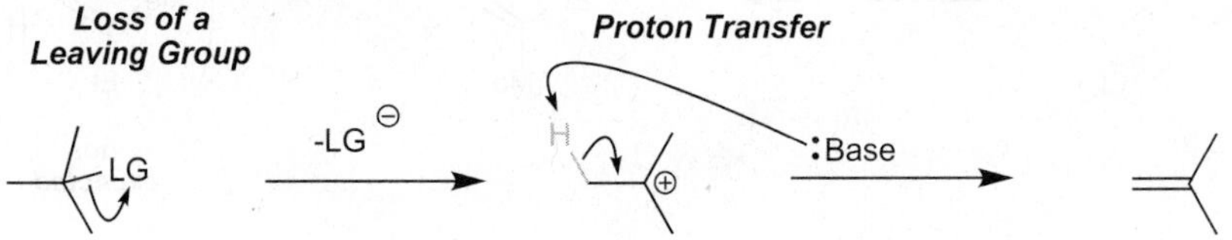

Loss of a Leaving Group   **Proton Transfer**

[reaction scheme showing LG loss, carbocation with H, and :Base giving alkene]

$-LG^{\ominus}$

The first step (loss of the leaving group) is the rate-determining step, much like we saw for S_N1 processes. The base does not participate in this step, and therefore, the concentration of the base does not affect the rate. Because this step involves only one chemical entity, it is said to be unimolecular. Unimolecular elimination reactions are called E1 reactions, where the "1" stands for "unimolecular."

Notice that the first step of an E1 process is identical to the first step of an S_N1 process. In each process, the first step involves loss of the leaving group to form a carbocation intermediate:

Loss of a Leaving Group   **Nucleophile**   [structure with Nuc]

$-LG^{\ominus}$

Substitution

Carbocation Intermediate

Base   [alkene structure]

Elimination

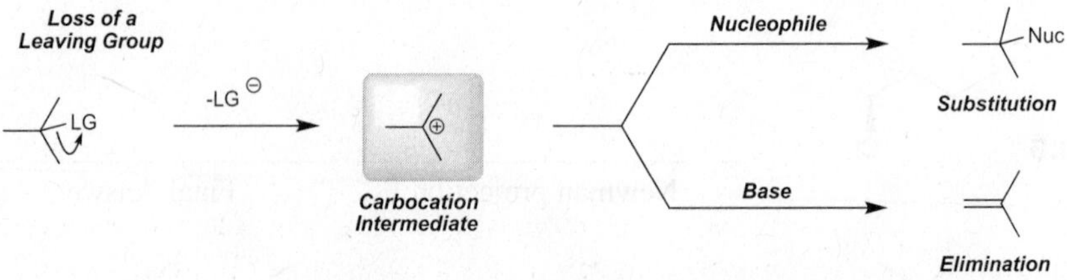

An E1 reaction is generally accompanied by a competing S_N1 reaction, and a mixture of products is generally obtained. At the end of this chapter, we will explore the main factors that affect the competition between substitution and elimination reactions.

For now, let's consider the effect of the substrate on the rate of an E1 process. The rate is found to be very sensitive to the nature of the starting alkyl halide, with tertiary halides reacting more readily than secondary halides; and primary halides generally do not undergo E1 reactions. This trend is identical to the trend we saw for S_N1 reactions, and the reason for the trend is the same as well. Specifically, the rate-determining step of the mechanism involves formation of a carbocation intermediate, so the rate of the reaction will be dependent on the stability of the carbocation (recall that tertiary carbocations are more stable than secondary carbocations).

In the previous chapter, we saw that an OH group is a terrible leaving group, and that an S_N1 reaction can only occur if the OH group is first protonated to give a better leaving group:

Bad
Leaving Group

Excellent
Leaving Group

The same is true with an E1 process. If the substrate is an alcohol, a strong acid will be required in order to protonate the OH group:

$$\text{OH} \xrightarrow[\text{heat}]{\text{H}_2\text{SO}_4} \quad + \quad \text{H}_2\text{O}$$

10.5   THE REGIOCHEMICAL OUTCOME OF AN E1 REACTION

E1 processes show a regiochemical preference for the Zaitsev product, just as we saw for E2 reactions. For example:

$$\text{OH} \xrightarrow[\text{heat}]{\text{H}_2\text{SO}_4}$$

Major   **Minor**

The more-substituted alkene (Zaitsev product) is the major product. However, there is one critical difference between the regiochemical outcomes of E1 and E2 reactions. Specifically, we have seen that the regiochemical outcome of an E2 reaction can often be controlled by carefully choosing the base (sterically hindered or not sterically hindered). In contrast, the regiochemical outcome of an E1 process *cannot* be controlled. The Zaitsev product will generally be obtained.

PROBLEMS Draw the major and minor products that are expected when each of the following substrates is heated in the presence of concentrated sulfuric acid to give an E1 reaction:

10.9

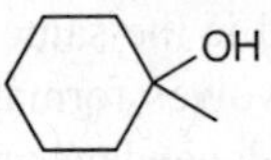

Major Product   Minor Product

10.10

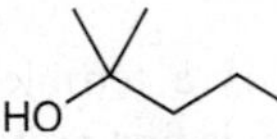

Major Product   Minor Product

10.11

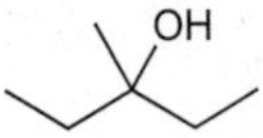

Major Product   Minor Product

10.6 THE STEREOCHEMICAL OUTCOME OF AN E1 REACTION

E1 reactions are not stereospecific. That is, they do not require anti-periplanarity in order for the reaction to occur. Nevertheless, E1 reactions are stereoselective. In other words, when *cis* and *trans* products are possible, we generally observe a preference for formation of the *trans* stereoisomer:

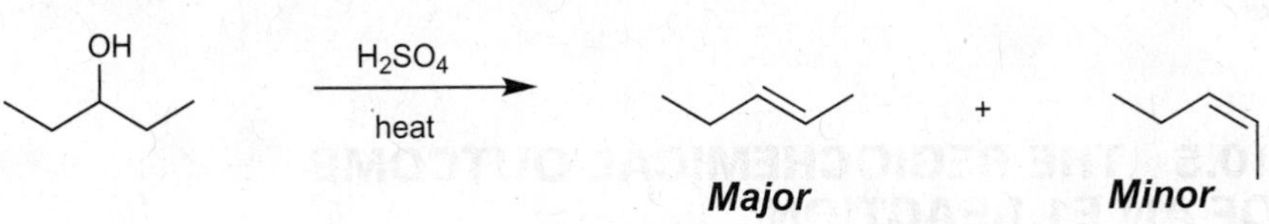

H_2SO_4

heat

Major   +   **Minor**

10.7 SUBSTITUTION VS. ELIMINATION

Substitution and elimination reactions are almost always in competition with each other. In order to predict the products of a reaction, you must determine which mechanism(s) win the competition. In some cases, there is one clear winner. For example, consider a case in which a tertiary alkyl halide is treated with a strong base, such as hydroxide:

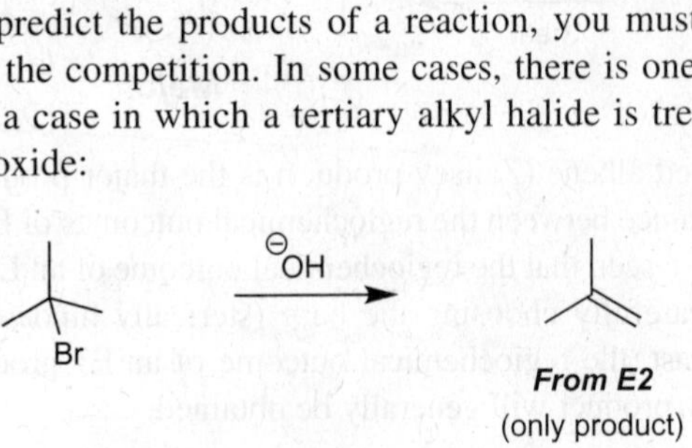

From E2
(only product)

In a case like this, E2 wins the competition, and no other mechanisms can successfully compete. Why not? An S_N2 process cannot occur at a reasonable rate because the substrate is tertiary (steric hindrance prevents an S_N2 from occurring). And unimolecular processes (E1 and S_N1) cannot compete because they are too slow. Recall that the rate-determining step for an E1 or S_N1 process is the loss of a leaving group to form a carbocation, which is a slow step. Therefore, E1 and S_N1 could only win the competition if the competing E2 process is extremely slow (when a weak base is used). However, when a strong base is used, E2 occurs rapidly, so E1 and S_N1 cannot compete.

Now let's consider a case where there is more than one winner, for example:

In this case, there are two winners! Don't fall into the trap of thinking that there must always be one clear winner. Sometimes there is, but sometimes, there are multiple products (perhaps even more than two). The goal is to predict *all* of the products, and to predict which products are major and which are minor. To accomplish this goal, you will need to perform the following three steps:

1. Determine the function of the reagent.

2. Analyze the substrate and determine the expected mechanism(s).

3. Consider regiochemical and stereochemical requirements.

The last three sections of this chapter are devoted to helping you become competent in performing all three steps. Let's begin with Step 1: Determining the function of the reagent.

10.8 DETERMINING THE FUNCTION OF THE REAGENT

We have seen earlier in this chapter that the main difference between substitution and elimination is the function of the reagent. A substitution reaction occurs when the reagent functions as a nucleophile, while an elimination reaction occurs when the reagent functions as a base. So the first step in any specific case is to determine whether the reagent is a strong or weak nucleophile, and whether it is a strong or weak base. Students generally assume that a strong base must also be a strong nucleophile, but this is not always true. It is possible for a reagent to be a weak nucleophile and a strong base. Similarly, it is possible for a reagent to be a strong nucleophile and a weak base. In other words, basicity and nucleophilicity do not always parallel each other. Let's begin by seeing when they *do* parallel each other.

When comparing atoms *in the same row* of the periodic table, basicity and nucleophilicity *do* parallel each other:

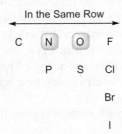

In the Same Row

For example, let's compare H_2N^- and HO^-. The difference between these two reagents is the identity of the atom bearing the charge (O *vs.* N). We already saw in Chapter 3 (when we saw the factors determining charge stability) that oxygen, being more electronegative than nitrogen, can stabilize a charge better than nitrogen can. Therefore, HO^- will be more stable than H_2N^-, so H_2N^- will be a stronger base.

As it turns out, H_2N^- will also be a stronger nucleophile than HO^-, because basicity and nucleophilicity parallel each other when comparing atoms in the same row of the periodic table.

When comparing atoms *in the same column* of the periodic table, basicity and nucleophilicity *do not* parallel each other:

In the Same Column

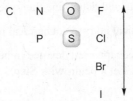

For example, let's compare HO^- and HS^-. Once again, the difference between these two reagents is the identity of the atom bearing the charge (O *vs.* S). We already saw in Chapter 3 that sulfur, being larger than oxygen, can stabilize a charge better than oxygen can (remember we saw that size is more important than electronegativity when comparing atoms in the same column). Therefore, HS^- will be more stable than HO^-, so HO^- will be a stronger base. Nevertheless, HS^- is a better nucleophile than HO^-. Why?

Recall that basicity and nucleophilicity are different concepts. Basicity measures stability of the charge (a thermodynamic argument), whereas nucleophilicity measures how fast a nucleophile attacks something (a kinetic argument). When you have a large atom, like sulfur, an interesting effect comes into play. As the sulfur atom approaches an electrophile (a compound with $\delta+$), the electron density within the sulfur atom gets polarized, meaning that the electron density can move around. This effect increases the force of attraction between the nucleophile and the electrophile, so the rate of attack is very fast. Since *nucleophilicity* is a measure of **how fast** the nucleophile attacks, this effect renders the sulfur atom very nucleophilic. As a result, HS$^-$ functions almost exclusively as a nucleophile and rarely functions as a base. The same is true for most of the halides (especially iodide), which function exclusively as nucleophiles. The halides are generally too weakly basic to function as bases. So, when you see one of these nucleophiles, you do not need to worry about elimination reactions – you will only get substitution reactions. It is very common to see the halides being used as nucleophiles, so it is very helpful to know that you do not need to worry about elimination reactions when you see a halide as the reagent.

Armed with the understanding that nucleophilicity and basicity are not the same concepts, we can now categorize reagents into the following four groups:

NUCLEOPHILE (ONLY)	BASE (ONLY)	STRONG NUC / STRONG BASE	WEAK NUC / WEAK BASE
Halides **Sulfur nucleophiles** Cl$^{\ominus}$ HS$^{\ominus}$ H$_2$S Br$^{\ominus}$ RS$^{\ominus}$ RSH I$^{\ominus}$	H:$^{\ominus}$ ($>$$-$$-O^{\ominus}$)	HO$^{\ominus}$ MeO$^{\ominus}$ $>$$-O^{\ominus}$ EtO$^{\ominus}$	H$_2$O MeOH EtOH

Let's quickly review each of these four categories. The first category contains reagents that function only as nucleophiles. They are strong nucleophiles because they are highly polarizable, but they are weak bases. When you see a reagent from this category, you should focus exclusively on substitution reactions (not elimination). Notice that sulfuric acid is NOT in this category (or any of the categories above). It is true that sulfuric acid contains sulfur, but the sulfur atom in sulfuric acid does not possess a lone pair, so it cannot function as a nucleophile. As its name implies, sulfuric acid functions only as an acid, so it is not listed in any of the four categories above.

The second category contains reagents that function only as bases; not as nucleophiles. The first reagent on this list is the hydride ion, usually shown as NaH, where Na$^+$ is the counter ion. The hydride ion of NaH is not a good nucleophile, despite the presence of a negative charge, because hydrogen is very small so it is not sufficiently polarizable. Nevertheless, the hydride ion is a very strong base. The use of a hydride ion as the reagent indicates that elimination will occur rather than substitution.

Notice that *tert*-butoxide appears in both the second and third categories. Technically, it is a strong nucleophile and a strong base, so it belongs in the third category. But practically, *tert*-butoxide is sterically hindered, which prevents it from functioning as a nucleophile in most cases. Therefore, it is often used as a base, to favor E2 over S$_N$2.

The third category contains reagents that are both strong nucleophiles and strong bases. These reagents include hydroxide (HO$^-$) and alkoxide ions (RO$^-$), and are generally used for bimolecular processes (S$_N$2 and E2).

The fourth and final category contains reagents that are weak nucleophiles and weak bases. These reagents include water (H_2O) and alcohols (ROH), and are generally used for unimolecular processes (S_N1 and E1).

In order to predict the products of a reaction, the first step is determining the identity and nature of the reagent. That is, you must analyze the reagent and determine the category to which it belongs. Let's get some practice with this critical skill.

PROBLEMS Identify the function of each of the following reagents. In each case, the reagent will fall into one of the following four categories:

(a) strong nucleophile and weak base

(b) weak nucleophile and strong base

(c) strong nucleophile and strong base

(d) weak nucleophile and weak base

10.12 _____
Function

10.13 _____
Function

10.14 HO$^{\ominus}$ _____
Function

10.15 Cl$^{\ominus}$ _____
Function

10.16 H^{O}H _____
Function

10.17 HS$^{\ominus}$ _____
Function

10.18 _____
Function

10.19 I$^{\ominus}$ _____
Function

10.9 IDENTIFYING THE MECHANISM(S)

We mentioned that there are three main steps for predicting the products of substitution and elimination reactions. In the previous section, we explored the first step (determining the function of the reagent). In this section, we now explore the second step of the process in which we analyze the substrate and identify which mechanism(s) operates.

As described in the previous section, there are four categories of reagents. For each category, we must explore the expected outcome with a primary, secondary, or tertiary substrate. All of the relevant information is summarized in the following flow chart. It is important to know this flow chart extremely well, but be careful not to memorize it. It is more important to "understand" the reasons for all of these outcomes. A proper understanding will prove to be far more useful on an exam than simply memorizing a set of rules.

Reagent	Substrate	Mechanism	Explanation

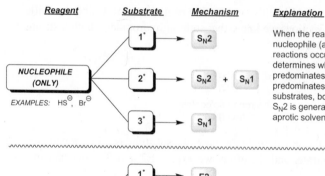

When the reagent functions exclusively as a nucleophile (and not as a base), only substitution reactions occur (not elimination). The substrate determines which mechanism operates. S_N2 predominates for primary substrates, and S_N1 predominates for tertiary substrates. For secondary substrates, both S_N2 and S_N1 can occur, although S_N2 is generally favored (especially when a polar aprotic solvent is used).

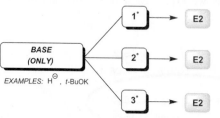

When the reagent functions exclusively as a base (and not as a nucleophile), only elimination reactions occur. Such reagents are generally strong bases, resulting in an E2 process. This mechanism is not sensitive to steric hindrance, and can occur for primary, secondary, and tertiary substrates

When the reagent is both a strong nucleophile and a strong base, bimolecular reactions (S_N2 and E2) are favored.

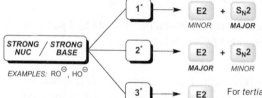

For *primary* substrates, S_N2 predominates over E2, unless *t*-BuOK is used as the reagent, in which case E2 predominates.

For *secondary* substrates, E2 predominates, because E2 is not sterically hindered, while S_N2 exhibits some steric hindrance.

For *tertiary* substrates, only E2 is observed, because the S_N2 pathway is too sterically hindered to occur.

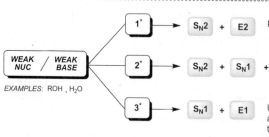

Not practical, because the reactions are too slow

Not practical, because the reactions are too slow, and too many products are formed. However, a secondary alcohol will undergo an E1 reaction when treated with sulfuric acid and heat.

Under these conditions, unimolecular reactions (S_N1 and E1) are favored. High temperature favors E1. A tertiary alcohol will undergo an E1 reaction when treated with sulfuric acid and heat.

The flow chart above can be used to determine which mechanism(s) operate for a specific case. Let's get some practice.

EXERCISE 10.20 Identify the mechanism(s) expected to occur when 3-bromopentane is treated with sodium hydroxide:

ANSWER Our first step is to identify the function of the reagent. Using the skills developed in the previous section, we know that sodium hydroxide is both a strong nucleophile and a strong base:

$$Na^{\oplus} \qquad :\overset{\ominus}{\underset{..}{O}}H$$

strong nucleophile
strong base

Our next step is to identify the substrate. In this case, the substrate is 3-bromopentane, which is a secondary substrate, and therefore, we expect E2 and S_N2 mechanisms to operate:

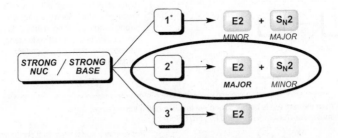

The E2 pathway is expected to provide the major product, because the S_N2 pathway is more sensitive to steric hindrance provided by secondary substrates.

PROBLEMS Identify the mechanism(s) expected to occur in each of the following cases. Do not worry about drawing the products yet. We will do that in the next section. For now, just identify which mechanisms are operating:

10.21
NaOEt
Br

10.22
H_2SO_4
heat
OH

10.23
H_2O
I

10.24
NaBr
Cl

10.25
NaBr
Cl

10.10   PREDICTING THE PRODUCTS

We mentioned that predicting the products of substitution and elimination reactions requires three discrete steps:

1. Determine the function of the reagent.
2. Analyze the substrate and determine the expected mechanism(s).
3. Consider any relevant regiochemical and stereochemical requirements.

In the previous two sections, we explored the first two steps of this process. In this section, we will explore the third and final step. After determining which mechanism(s) are expected to operate, the final step is to consider the regiochemical and stereochemical outcomes for each of the expected mechanisms. The following table provides a summary of guidelines that must be followed when drawing products.

	Regiochemical Outcome	**Stereochemical Outcome**
S_N2	The nucleophile attacks the α position, where the leaving group is connected.	The nucleophile replaces the leaving group with inversion of configuration.
S_N1	The nucleophile attacks the carbocation, which is generally where the leaving group was originally connected, unless a carbocation rearrangement took place.	The nucleophile replaces the leaving group with racemization.
E2	The Zaitsev product is generally favored over the Hofmann product, unless a sterically hindered base is used, in which case the Hofmann product will be favored.	This process is both stereoselective and stereospecific. When applicable, a *trans* disubstituted alkene will be favored over a *cis* disubstituted alkene. When the β position of the substrate has only one proton, the stereoisomeric alkene resulting from antiperiplanar elimination will be obtained (exclusively, in most cases).
E1	The Zaitsev product is always favored over the Hofmann product.	The process is stereoselective. When applicable, a *trans* disubstituted alkene will be favored over a *cis* disubstituted alkene.

The table above does not contain any new information. All of the information can be found in this chapter and in the previous chapter. The table is meant only as a summary of all of the relevant information, so that it is easily accessible in one location. Let's get some practice applying these guidelines.

EXERCISE 10.26 Predict the product(s) of the following reaction, and identify the major and minor products:

Answer In order to draw the products, we must follow these three steps:

1. Determine the function of the reagent.

2. Analyze the substrate and determine the expected mechanism(s).

3. Consider any relevant regiochemical and stereochemical requirements.

We begin by analyzing the reagent. The methoxide ion is both a strong base and a strong nucleophile. Next, we move on to Step 2 and we analyze the substrate. In this case, the substrate is secondary, so we would expect E2 and S_N2 pathways to compete with each other:

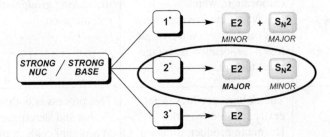

We expect the E2 pathway to predominate, because it is less sensitive to steric hindrance than the S_N2 pathway. Therefore, we would expect the major product(s) to be generated via an E2 process, and the minor product(s) to be generated via an S_N2 process. In order to draw the products, we must complete the third and final step. That is, we must consider the regiochemical and stereochemical outcomes for both the E2 and S_N2 processes. Let's begin with the E2 process.

For the regiochemical outcome, we expect the Zaitsev product to be the major product, because the reaction does not utilize a sterically hindered base:

Br
|
NaOMe
→
Major   +   *Minor*

Next, look at the stereochemistry. The E2 process is stereoselective, so we expect *cis* and *trans* isomers, with a predominance of the *trans* isomer:

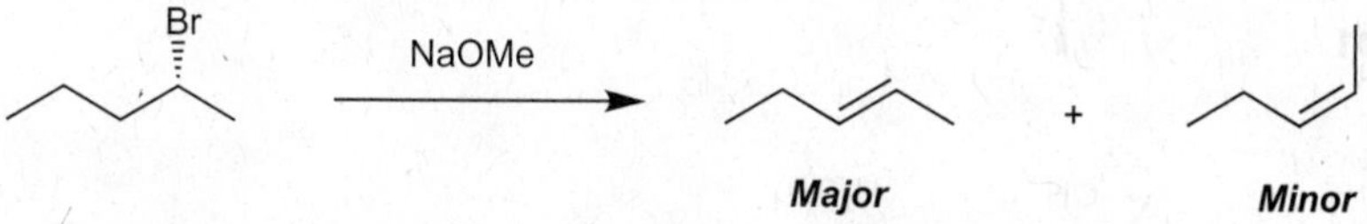

The E2 process is not only stereoselective, but it is also stereospecific. However, in this case, the β position has more than one proton, so the stereospecificity of this reaction is not relevant.

Now consider the S$_N$2 product. This case involves a stereocenter, so we expect inversion of configuration:

In summary, we expect the following products:

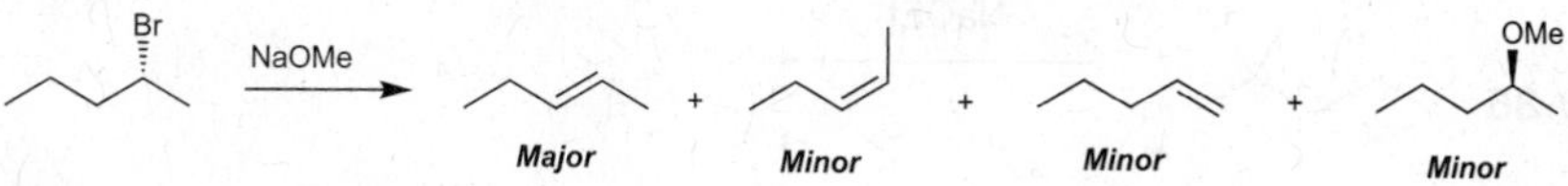

PROBLEMS   Identify the major and minor product(s) that are expected for each of the following reactions.

10.27

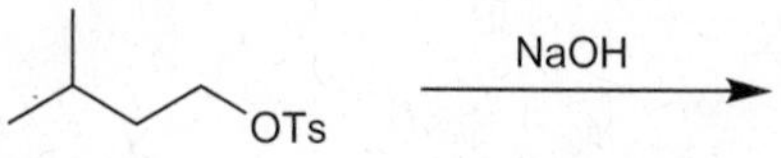

NaOH

10.28

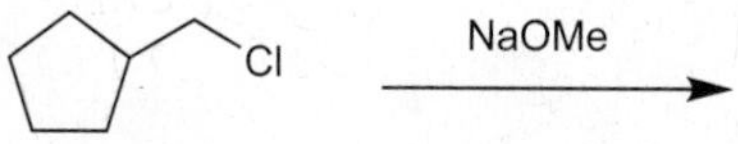

NaOMe

10.29

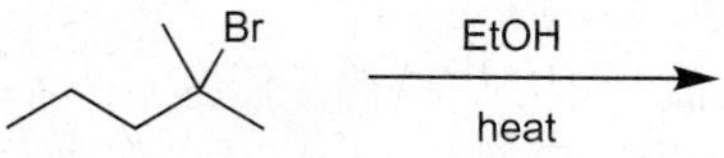

EtOH
heat

10.30

NaOEt

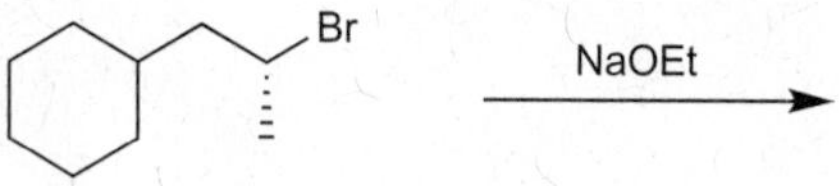

10.31 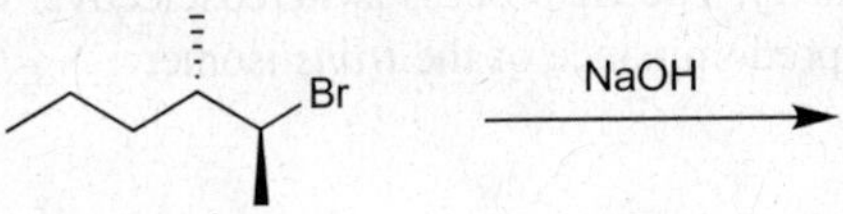
Br
NaOH

10.32 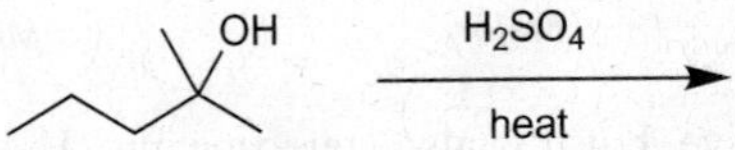
OH
H_2SO_4
heat

10.33 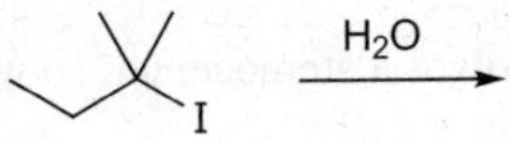
I
H_2O

10.34 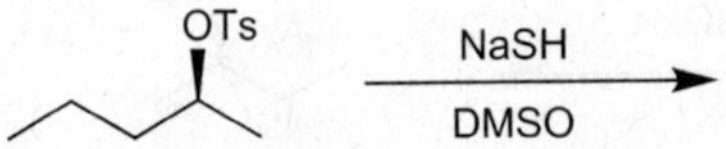
OTs
NaSH
DMSO

10.35 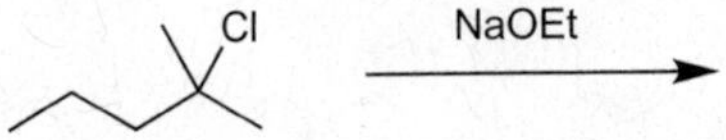
Cl
NaOEt

10.36 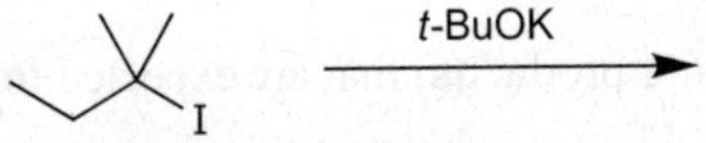
I
t-BuOK

10.37 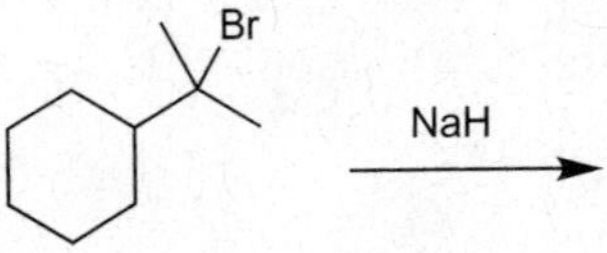
Br
NaH

10.38 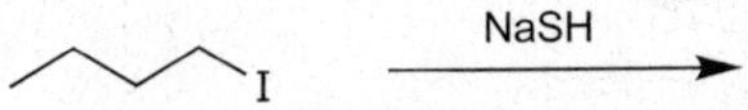
I
NaSH

10.39 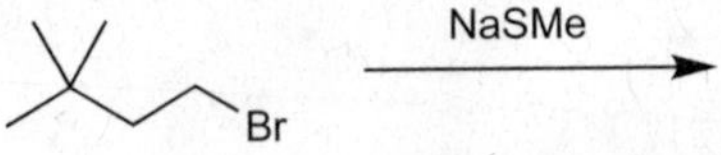
Br
NaSMe

ADDITION REACTIONS

Addition reactions are characterized by two groups adding across a double bond:

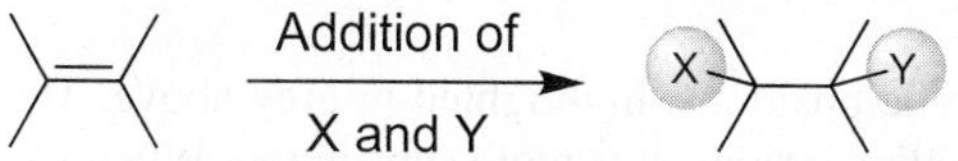

In the process, the double bond is destroyed, and we say that the two groups (X and Y) have "added" across the double bond. In this chapter, we will see many addition reactions, and we will focus on the following three types of problems: (1) predicting the products of a reaction, (2) proposing a mechanism, and (3) proposing a synthesis. In order to gain mastery over these types of problems, you must first become comfortable with some crucial terminology. We will focus on this terminology now, before learning any reactions.

11.1 TERMINOLOGY DESCRIBING REGIOCHEMISTRY

When adding two different groups across an unsymmetrical alkene, there is special terminology describing the *regiochemistry* of the addition. For example, suppose you are adding H and Br across an alkene. Regiochemistry refers to the positioning of the H and the Br in the product: which side gets the H and which side gets the Br?

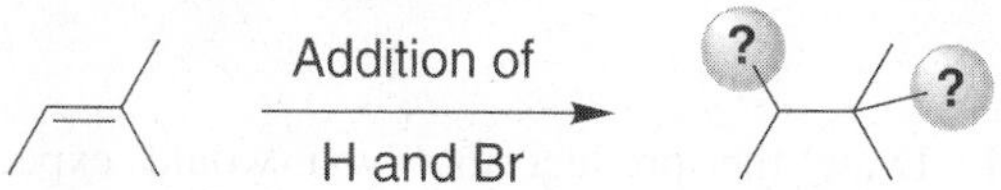

Regiochemistry is only relevant when adding two *different* groups (such as H and Br). However, when adding two of the same group (such as Br and Br), regiochemistry becomes irrelevant:

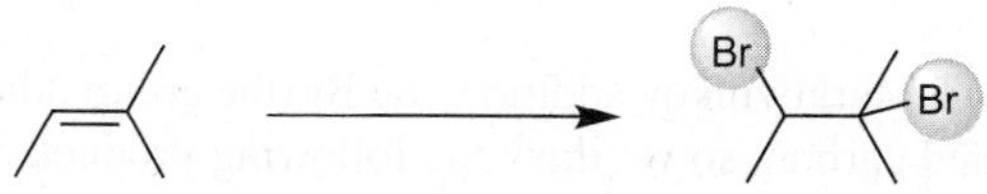

Similarly, when adding two different groups across a *symmetrical* alkene, regiochemistry is also irrelevant:

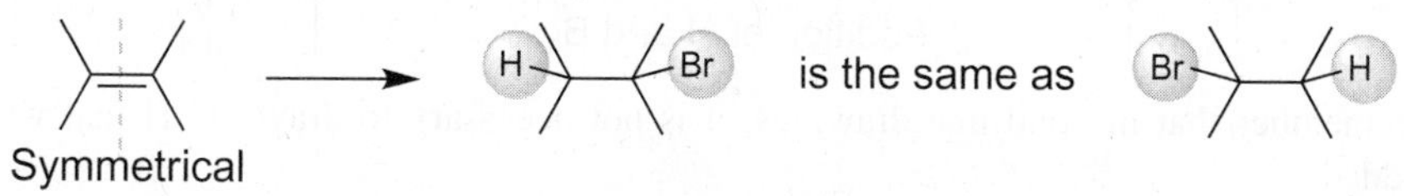

Symmetrical

245

The bottom line is: regiochemistry is only relevant when adding two *different groups* across an *unsymmetrical* alkene.

As we learn addition reactions, we will be using two important terms to describe the regiochemistry: *Markovnikov* and *anti-Markovnikov*. To use these terms properly, we must be able to recognize which carbon is more substituted. Consider the following example:

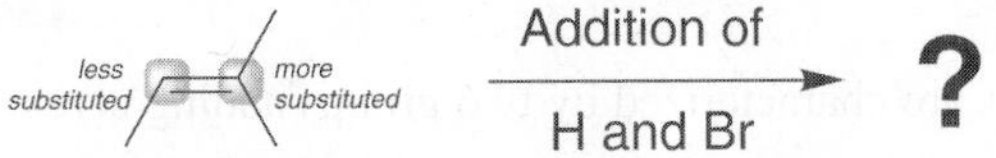

There are two vinylic positions, highlighted in grey above. The vinylic position on the right has more alkyl groups; it is more substituted. When Br ends up on the more substituted carbon, we call it a *Markovnikov addition:*

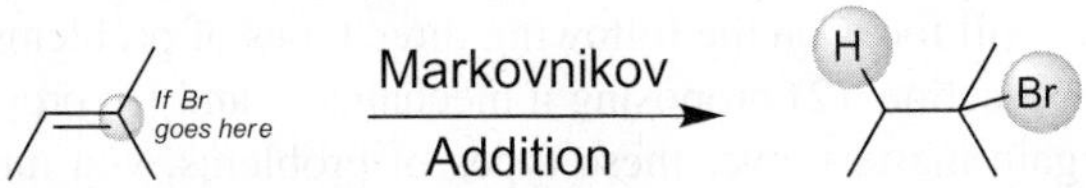

When Br ends up on the less substituted carbon, we call it an *anti-Markovnikov addition:*

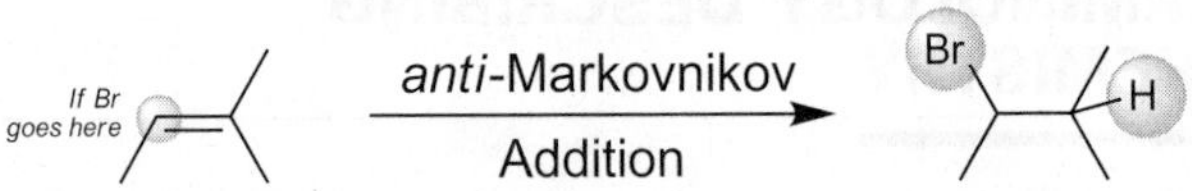

When we explore the mechanisms of addition reactions, we will see why some reactions proceed through a Markovnikov addition while others proceed through an *anti*-Markovnikov addition. For now, let's make sure that we are comfortable using the terms.

EXERCISE 11.1 Draw the product that you would expect from an *anti*-Markovnikov addition of H and Br across the following alkene:

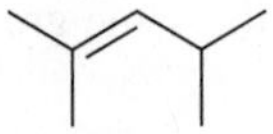

Answer In an *anti*-Markovnikov addition, the Br (the group other than H) ends up at the less substituted carbon, so we draw the following product:

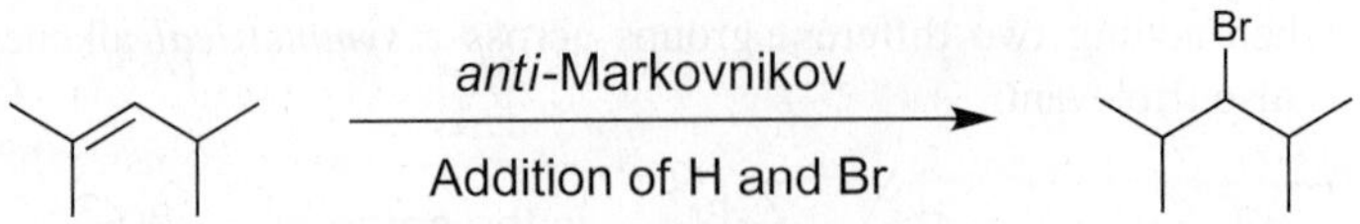

Remember that in bond-line drawings, it is not necessary to draw the H that was added.

PROBLEMS   In each of the following cases, use the information provided to draw the product that you expect.

11.2

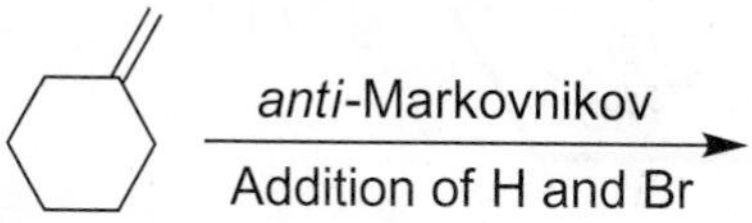

anti-Markovnikov
Addition of H and Br

11.3

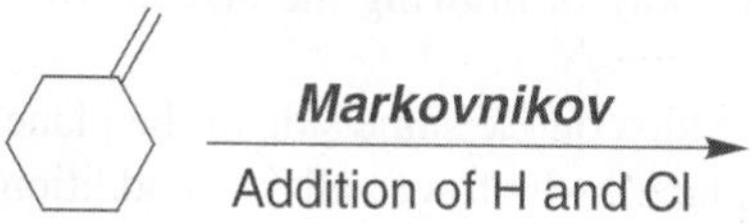

Markovnikov
Addition of H and Cl

11.4

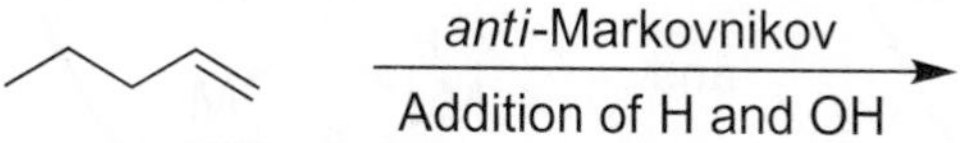

anti-Markovnikov
Addition of H and OH

11.5

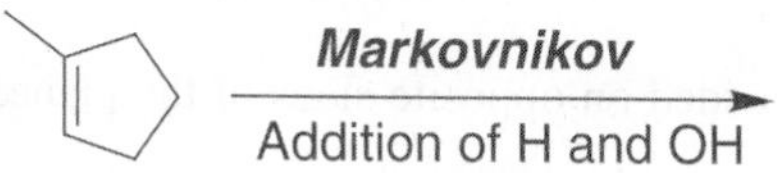

Markovnikov
Addition of H and OH

11.2   TERMINOLOGY DESCRIBING STEREOCHEMISTRY

In addition to the regiochemistry, there is also special terminology used to describe the stereochemistry of a reaction. As an example, consider the following simple alkene:

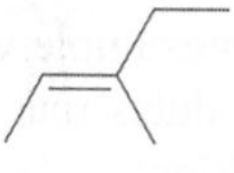

Suppose that we have an *anti*-Markovnikov addition of H and OH across this alkene:

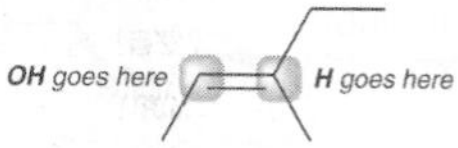

OH goes here   *H goes here*

We know which two groups are adding to the double bond, and we know the regiochemistry of the addition. But in order to draw the products correctly, we also need to know the stereochemistry of the reaction. To better explain this, we will redraw the alkene in a different way.

The *vinylic* carbon atoms, highlighted above, are both sp^2 hybridized, and therefore trigonal planar. As a result, all four groups (connected to the vinylic positions) are in one plane. In order to discuss stereochemistry, we will rotate the molecule so that the plane is coming in and out of the page:

This is an unusual way to draw an alkene (where all bonds are shown as wedges and dashes, rather than straight lines), but this way of drawing the alkene will make it easier to explore stereochemistry.

We can imagine both groups being added on the same side of the plane (either from above the plane or from below the plane), which we call a *syn* addition:

Or, we can imagine both groups being added on opposite sides of the plane, which we call an *anti* addition:

Note: Do not confuse the term *"anti"* with the term *"anti*-Markovnikov." The term *"anti"* describes the stereochemistry, while the term *"anti*-Markovnikov" describes the regiochemistry. It is possible for an *anti*-Markovnikov reaction to be a *syn* addition. In fact, we will see such an example very soon.

We see that there are two products that arise from a *syn* addition, and two products that arise from an *anti* addition:

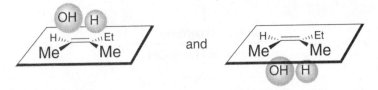

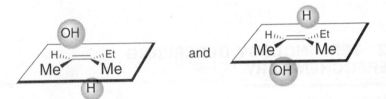

In total, there are four possible products (two pairs of enantiomers). The two products of a *syn* addition represent one pair of enantiomers. And the two products from an *anti* addition represent the other pair of enantiomers.

Some reactions are not stereospecific, and we might expect all four possible products (both pairs of enantiomers). Other reactions *are* stereospecific—we might predominantly get the two enantiomers from a *syn* addition, or we might predominantly get the two enantiomers from an *anti* addition. It is important to know which reactions occur through an *anti* addition, which reactions occur through a *syn* addition, and which reactions are not stereospecific at all. As we go through each reaction in this chapter, we will look closely at the mechanism for each reaction, because the mechanism will always explain the stereochemistry of the reaction. For now, let's make sure that we know the terminology. We will practice drawing the products when all of the information has been provided (which two groups to add, the regiochemistry, and the stereochemistry).

EXERCISE 11.6   Consider the following alkene:

Draw the products that you would expect when adding H and OH in the following way:

   Regiochemistry = *anti*-Markovnikov

   Stereochemistry = *syn* addition

Answer   We begin by looking at the *regio*chemistry. The reaction is *anti*-Markovnikov, which means that the OH group will be positioned at the less substituted carbon:

Next, we look at the *stereo*chemistry. The reaction is a *syn* addition, which means that the H and OH both add on the same side of the double bond. In order to see this more clearly, we rotate the molecule so that the plane of the double bond is coming out of the page, and we draw the pair of enantiomers that we expect from a *syn* addition:

Our products are a pair of enantiomers, so we can record our answer more quickly in the following way:

PROBLEMS For each of the following problems, predict the products using the information provided:

11.7
Addition of OH and OH
anti addition

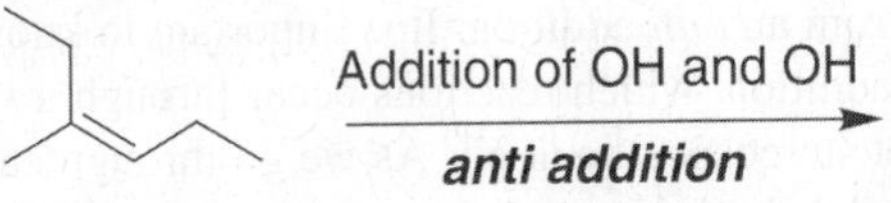

11.8
Addition of H and OH
anti-Markovnikov
syn addition

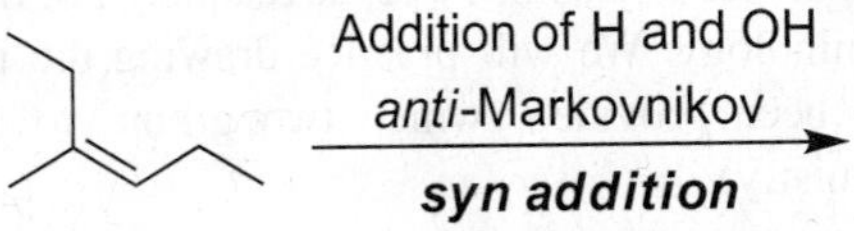

11.9
Addition of OH and OH
syn addition

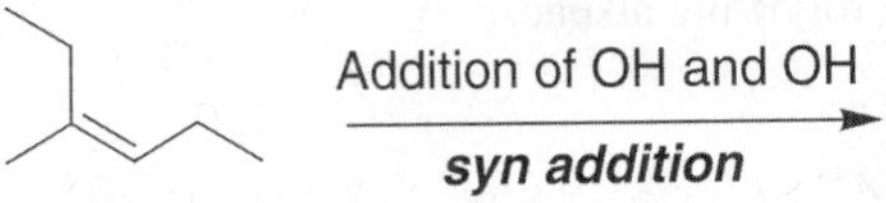

11.10
Addition of H and OH
Markovnikov
not stereospecific

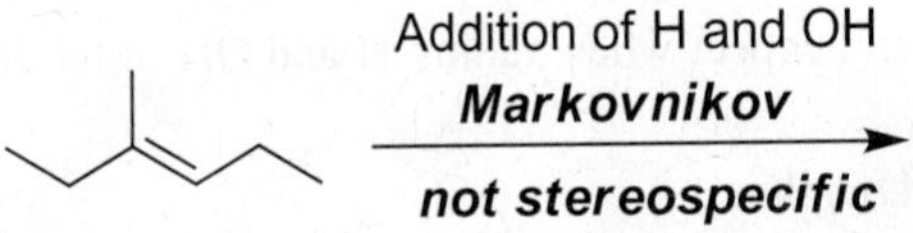

The examples we have seen so far have been acyclic alkenes (not containing a ring). When we add across cyclic alkenes, the products are easier to draw because we don't have to rotate and redraw the alkene before starting. Let's see an example:

EXERCISE 11.11 Predict the products using the following information:

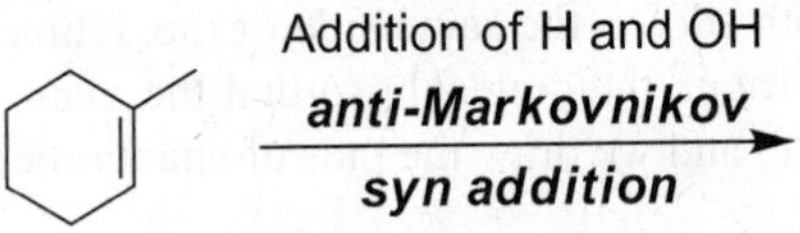

Addition of H and OH
anti-Markovnikov
syn addition

Answer We begin by looking at the regiochemistry. The reaction is *anti*-Markovnikov, which means that the OH group will be positioned at the less substituted carbon:

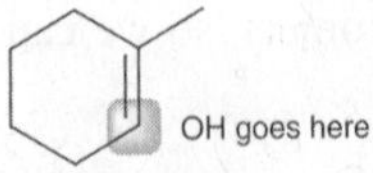

OH goes here

Next, we look at the stereochemistry. The reaction is a *syn* addition, which means that the H and OH both add on the same side of the double bond. Since the alkene is a cyclic compound, the products are easy to draw (without having to rotate the alkene first). We simply place the groups on wedges and dashes, like this:

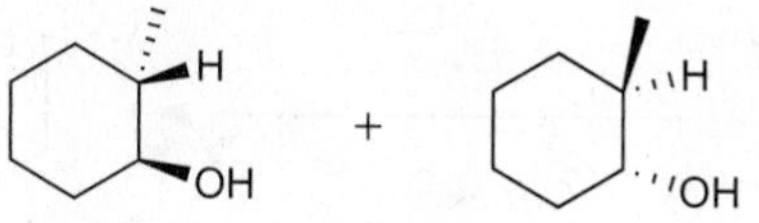

Our two products represent a pair of enantiomers, so we can record our answer more quickly in the following way:

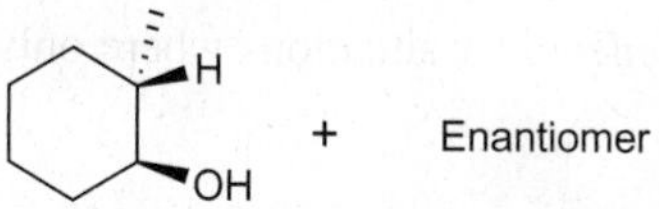

+ Enantiomer

PROBLEMS For each of the following problems, draw the products using the information provided:

11.12

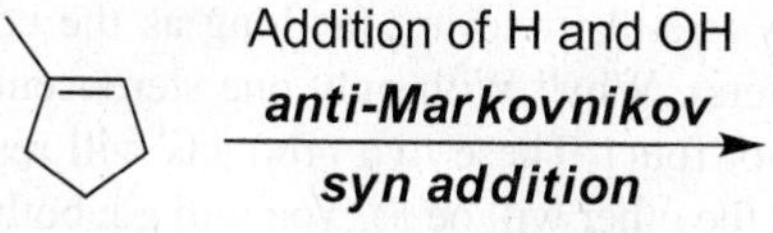

Addition of H and OH
anti-Markovnikov
syn addition

11.13

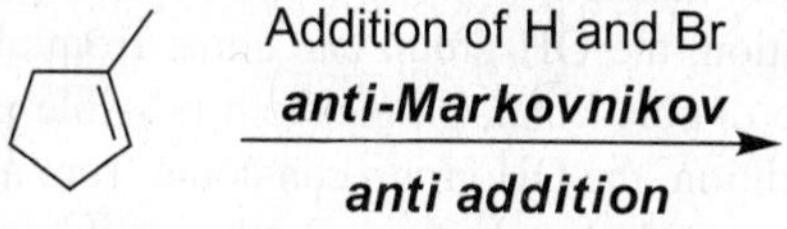

Addition of H and Br
anti-Markovnikov
anti addition

11.14

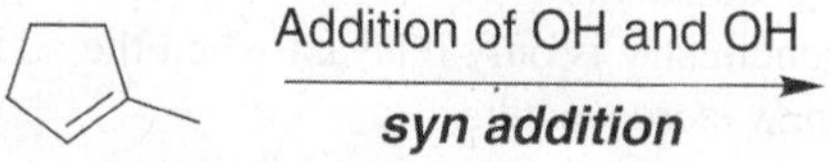

Addition of OH and OH
syn addition

11.15

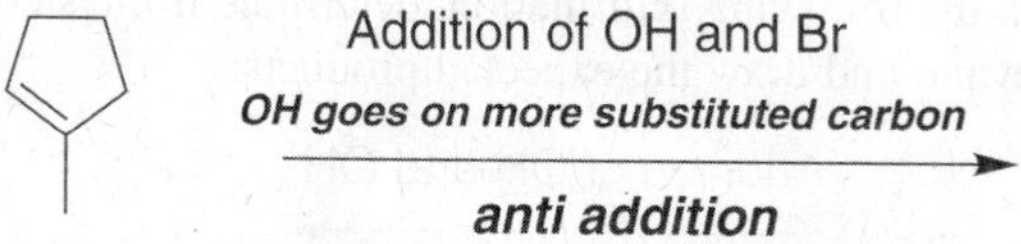

Addition of OH and Br
OH goes on more substituted carbon
anti addition

In all of the examples we have seen so far, we were creating *two* stereocenters:

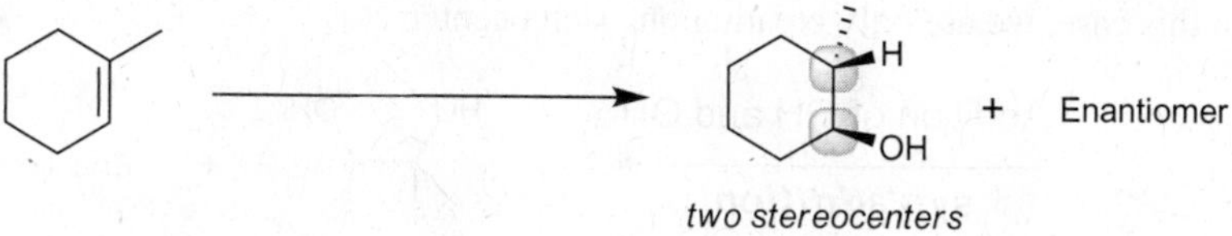

+ Enantiomer

two stereocenters

However, you may encounter examples where no stereocenters are being formed. For example,

no stereocenters

In situations like this, the stereochemistry is irrelevant. You will only get one product (no stereoisomers).

Similarly, you may encounter situations where only *one* stereocenter is being formed. For example,

only one stereocenter

In cases like this, the stereochemistry is still irrelevant (as long as the compound does not possess any other stereocenters). Why? With only one stereocenter, there will only be two possible products (not four). These two products will represent a pair of enantiomers (one will be *R* and the other will be *S*). You will get both of these products whether the reaction proceeds through a *syn* addition *or* through an *anti* addition. If the reaction is a *syn* addition, the OH group can come from above the plane or from below the plane of the double bond, giving both possible products. Similarly, if the reaction is an *anti* addition, the OH group can come from above the plane or from below the plane of the double bond, giving both possible products. Either way, we get the two possible products.

The bottom line is: the stereochemistry is only relevant when the addition reaction involves the creation of *two* new stereocenters.

EXERCISE 11.16   Given the following information, determine if the stereochemistry of the reaction is relevant, and draw the expected products:

Addition of OH and OH

syn addition

Answer   We begin by looking at the regiochemistry. We are not told what the regiochemistry is, because it is irrelevant (we are adding two groups of the same kind: OH and OH). Next, we look at the stereochemistry. The reaction is a *syn* addition. But in this case, we are only creating one stereocenter:

Addition of OH and OH

syn addition

Therefore, the fact that the reaction proceeds through a *syn* addition is not important for predicting the products. If the reaction had been an *anti* addition, we would have obtained the same products. In fact, if the reaction had not been stereospecific at all, we still would have obtained the same two products (the pair of enantiomers above).

PROBLEMS For each of the following problems, draw the products using the information provided:

11.17

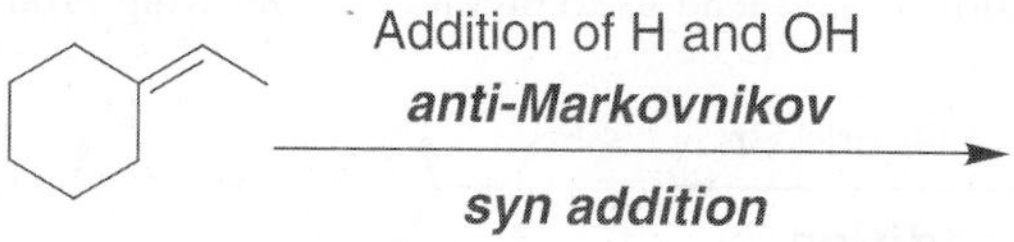

Addition of H and OH
anti-Markovnikov
syn addition

11.18

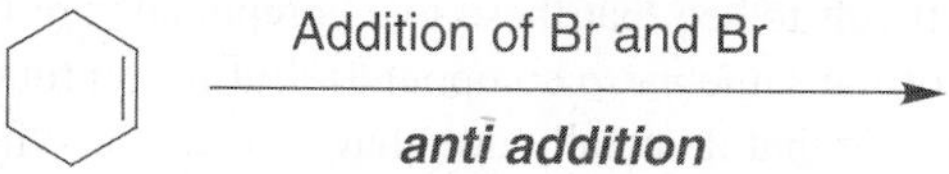

Addition of Br and Br
anti addition

11.19

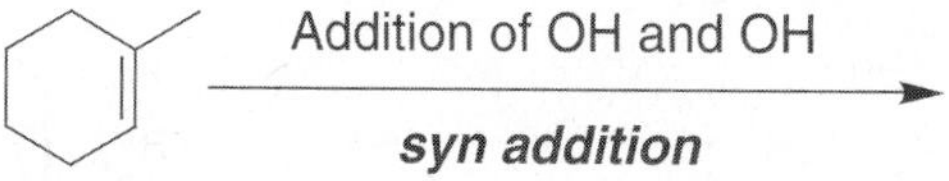

Addition of OH and OH
syn addition

11.20

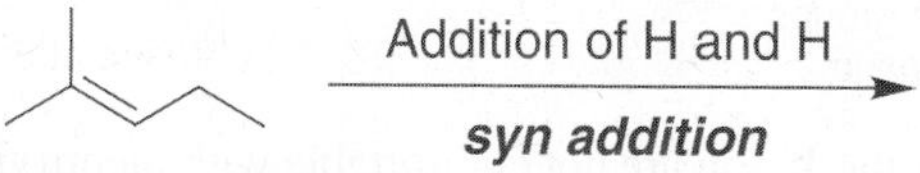

Addition of H and H
syn addition

We are almost ready to begin learning the actual reactions. But first, we must explore one more subtlety associated with the stereochemistry of addition reactions. Consider the following example:

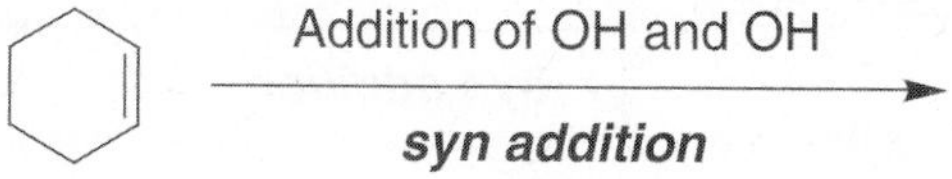

Addition of OH and OH
syn addition

An analysis of the information given should lead us to draw the following products:

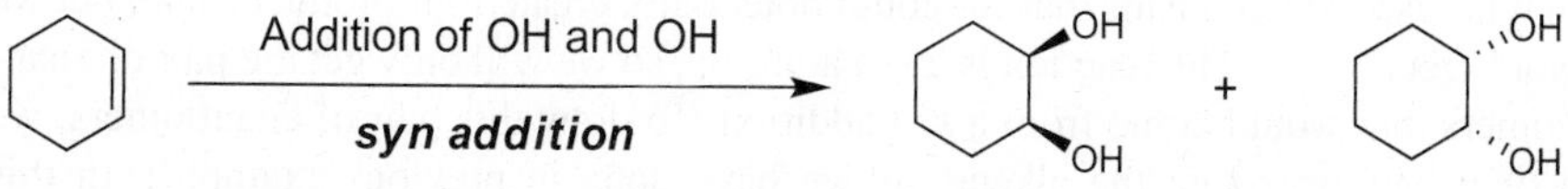

Addition of OH and OH
syn addition

OH
OH

+

OH
OH

However, there are not two products here. Look closely and you will see that the two drawings above are actually the same compound. This compound is a *meso* compound, because it has an internal plane of symmetry. In a case like this, there is actually only one product. You can either draw the product with both OH groups on wedges, or you can draw it with both OH groups on dashes. Either way, you are drawing the same product. Just make sure not to draw *both* drawings, because that would imply that you don't recognize that it is a *meso* compound.

Here is another case where it is not so simple to see:

Addition of Br and Br

anti addition

An analysis of the information should lead us to predict the following products:

Addition of Br and Br

anti addition

At first glance, it might be difficult to see that these two compounds are the same (they represent two drawings of the same *meso* compound). But if you rotate about the C—C single bond, you can see that it actually does have a plane of symmetry:

Rotate about this C-C single bond

and these three groups
get rotated in space...

Plane of Symmetry

This is a subtle but important point. If you are not comfortable with identifying *meso* compounds, you should go back and review *meso* compounds in Chapter 7.

EXERCISE 11.21 Given the following information, draw the expected products:

Addition of OH and OH

syn addition

Answer We begin by looking at the regiochemistry. In this case, we are adding two groups of the same kind (OH and OH), so the regiochemistry is irrelevant.

Next, we look at the stereochemistry. We are creating two new stereocenters in this case, which means that we could potentially create four products here, but we won't get all four. The reaction is a *syn* addition, so we will only get the pair of enantiomers that would come from a *syn* addition. To draw this pair of enantiomers, we do not have to rotate the alkene, as we have done in previous examples. In this example, it is simple enough to draw the products without rotating the alkene (sometimes, it will be simpler to do it this way):

But wait! As a final step, we must determine whether the products are actually a pair of enantiomers or whether they are just two different ways to draw one *meso* compound. We look for a plane of symmetry, and in this case, we do have a plane of symmetry. Therefore, we don't draw both drawings above as our answer. We choose one drawing (either one). So our answer would look like this:

meso

PROBLEMS For each of the following, predict the products using the information provided. (Some products might be *meso* compounds, and others might not—be careful.)

11.22

Addition of OH and OH

anti addition

11.23

Addition of OH and OH

anti addition

11.24

Addition of Br and Br

anti addition

11.25

Addition of Br and Br

anti addition

Until now, we have learned the basic terminology that you will need in order to predict products of addition reactions. To summarize, there are three pieces of information that you must have in order to predict products:

1. Which two groups are being added across the double bond (X and Y)?

2. What is the regiochemistry? (Markovnikov or *anti*-Markovnikov)

3. What is the stereochemistry? (*syn* or *anti*)

With these three pieces of information, you should be able to predict products with ease. Until now, you have been given all three pieces of information in each problem. However, as we progress through this chapter, this information will not be given to you. Rather, you will have to look at the reagents being used, and you will have to determine all three pieces of information on your own. That might sound like it involves a lot of memorization. Not so. We will soon see that the mechanism of each reaction contains all three pieces of information that you need. By understanding the mechanism for each reaction, you will "know" all three pieces of information about each reaction. We will focus on understanding, rather than memorization.

11.3 ADDING H AND H

It is possible to add H and H across an alkene. Here are two examples:

In this type of reaction, called *hydrogenation,* the regiochemistry will always be irrelevant, regardless of what alkene we use (we are adding two of the same group). However, we do need to explore the stereochemistry of hydrogenation reactions. In order to do this, let's take a close look at how the reaction takes place.

Notice the reagents that we use to accomplish a hydrogenation reaction (H_2 and a metal catalyst). A variety of metal catalysts can be used, such as Pt, Pd, or Ni. The hydrogen molecules (H_2) interact with the surface of the metal catalyst, effectively breaking the H—H bonds:

This forms individual hydrogen atoms adsorbed to the surface of the metal. These hydrogen atoms are now available for addition across the alkene. The addition reaction begins when the alkene coordinates with the metal surface:

Surface chemistry then allows for the following two steps, effectively adding H and H across the alkene:

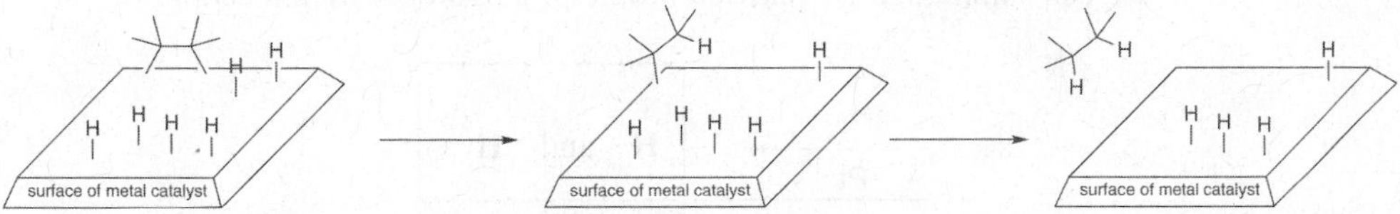

Notice that the alkene grabs both hydrogen atoms *on the same side of the alkene.* Therefore, we get a *syn* addition.

The requirement for *syn* addition can be seen in the following example:

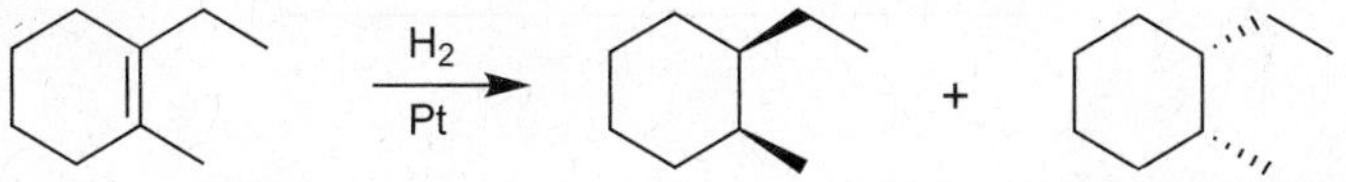

The hydrogen atoms that were added are not explicitly shown, but remember that H's don't need to be drawn in bond-line drawings. You should be able to see them, even though they are not drawn.

In the example above, we are creating two new stereocenters. So, theoretically, we could imagine four possible products (two pairs of enantiomers):

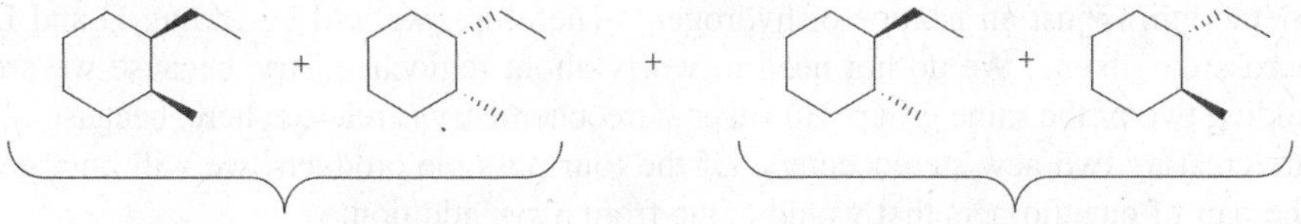

Pair of enantiomers   Pair of enantiomers

But we don't get all four products. We only get the pair of enantiomers that come from a *syn* addition (above left):

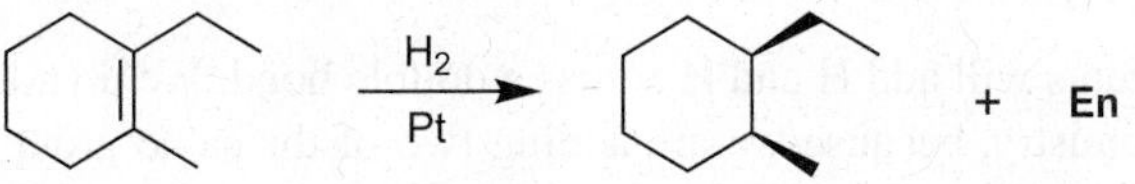

But be careful—make sure to be on the lookout for *meso* compounds. Consider this example:

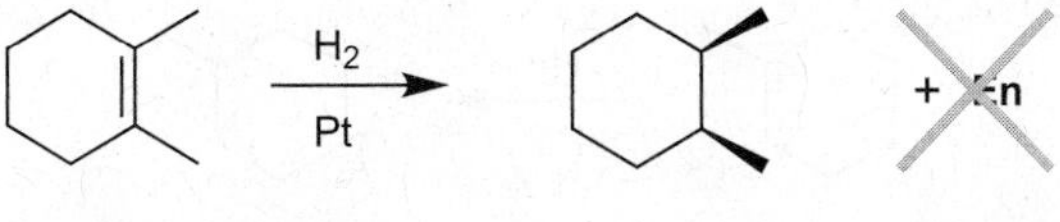

In this example, we would not write "+ Enantiomer," because the product is a *meso* compound.

Now we can summarize the reaction profile of a hydrogenation reaction:

$\xrightarrow[\text{Pt}]{\text{H}_2}$	**H** and **H**
Regiochem	not relevant
Stereochem	*syn*

EXERCISE 11.26 Predict the products for each of the following reactions:

(a)

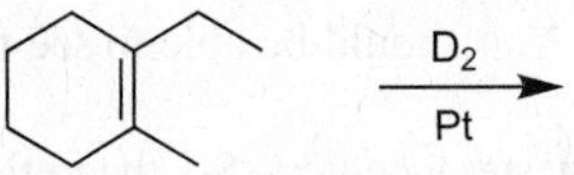

$$\xrightarrow[\text{Pt}]{\text{D}_2}$$

(b)

$$\xrightarrow[\text{Pd}]{\text{H}_2}$$

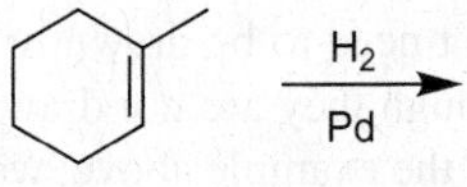

Answer **(a)** Just as we can hydrogenate an alkene, we can also deuterate an alkene (deuterium is just an isotope of hydrogen). Therefore, we will be adding D and D across the alkene. We do not need to worry about regiochemistry, because we are adding two of the same group. However stereochemistry *is* relevant here, because we are creating two new stereocenters. Of the four possible products, we will only get the pair of enantiomers that would come from a *syn* addition:

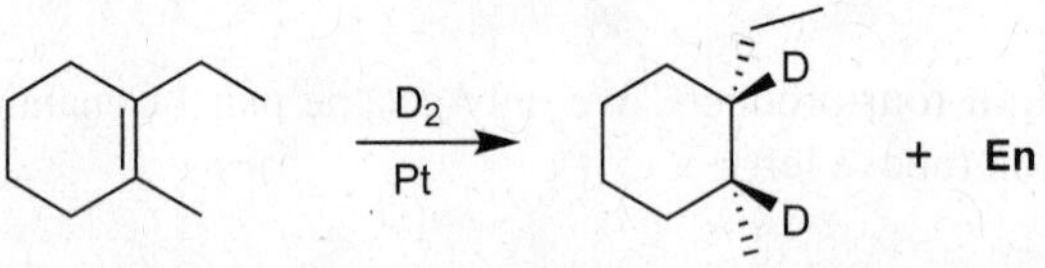

$$\xrightarrow[\text{Pt}]{\text{D}_2}$$

+ En

(b) These reagents will add H and H across a double bond. We do not need to worry about regiochemistry, because we are adding two of the same group. To determine whether stereochemistry is relevant, we must ask whether we are creating two new stereocenters. In this example, we are *not* creating two new stereocenters. In fact, we are not even creating one stereocenter. Therefore, stereochemistry is irrelevant in this example. There will only be one product here:

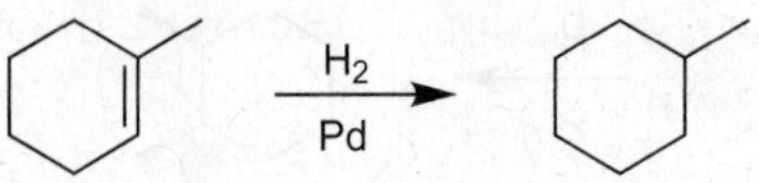

$$\xrightarrow[\text{Pd}]{\text{H}_2}$$

PROBLEMS Predict the products for each of the following reactions. In each example, make sure to determine whether or not you are forming two stereocenters. If not, then the stereochemistry will be irrelevant.

11.27 $\xrightarrow[\text{Ni}]{\text{H}_2}$

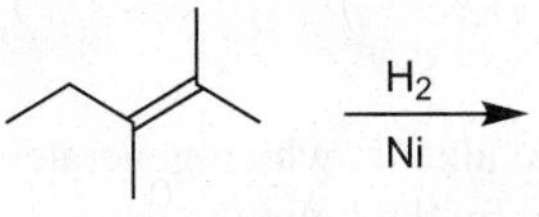

11.28 $\xrightarrow[\text{Pd}]{\text{H}_2}$

11.29 $\xrightarrow[\text{Pt}]{\text{H}_2}$

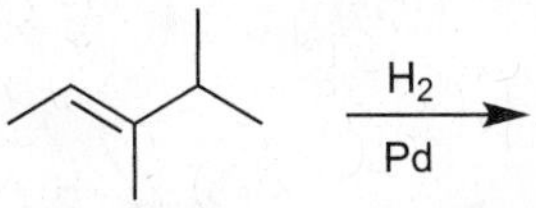

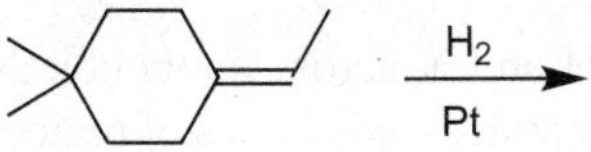

11.30 $\xrightarrow[\text{Pt}]{\text{H}_2}$

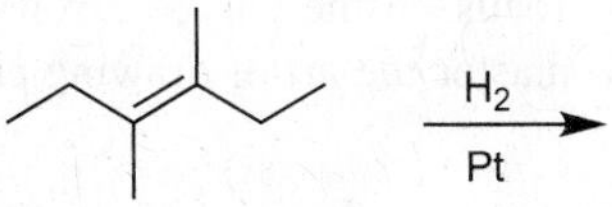

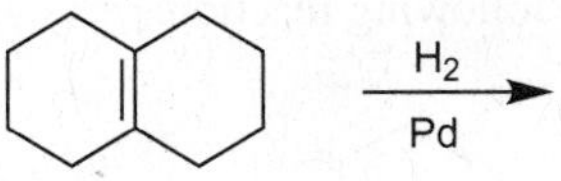

11.31 $\xrightarrow[\text{Pd}]{\text{H}_2}$

11.32 $\xrightarrow[\text{Pd}]{\text{H}_2}$

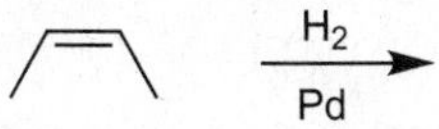

11.4   ADDING H AND X, MARKOVNIKOV

We will now explore the details of adding a hydrogen halide (HX) across a double bond. Here are two examples:

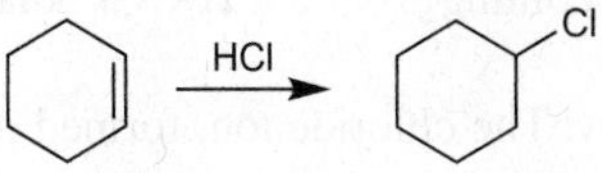

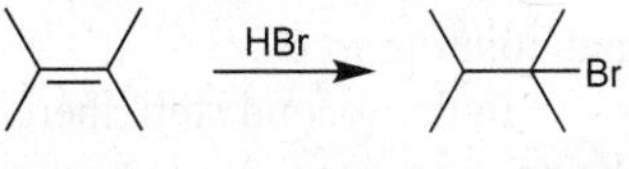

In order to understand the regiochemistry and stereochemistry of HX addition, we must analyze the proposed mechanism. When adding HX across a double bond, there are two key steps involved in the mechanism:

Step 1: *Proton Transfer:*

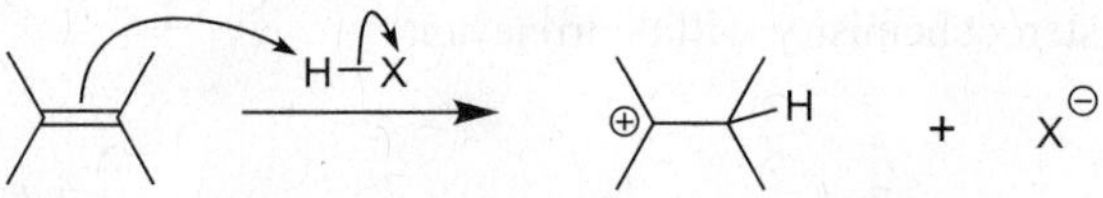

In this step, a proton is transferred to the alkene, which generates a carbocation. This carbocation is then attacked by the halide in step 2:

Step 2: *Nucleophilic Attack:*

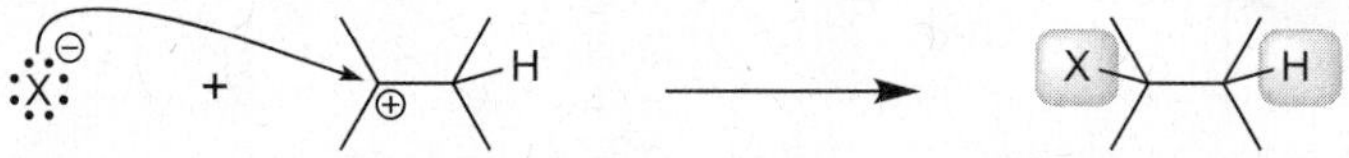

The overall result is the addition of H and X across the double bond. We have specifically used a starting alkene that avoids issues of regiochemistry or stereochemistry; we will soon see other examples in which we must explore both of those issues. For now, focus on the curved arrows used in both steps. It is absolutely critical to master the art of drawing curved arrows properly. Let's quickly practice:

EXERCISE 11.33 Draw a mechanism for the following reaction:

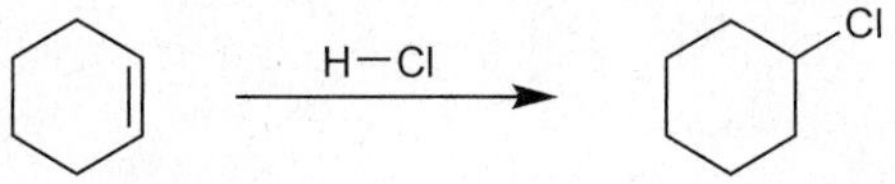

Answer In the first step, there are two curved arrows:

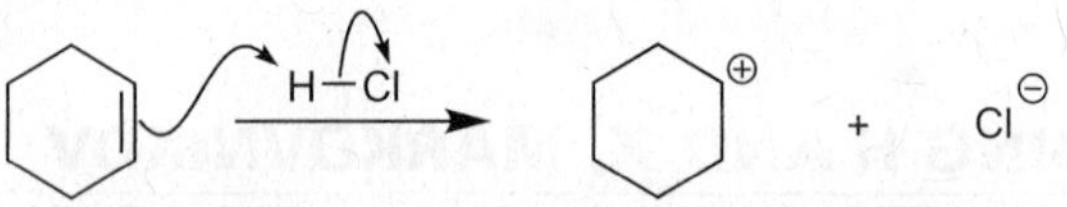

One curved arrow is drawn coming *from the alkene* and pointing *to the proton* (take special notice of this arrow, as it is a *very* common mistake to draw this arrow in the wrong direction). The second curved arrow is drawn coming *from* the H—Cl bond and pointing *to* Cl.

In the second step, there is just one curved arrow. The chloride ion, formed in the previous step, now attacks the carbocation:

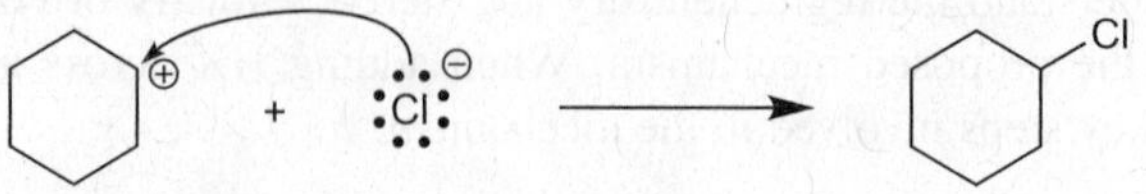

PROBLEMS   Draw a mechanism for each of the following reactions:

11.34

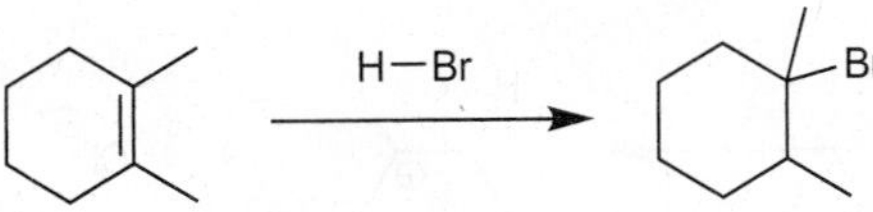

H—Br

11.35

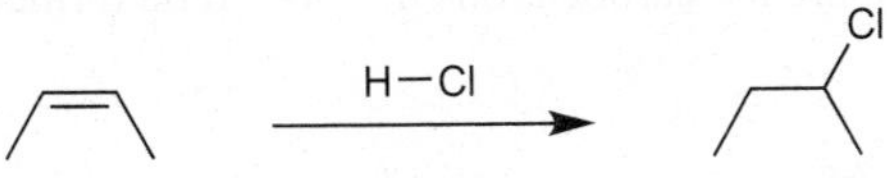

H—Cl

11.36

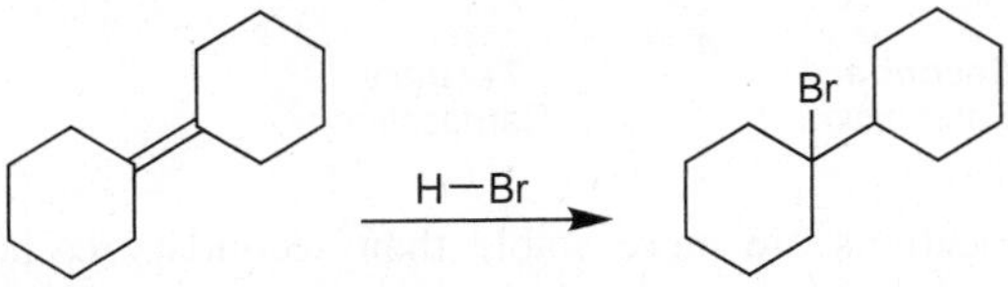

H—Br

11.37

H—I

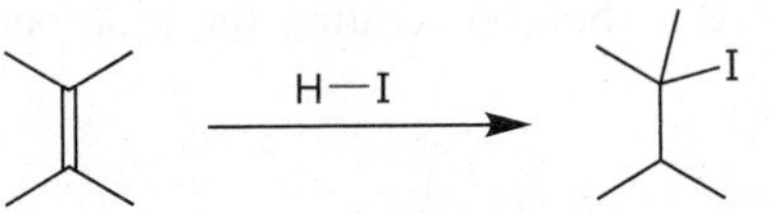

All of the examples above were symmetrical alkenes, so regiochemistry was not relevant. Now let's consider a case where regiochemistry is relevant. With an unsymmetrical alkene, we must decide where to put the H and where to put the X. For example,

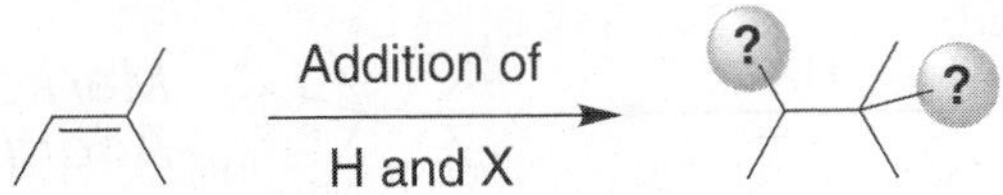

Addition of

H and X

In other words, we must determine whether the reaction is a Markovnikov addition or an *anti*-Markovnikov addition. As promised, the answer to this question is contained in the mechanism. In the first step of the mechanism, a proton was transferred to the alkene, to form a carbocation. When starting with an unsymmetrical alkene, we are confronted with two possible carbocations that can form (depending on where we place the proton):

Do we transfer the proton to the more substituted carbon:

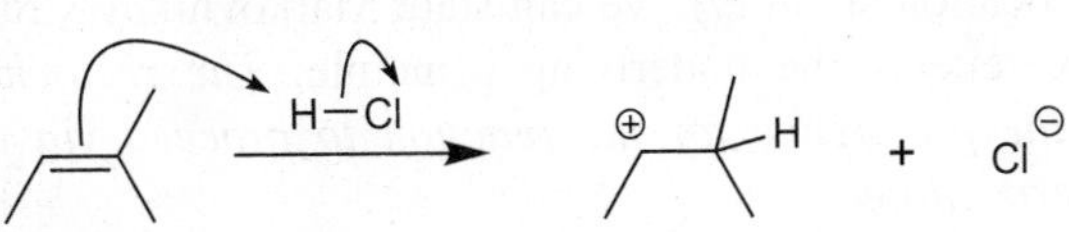

H—Cl   +   Cl⁻

Or do we transfer the proton to the less substituted carbon:

To answer this question, we compare the carbocations that would be formed in each scenario:

Secondary
Carbocation

Tertiary
Carbocation

Recall that tertiary carbocations are more stable than secondary carbocations. When given the choice, we expect the alkene to accept the proton in such a way as to form the more stable carbocation intermediate. In order to accomplish this, the proton must add to the less substituted carbon, generating the more substituted carbocation:

proton goes here   *carbocation ends up here*

The last step of the mechanism involves the halide attacking the carbocation. As a result, the halide will end up on the more substituted carbon (where the carbocation was). Therefore, this reaction is said to follow Markovnikov's rule:

HX   *Markovnikov Addition*

X ends up on more substituted carbon

As we saw in the previous section, Markovnikov's rule tells us to place the H on the less substituted carbon, and to place the X on the more substituted carbon. The rule is named after Vladimir Markovnikov, a Russian chemist, who first showed the regiochemical preference of HBr additions to alkenes. When Markovnikov recognized this pattern in the late 19th century, he stated the rule in terms of the placement *of the proton* (specifically, that the proton will end up on the less substituted carbon atom). Now that we understand the reason for the regiochemical preference (carbocation stability), we can state Markovnikov's rule in a way that more accurately reflects the underlying principle: *The regiochemistry will be determined by the preference for the reaction to proceed via the more stable carbocation intermediate.*

Notice that the regiochemistry of this reaction is explained by the mechanism. Don't try to memorize that the regiochemistry of this reaction is Markovnikov. Rather, try to "understand why" it must be that way.

In any reaction, the mechanism should explain not only the regiochemistry, but the stereochemistry as well. In this particular reaction (addition of H—X across alkenes), the stereochemistry is generally not relevant. Recall from the previous section that we need to consider stereochemistry (*syn* vs. *anti*) *only* in cases where the reaction generates *two* new stereocenters. If only one stereocenter is formed, then we expect a pair of enantiomers (racemic mixture), regardless of whether the reaction was *anti* or *syn*. You will probably not see an example where two new stereocenters are formed, because the stereochemical outcome in such a case is complex and is beyond the scope of our conversation.

The details of this reaction can now be summarized with the following chart:

$\xrightarrow{\text{HX}}$	**H** and **X**
*Regio*chem	Markovnikov
*Stereo*chem	Beyond the scope of this course

EXERCISE 11.38   Predict the product of the following reaction:

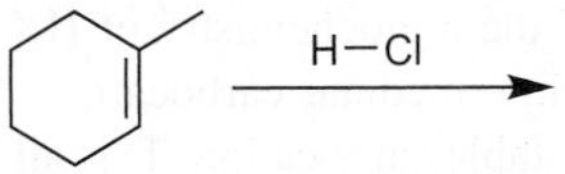

$$\text{H—Cl} \longrightarrow$$

Answer   We begin by focusing on the regiochemistry. This alkene is unsymmetrical, so we must decide where to place the H and where to place the Cl. To do this, we must identify which carbon is more substituted:

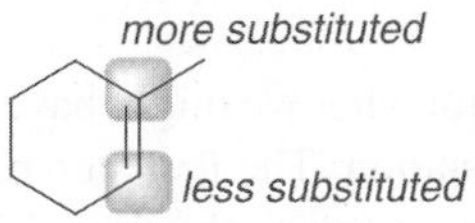

more substituted

less substituted

This reaction proceeds according to Markovnikov's rule, which tells us to place the H on the less substituted carbon, and to place the Cl on the more substituted carbon:

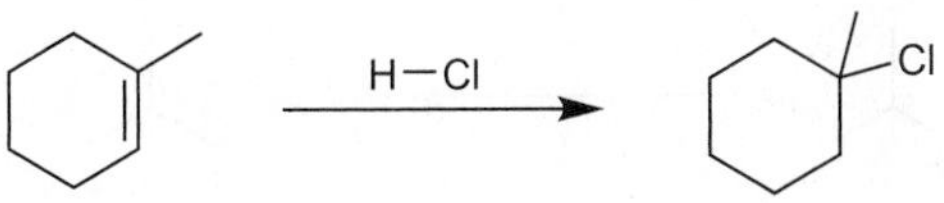

We do not need to think about stereochemistry here, because the product does not contain two stereocenters (in fact, it doesn't even have one stereocenter). Therefore, stereochemistry is irrelevant in this example. As we have said, the stereochemistry of this reaction (H—X addition) will generally not be relevant in the problems that you will encounter.

PROBLEMS Predict the products for each of the following reactions. After finishing each problem, also try to draw the mechanism so that you can "see" exactly why the reaction proceeds through Markovnikov's rule.

11.39
H—Br

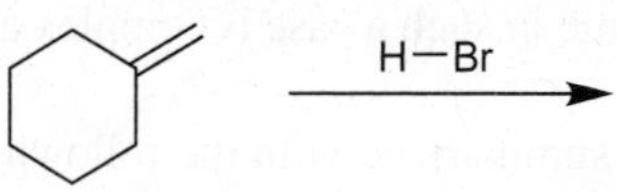

11.40
H—Cl

11.41
H—Br

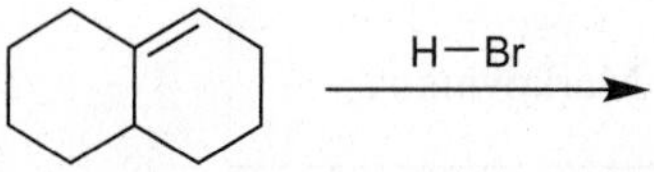

11.42
H—I

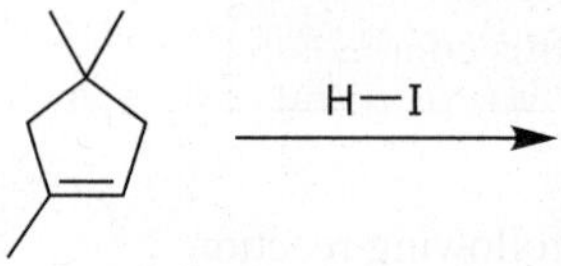

In order to understand the regiochemistry of HX additions to alkenes, we focused our attention on the intermediate carbocation. We argued that the reaction would proceed via the more stable carbocation. This all-important principle will also help explain why some reactions will involve a rearrangement. For example, consider the following reaction:

H—Cl

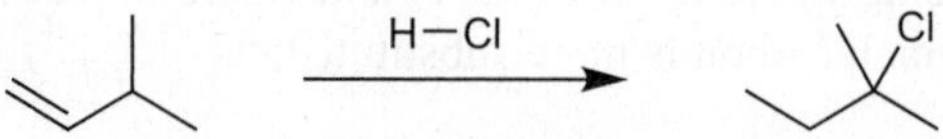

At first glance, the product is not what we might have expected. Once again, we turn to the mechanism for an explanation. The first step of the mechanism is identical to what we have seen so far—we protonate the double bond to produce the more stable carbocation (secondary, rather than primary):

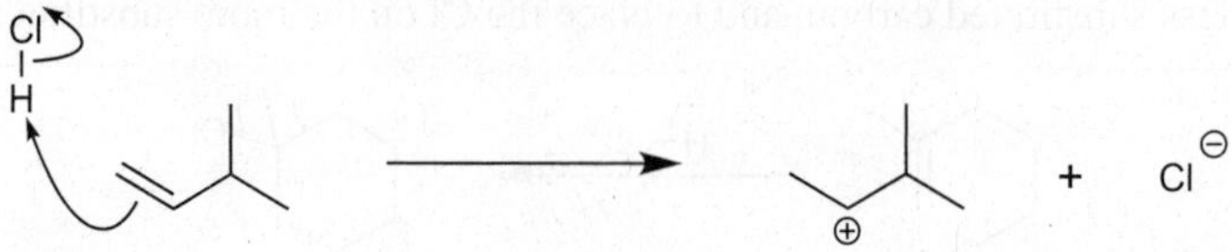

Now the carbocation is ready to be attacked by chloride. However, there is something else that can happen first (before chloride has a chance to attack): A hydride shift can produce a more stable carbocation:

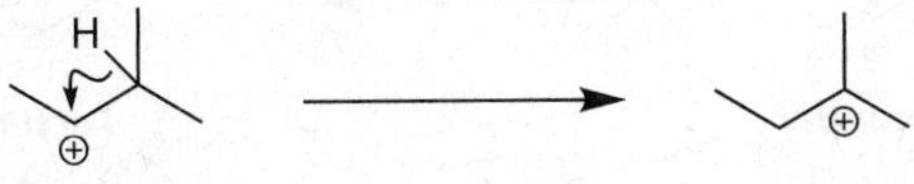

Secondary   **Tertiary**

This tertiary carbocation can now be attacked by chloride to give the product:

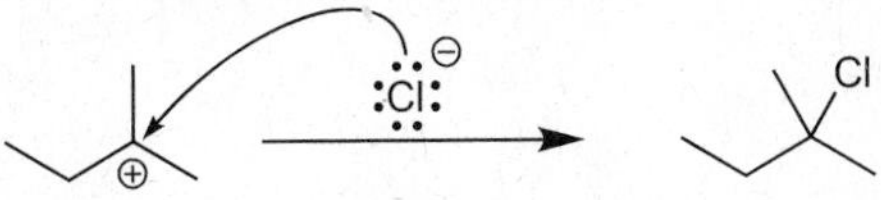

Clearly, we must be able to predict when to expect a carbocation rearrangement. There are two common ways for a carbocation to rearrange: either through a hydride shift or through a methyl shift. Your textbook will have examples of each. Carbocation rearrangements are possible for any reaction that involves an intermediate carbocation (not just for addition of HX across an alkene). Later in this chapter, we will see other addition reactions that also proceed through carbocation intermediates. In those cases, you will be expected to know that there will be a possibility for carbocation rearrangements.

Let's get some practice.

EXERCISE 11.43   Draw a mechanism for the following reaction:

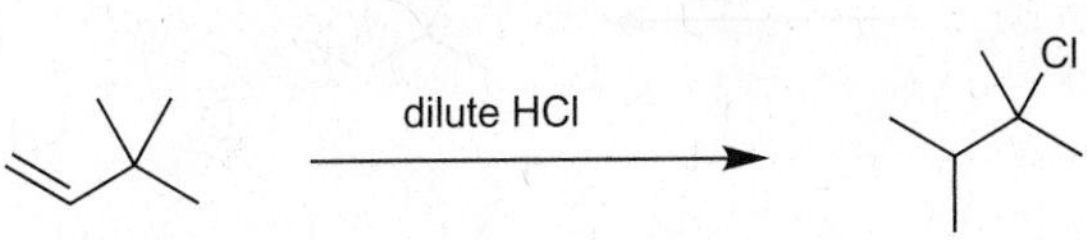

Answer   In the first step, we protonate the alkene:

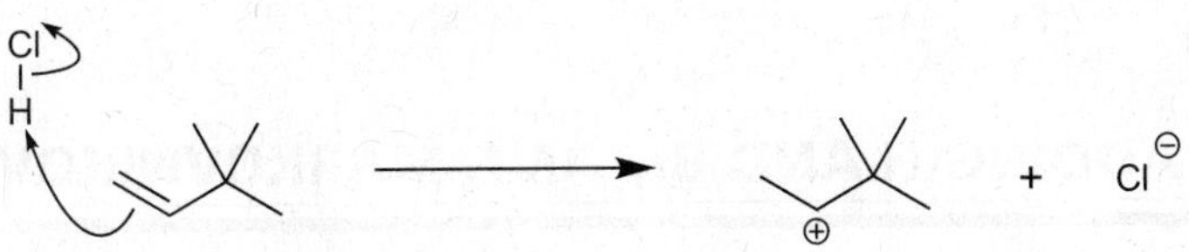

There were two ways that we could have protonated, and we chose the way that would produce the secondary carbocation (rather than producing a primary carbocation). Before we simply attack with the halide to end the reaction, we consider whether a rearrangement can take place. In this case, a methyl shift will produce a more stable, tertiary carbocation:

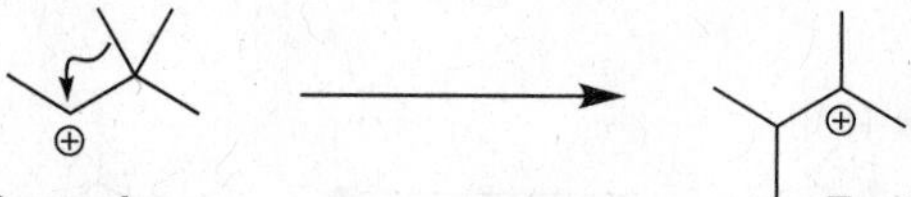

Secondary   **Tertiary**

Finally, the chloride ion now attacks the tertiary carbocation to give our product:

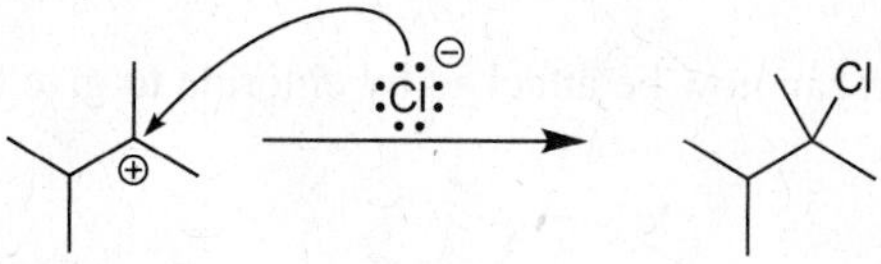

PROBLEMS   Draw the mechanism for each of the following reactions:

11.44

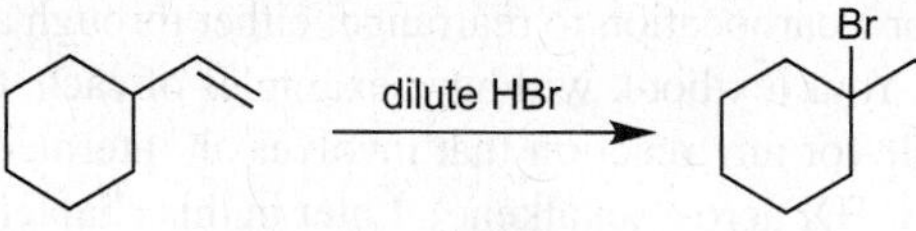

dilute HBr

11.45

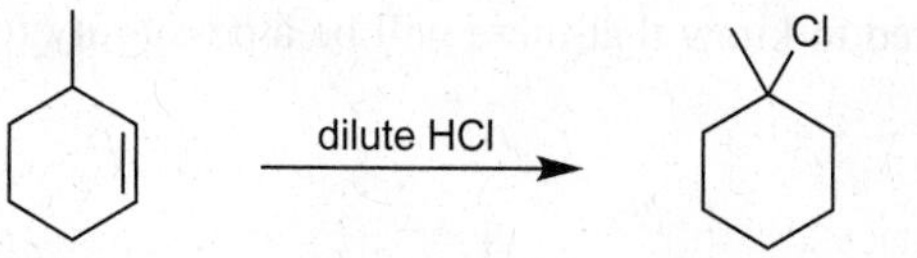

dilute HCl

11.46

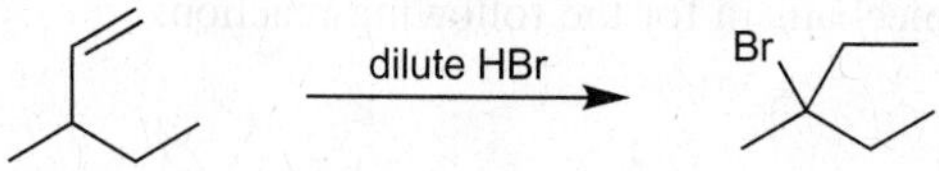

dilute HBr

11.47

dilute HCl

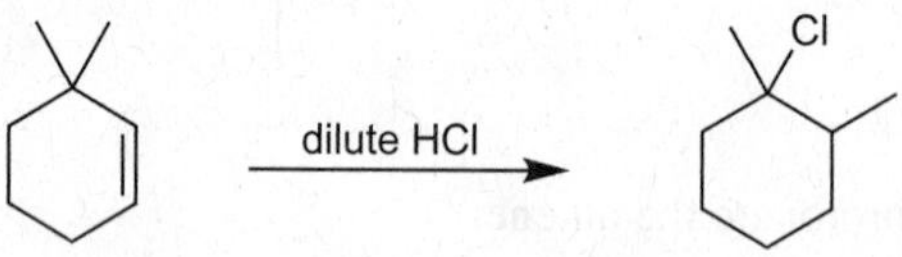

11.5   ADDING H AND Br, *ANTI*-MARKOVNIKOV

In the previous section, we saw how to add H and X, placing X at the more substituted carbon (Markovnikov addition). There is another reaction that will allow us to add H and X *anti*-Markovnikov, but it only works well with HBr (not any other H—X).

If we use HBr in the presence of peroxides (ROOR), Br ends up on the less substituted carbon:

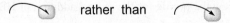

Why does the presence of peroxides cause the addition to be *anti*-Markovnikov? In order to understand the answer to this question, we will need to explore the mechanism in detail. This reaction follows a mechanism that involves radical intermediates (such as Br$^{\bullet}$), rather than ionic intermediates (such as Br^{-}). Peroxides are used to generate bromine radicals, in the following way:

The O—O bond of the peroxide is easily broken in the presence of light (*hv*) or heat. When this happens, the bond is broken homolytically, which means that two radicals are formed:

$$RO{-}OR \xrightarrow{h\nu} RO^{\bullet} \quad {}^{\bullet}OR$$

Each of these RO radicals can then abstract a hydrogen atom from HBr, to form the reactive intermediate (Br$^{\bullet}$):

$$RO^{\bullet} \quad H{-}Br \longrightarrow ROH + \boxed{Br^{\bullet}}$$

In your mind, you can compare the step above to a proton transfer. But there is one important difference. In a proton transfer, we are transferring H^{+} (a proton is the nucleus of a hydrogen atom, *without* its corresponding electron) from one place to another, via an ionic process. But here, we are transferring an H$^{\bullet}$ (an entire hydrogen atom: proton *and* electron), and therefore, we are dealing with a radical process.

Now that Br$^{\bullet}$ has formed, it can attack the alkene, like this:

Notice that we have been using one-headed curved arrows exclusively:

rather than

These arrows (called fishhook arrows) are the hallmark of radical reactions. We use fishhook arrows in radical mechanisms, because they indicate the movement of only *one* electron, rather than *two* electrons (by contrast, two-headed curved arrows are used in ionic mechanisms to show the movement of two electrons).

In the step above, Br$^{\bullet}$ attacked the alkene at the less substituted carbon, in order to form the more substituted carbon radical (C$^{\bullet}$). Tertiary radicals are more stable than secondary radicals, for the same reason that tertiary carbocations are more stable than secondary carbocations. Just as alkyl groups donate electron density to stabilize a neighboring, *empty p*-orbital, so too, alkyl groups can stabilize a neighboring, *partially filled* orbital. This preference for forming a tertiary radical (rather than a secondary radical) dictates that Br$^•$ will attack the *less substituted* carbon. This explains the observed *anti*-Markovnikov regiochemistry.

As a final step, the carbon radical then abstracts a hydrogen atom from HBr to give the product:

As a side product of this reaction, we regenerate another Br$^•$, which can go and react with another alkene. We call this a chain reaction, and the reaction occurs very rapidly. In fact, when peroxides are present (to jump-start this chain process), the reaction occurs much more rapidly than the competing ionic addition of HBr that we saw in the previous section.

Compare the intermediate of this radical mechanism with the intermediate of an ionic mechanism:

Ionic Mechanism   Radical Mechanism

Tertiary carbocation   Tertiary radical
intermediate   intermediate

In both mechanisms, the regiochemistry is determined by a preference for forming the most stable intermediate possible. For example, in the ionic mechanism, H$^+$ adds *to produce a tertiary carbocation,* rather than a secondary carbocation. Similarly, in the radical mechanism, Br$^•$ adds *to produce a tertiary radical,* rather than a secondary radical. In this respect, the two reactions are very similar. But take special notice of the fundamental difference. In the ionic mechanism, the *proton* comes on first. However, in the radical mechanism, the *bromine* comes on first. This critical difference explains why an ionic mechanism gives a Markovnikov addition while a radical mechanism gives an *anti*-Markovnikov addition.

Now let's review the profile for the radical addition of HBr:

	Br and **H**
$\xrightarrow{\text{HBr}}$ ROOR	
***Regio*chem**	*anti*-Markovnikov
***Stereo*chem**	Beyond the scope of this course

Until now we have focused on the regiochemistry of this reaction. We did not explore the stereochemistry, because it is beyond the scope of the course. In situations where two stereocenters are formed, the results are dependent on the starting alkene and on the temperature. Therefore, we will only present problems where no stereocenters are formed, or where only one stereocenter is formed.

EXERCISE 11.48 Predict the product for the following reaction:

Answer HBr indicates that we will be adding H and Br across the double bond. The presence of peroxides indicates that the regiochemistry will be *anti*-Markovnikov. To determine whether stereochemistry is relevant in this particular case, we need to look at whether we are creating two new stereocenters. When we place the Br on the less substituted carbon (and the H on the more substituted carbon), we will only be creating one new stereocenter. With only one stereocenter, there are not four possible stereoisomers but just two possible products (a pair of enantiomers). And we will get this pair of enantiomers regardless of whether the reaction was *syn* or *anti*:

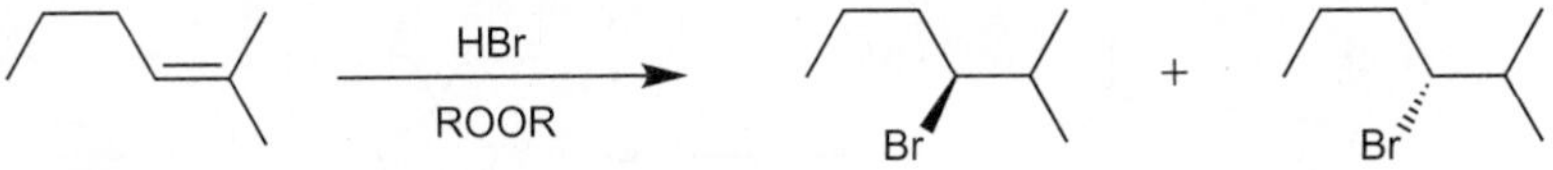

PROBLEMS Predict the products for each of the following reactions:

11.49

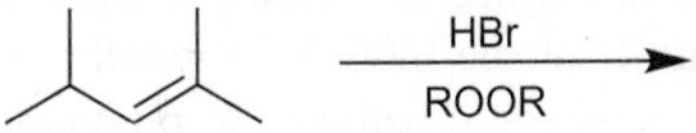

11.50

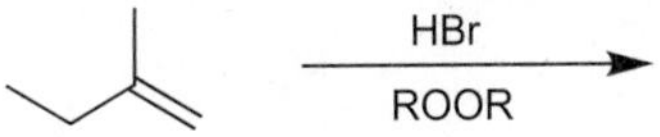

11.51

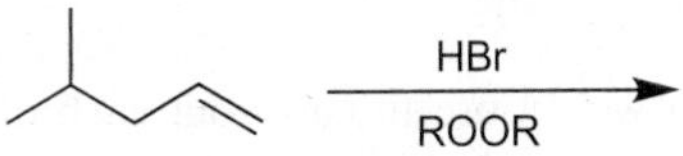

11.52

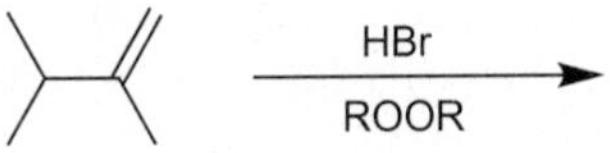

We have now seen two pathways for adding HBr across a double bond: the ionic pathway (which gives Markovnikov addition) and the radical pathway (which gives *anti*-Markovnikov addition). Both pathways are actually in competition with each other. However, the radical reaction is a much faster reaction. Therefore, we can control the regiochemistry of addition *by carefully choosing the conditions*. If we use a radical initiator, like ROOR, then the radical pathway will predominate, and we will see an *anti*-Markovnikov addition. If we do *not* use a radical initiator, then the ionic pathway will predominate, and we will see a Markovnikov addition:

Alkene

HBr ⟶ ionic pathway (*Markovnikov Addition*)

HBr / **ROOR** ⟶ radical pathway (*anti-Markovnikov Addition*)

Let's get a bit of practice with choosing the appropriate conditions for addition of HBr.

EXERCISE 11.53 In the following hydrobromination reaction, determine whether or not you should use peroxides:

Answer In order to determine whether or not to use peroxides, we must decide whether the desired transformation represents a Markovnikov addition or an *anti*-Markovnikov addition. When we compare the starting alkene above with the desired product, we see that we need to place the Br at the more substituted carbon (i.e., Markovnikov addition). Therefore, we need an ionic pathway to predominate, and we should *not* use peroxides. We just use HBr:

PROBLEMS Identify what reagents you would use to carry out each of the following transformations:

11.54

11.55

11.56

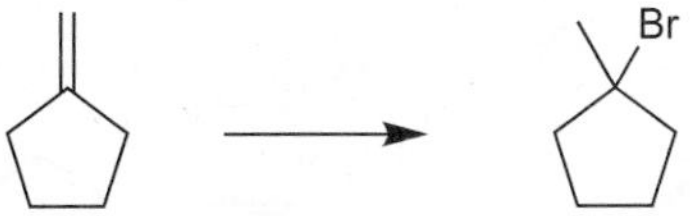

11.57

11.6 ADDING H AND OH, MARKOVNIKOV

Over the next two sections, we will learn how to add H and OH across a double bond. The process of adding water across an alkene is called *hydration,* and it can be achieved through a Markovnikov addition or through an *anti*-Markovnikov addition. We just need to carefully choose our reagents. In this section, we will explore the reagents that give a Markovnikov addition of water. Then, in the next section, we will explore reagents that give an *anti*-Markovnikov addition of water.

Consider the following reaction:

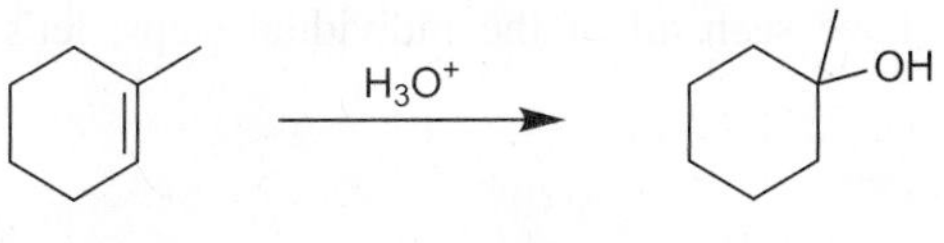

Careful comparison of the starting material and product will reveal that we have performed a hydration, with Markovnikov regiochemistry. Notice the reagent that we used (H_3O^+). Essentially, this is water (H_2O) and an acid source (such as sulfuric acid). There are many ways to show this reagent. Sometimes, it is written as H_3O^+ (as above), while at other times, it might be written like this: H_2O , H^+. You might even see it like this, with brackets around the acid:

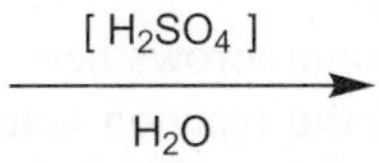

$$\xrightarrow[\text{H}_2\text{O}]{[\text{H}_2\text{SO}_4]}$$

These brackets indicate that H^+ is not consumed in the reaction. In other words, H^+ is a catalyst, and therefore, we call this reaction an *acid-catalyzed hydration*. In order to understand why this reaction proceeds via a Markovnikov addition, we turn our attention to the mechanism. The proposed mechanism of an acid-catalyzed hydration is similar to the mechanism of addition of HX (the ionic pathway). Compare these two mechanisms:

In each mechanism above, the first step involves protonation of the alkene to form a carbocation. Then, in both cases, a nucleophile (either X⁻ or H_2O) attacks the carbocation to give a product. The difference between these two reactions is in the nature of the product. The first reaction above (hydrohalogenation) gives a product that is neutral (no charge). However, the second reaction above (hydration) produced a charged species. Therefore, one more step is necessary at the end of the hydration reaction—we must get rid of the positive charge. To do this, we simply deprotonate:

Notice what reagent we use to pull off the proton. We use H_2O, rather than using a hydroxide ion (HO⁻). To understand why, remember that we are in acidic conditions; there really aren't many hydroxide ions floating around. But there is plenty of water, and a mechanism must always be consistent with the conditions that are present.

Now that we have seen all of the individual steps, let's look at the entire mechanism:

Notice that we are using equilibrium arrows here (⇌ rather than ⟶). These equilibrium arrows indicate that the reaction actually goes in both directions. In fact, the reverse path (starting from the alcohol and ending with the alkene) is a reaction that we have already studied. It is just an E1 reaction (follow the sequence above from the end to the beginning, and convince yourself that it is the E1 process). The truth is that most reactions represent equilibrium processes, however, organic chemists (generally) only make an effort to draw equilibrium arrows in situations where the equilibrium can be easily manipulated (allowing us to control which products we get). This reaction is one of those situations. By carefully controlling the amount of water present (using either concentrated acid or dilute acid), we can favor one side of the equilibrium or the other:

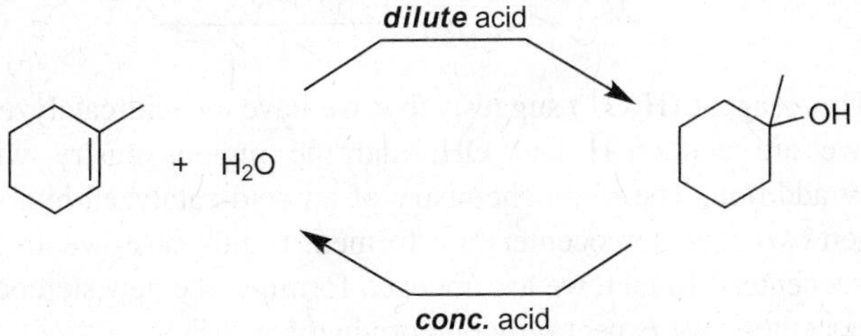

We are exploiting Le Chatelier's principle, which tells us that the equilibrium can be pushed toward one side or the other by removing or adding reagents. Imagine that you have the system shown above (alkene + water on the left side; alcohol on the right side), and this system has reached equilibrium. Then you add water. The concentrations would no longer be at equilibrium, and the system would have to adjust to re-establish new equilibrium concentrations. The end result: adding water would cause more alkene to turn into alcohol. Therefore, we would use dilute acid (which is mostly water) to favor the alcohol. If we wanted to favor the alkene, then we would want to remove water, which would push the equilibrium toward the left. Therefore, if we want to form the alkene, we would use concentrated acid (which is mostly acid and very little water).

Once again, we see that carefully choosing reaction conditions can greatly affect the outcome of a reaction.

We have already explained the regiochemistry of acid-catalyzed hydration; there is a strong preference for Markovnikov addition. But what about the stereochemistry?

The stereochemistry of acid-catalyzed hydration is very similar to the stereochemistry of ionic addition of HX (this should make sense, as we have already seen that the mechanisms for each of these reactions are identical). If only one stereocenter is formed, then we expect a pair of enantiomers (racemic mixture), regardless of whether the reaction was *anti* or *syn*. You will probably not see an example where two new stereocenters are formed, because the stereochemical outcome in such a case is complex and is beyond the scope of our conversation.

Now we can summarize the profile for acid-catalyzed hydration:

H_3O^+	**H** and **OH**
***Regio*chem**	Markovnikov
***Stereo*chem**	Beyond the scope of this course

EXERCISE 11.58 Predict the product for the following reaction, and then propose a mechanism for formation of that product:

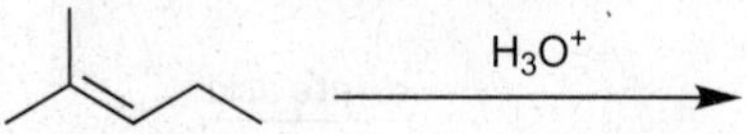

Answer This reagent (H_3O^+) suggests that we have an acid-catalyzed hydration. Therefore, we are adding H and OH, and the regiochemistry will follow a Markovnikov addition. The stereochemistry of an acid-catalyzed hydration is only complex when two new stereocenters are formed. In this case, we are not forming two new stereocenters. In fact, we are not even forming one new stereocenter. Without any stereocenters, we expect only one product:

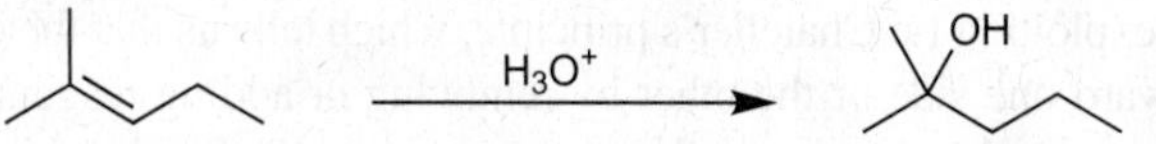

The mechanism of the reaction will have three steps: (1) protonate the alkene to form a carbocation, (2) water attacks the carbocation, and (3) deprotonate to form the product:

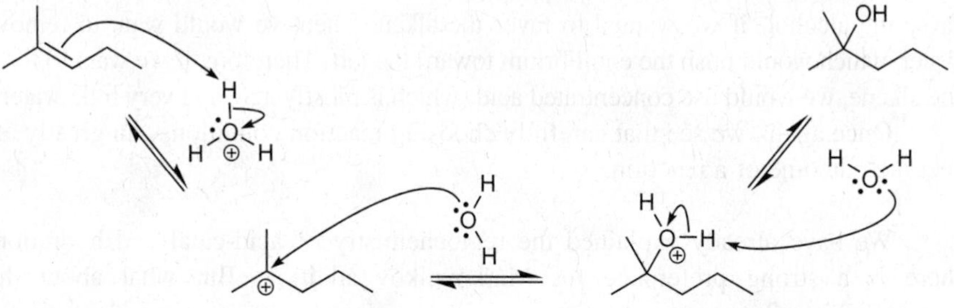

PROBLEMS For each of the following reactions, predict the expected product, and propose a plausible mechanism for formation of the product:

11.59

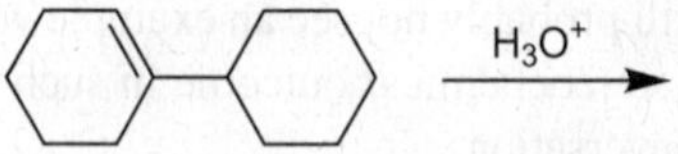

11.60

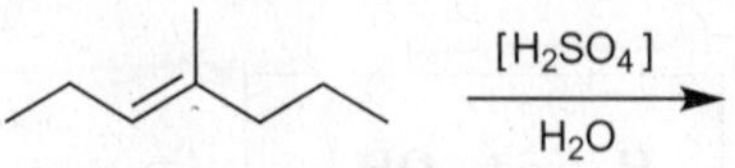

11.61

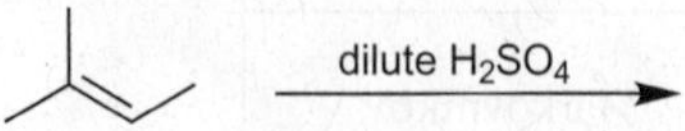

11.62

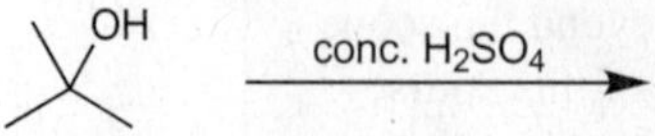

11.7 ADDING H AND OH, *ANTI*-MARKOVNIKOV

In the previous section, we saw how to perform a Markovnikov addition of H and OH across a double bond. In this section, we will learn how to perform an *anti-Markovnikov* addition of H and OH, for example:

A quick glance at the products indicates that we are adding H and OH across the alkene. Let's take a closer look and carefully analyze the regiochemistry *and* stereochemistry of this reaction. The OH is ending up on the less substituted carbon, and therefore, the regiochemistry represents an *anti*-Markovnikov addition. But what about the stereochemistry? Are we seeing a *syn* addition here, or is this *anti* addition?

Be careful. The example above represents somewhat of an optical illusion. The products seem to suggest an *anti* addition (the methyl and the OH are *trans* to each other). But think about what we added in this reaction: we did not add OH and a *methyl* group. The methyl group was already there. Rather, we added OH and H. The H that we added is not shown in the product above (because it does not have to be drawn in a bond-line drawing). If you draw that H on the compound above, you will see that it is on a dash—therefore, this was a *syn* addition of H and OH.

To recap, the reaction above is an addition of H and OH, with *anti*-Markovnikov regiochemistry, and *syn* stereochemistry. Now we have three important questions to answer:

1. How do these reagents (BH_3, etc.) cause an addition of H and OH?
2. Why *anti*-Markovnikov?
3. Why *syn*?

The answers to all three of these questions are encapsulated in the mechanism as usual. In order to explore the accepted mechanism, we must first acquaint ourselves with the reagents. In the first step, the reagents are BH_3 and THF. The former is called borane. The element boron uses its three valence electrons to comfortably form three bonds:

However, in this structure, boron does not have an octet. It has an empty *p* orbital, (very similar to a carbocation, except there is no positive charge here). Therefore, borane is very reactive. In fact, it reacts with itself to give dimeric structures, called diborane:

$$2\ BH_3 \rightleftharpoons B_2H_6$$

The empty p orbital in borane can be somewhat stabilized if we use a solvent (like THF) that can donate electron density into the empty p orbital of boron:

THF

This solvent is called tetrahydrofuran, or THF for short. Even though it somewhat stabilizes the empty p orbital on the boron atom in BH_3, nevertheless the boron atom is very eager to look for any other sources of electron density that it can find. It is an electrophile—it is scavenging for sites of high electron density to fill its empty orbital. A pi bond is a site of high electron density, and therefore, a pi bond can attack borane. In fact, this is the first step of our mechanism. A pi bond attacks the empty p orbital of boron, which triggers a *simultaneous* hydride shift:

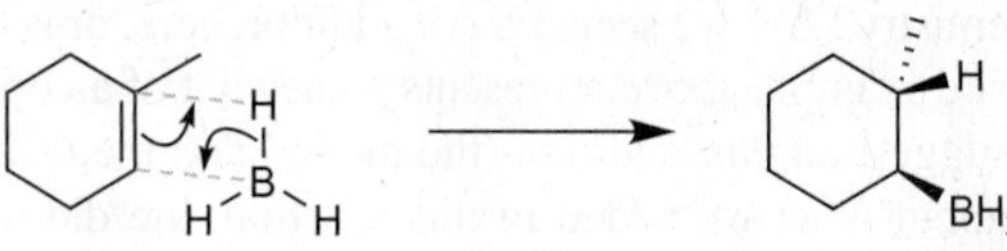

Notice that it all happens in one concerted process (via a four-membered transition state). Let's take a close look at this first step, and consider the regiochemistry and stereochemistry.

For the regiochemistry, we notice that the boron ends up on the less substituted carbon (and that is where the OH group will ultimately end up). Now we can understand one of the sources of this regiochemical preference. We are adding H and BH_2 across the double bond. BH_2 is bigger and bulkier than H, so it will have an easier time fitting over the less substituted carbon (the less sterically hindered position). Therefore, we get an *anti*-Markovnikov addition.

The stereochemical preference (for *syn* addition) can now also be understood. The step above represents a concerted process. Both BH_2 and H are adding simultaneously, so they must end up on the same face of the alkene. In other words, the reaction must be a *syn* addition.

The structure shown above still has two remaining B—H bonds (look at the BH_2 group), and so the reaction can occur again with those B—H bonds. In other words, one molecule of BH_3 can react with three molecules of alkene to give a trialkylborane:

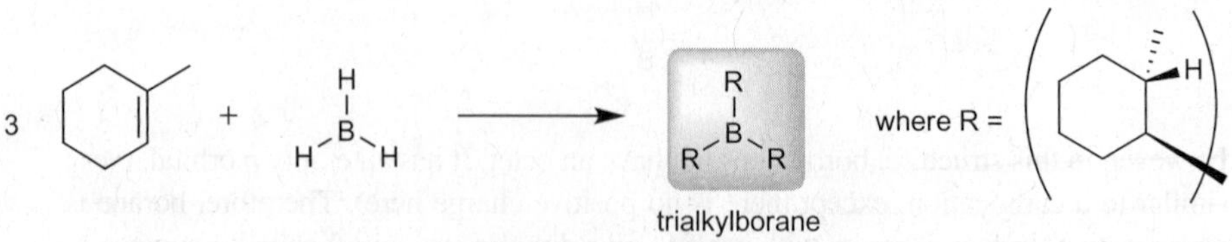

trialkylborane

What we have done until now (formation of the trialkylborane above) is called hydroboration, which occurs when you mix an alkene with BH_3 in THF. Now, we move on to the next set of reagents, which accomplish an oxidation reaction: H_2O_2 and hydroxide. These reagents give us an oxidation reaction:

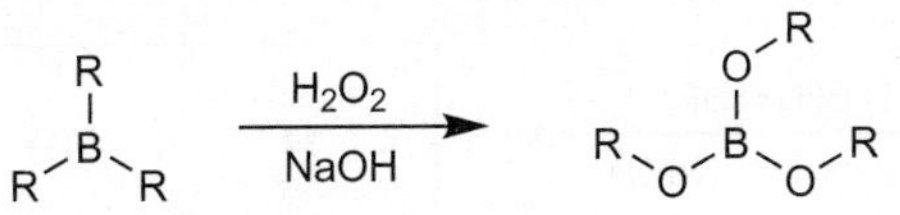

How does oxygen insert in between the B—R bonds? Let's take a closer look at the reagents—a hydroxide ion can deprotonate hydrogen peroxide to form a hydroperoxide anion:

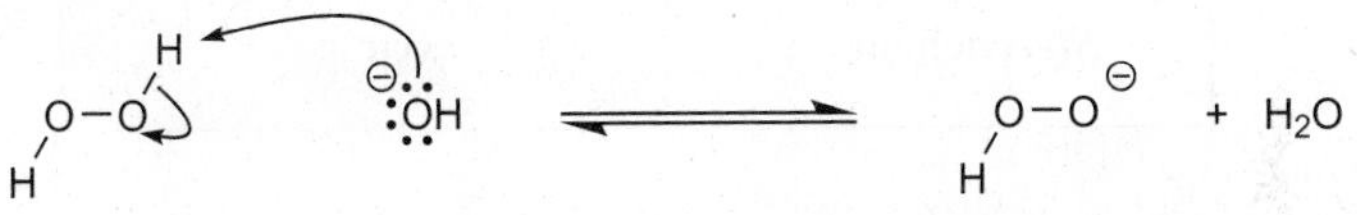

This hydroperoxide anion can attack the trialkylborane (remember that the boron atom still has an empty p orbital, and therefore, it is still scavenging for electron density):

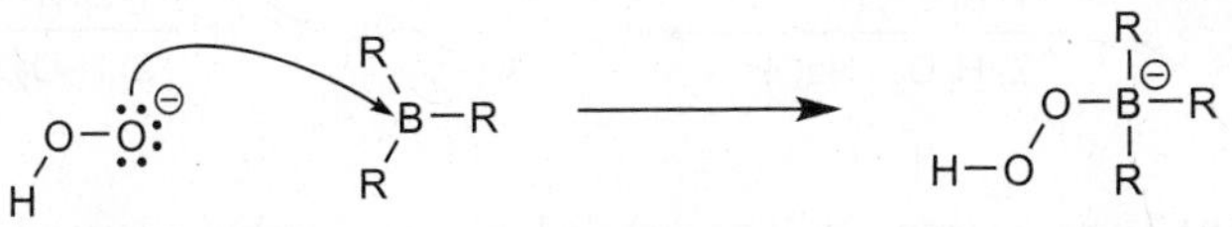

Now is where it gets interesting. One of the alkyl groups migrates over (an alkyl shift) to kick off hydroxide:

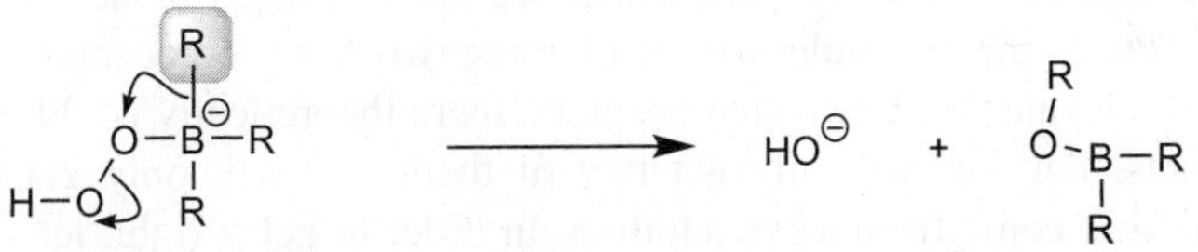

Notice the overall result. When R migrates over, the net result is to place oxygen in between B and R. Focus your attention on the stereocenter of the alkyl group (R). As R migrates, the configuration of the stereocenter is unaffected by the migration. In other words, the configuration of the stereocenter is preserved. This happens to all three B—R bonds:

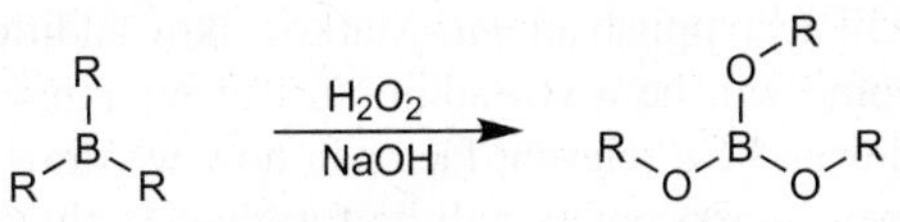

The final step involves removing the OR groups from B, which happens like this:

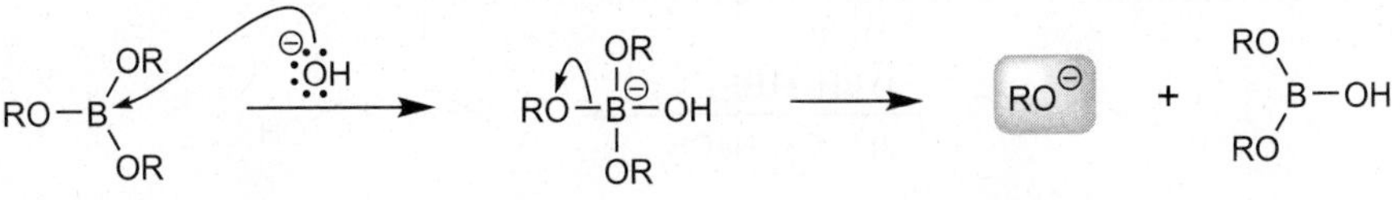

RO^- then removes a proton from water, and the final product is an alcohol. Overall, we have a two-step synthesis for converting an alkene into an alcohol. This two-step synthesis is called *hydroboration-oxidation*. Let's now summarize the profile of this two-step process:

1) BH$_3$•THF 2) H$_2$O$_2$, NaOH →	**H** and **OH**
*Regio*chem	*anti*-Markovnikov
*Stereo*chem	*syn*

EXERCISE 11.63 Predict the products for each of the following reactions:

(a)
1) BH$_3$•THF
2) H$_2$O$_2$, NaOH

(b)
1) BH$_3$•THF
2) H$_2$O$_2$, NaOH

Answer **(a)** These reagents will accomplish an *anti*-Markovnikov addition of OH and H. The stereochemical outcome will be a *syn* addition. But we must first decide whether stereochemistry will even be a relevant factor in how we draw our products. To do that, remember that we must ask if we are creating *two* new stereocenters in this reaction. In this example, we *are* creating two new stereocenters. So, stereochemistry *is* relevant. With two stereocenters, there theoretically could be four possible products, but we will only get two of them; we will only get the pair of enantiomers that come from a *syn* addition. In order to get it right, let's redraw the alkene (as we have done many times earlier), and add OH and H like this:

(b) These reagents will accomplish an *anti*-Markovnikov addition of OH and H. The stereochemical outcome will be a *syn* addition. But we must first decide whether stereochemistry will even be a relevant factor in how we draw our products. To do this, we ask if *two* new stereocenters will be forming. In this example, we are not creating two new stereocenters. In fact, we are not even creating one stereocenter. So, stereochemistry is not relevant for us in this problem:

1) BH$_3$•THF
2) H$_2$O$_2$, NaOH → OH

Whenever stereochemistry is irrelevant (i.e., whenever we are not creating two new stereocenters) the problem becomes a bit easier to solve.

PROBLEMS   Predict the products of the following reactions:

11.64
$$\text{1) BH}_3 \cdot \text{THF}$$
$$\text{2) H}_2\text{O}_2 \text{ , NaOH}$$

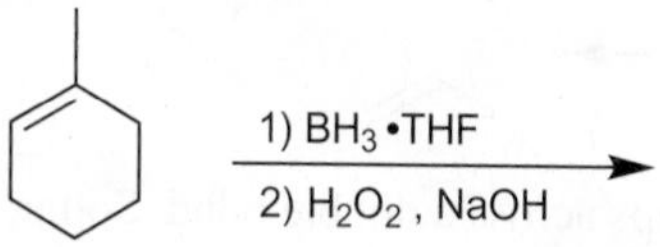

11.65
$$\text{1) BH}_3 \cdot \text{THF}$$
$$\text{2) H}_2\text{O}_2 \text{ , NaOH}$$

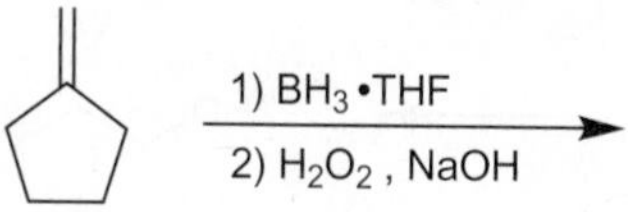

11.66
$$\text{1) BH}_3 \cdot \text{THF}$$
$$\text{2) H}_2\text{O}_2 \text{ , NaOH}$$

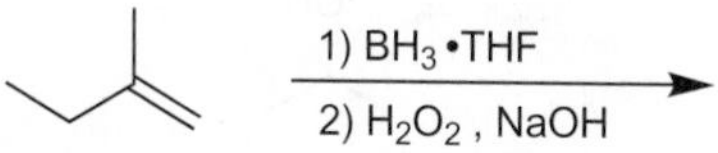

11.67
$$\text{1) BH}_3 \cdot \text{THF}$$
$$\text{2) H}_2\text{O}_2 \text{ , NaOH}$$

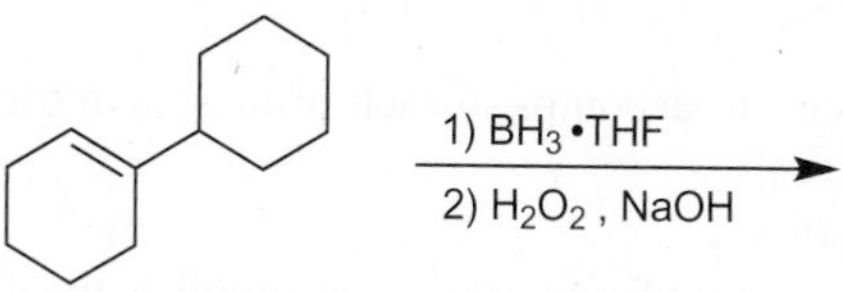

11.68
$$\text{1) BH}_3 \cdot \text{THF}$$
$$\text{2) H}_2\text{O}_2 \text{ , NaOH}$$

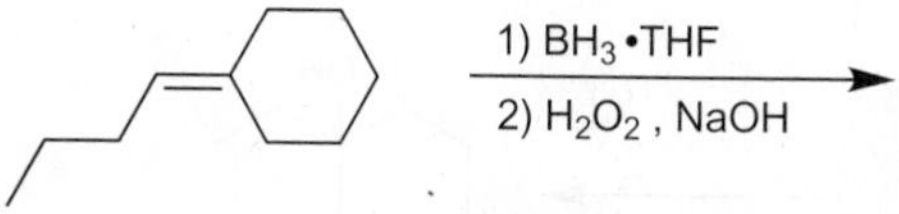

11.69
$$\text{1) BH}_3 \cdot \text{THF}$$
$$\text{2) H}_2\text{O}_2 \text{ , NaOH}$$

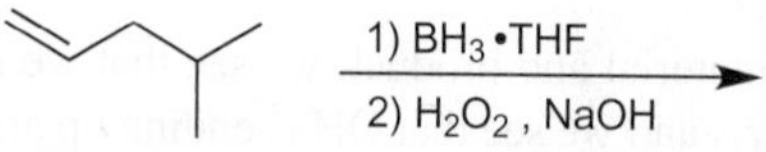

11.8   SYNTHESIS TECHNIQUES

11.8A   One-Step Syntheses

In order to begin practicing synthesis problems, it is absolutely essential that you master all of the individual reactions that we have seen so far. You must learn how to walk before you can start to run. Therefore, we will first focus on one-step synthesis problems. Once you feel comfortable with the individual reactions, then we can start stringing them together in various sequences to form synthesis problems.

Until now, we have seen substitution reactions (S_N1 and S_N2), elimination reactions (E1 and E2), and five addition reactions. Let's quickly review what these reactions can accomplish. *Substitution* allows us to interconvert groups:

Elimination allows us to form alkenes:

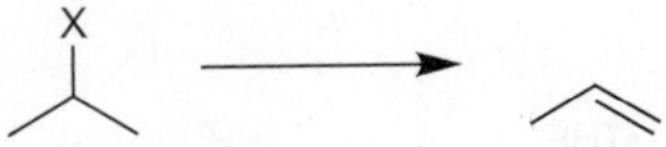

Addition reactions allow us to add two groups across a double bond. So far, we have seen the following five addition reactions:

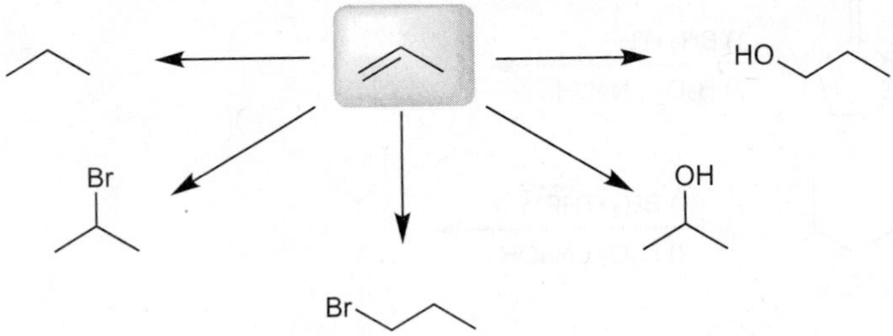

Can you fill in the reagents necessary to accomplish each of these five transformations? Try it. . . .

EXERCISE 11.70 What reagents would you use to accomplish the following transformation?

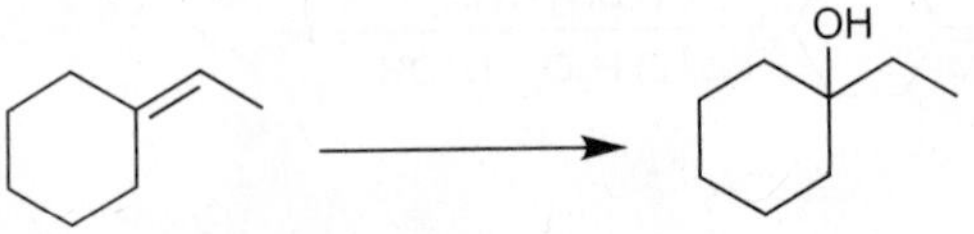

Answer If we compare the starting material and product, we see that we must add H and OH. We look at the regiochemistry, and we see that OH is ending up at the more substituted carbon—so we need a Markovnikov addition. Then, we look at the stereochemistry and we see that we are *not* creating two stereocenters in this reaction (in fact, we are not even creating one stereocenter). Therefore, the stereochemistry of the reaction will be irrelevant. So we need to choose reagents that will give a Markovnikov addition of H and OH. We can accomplish this with an acid-catalyzed hydration:

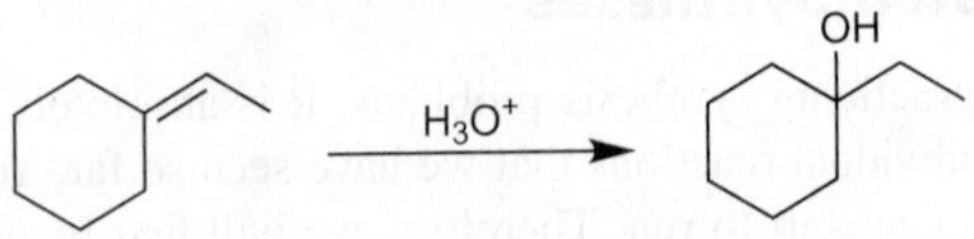

PROBLEMS What reagents would you use to accomplish each of the following transformations?

11.71

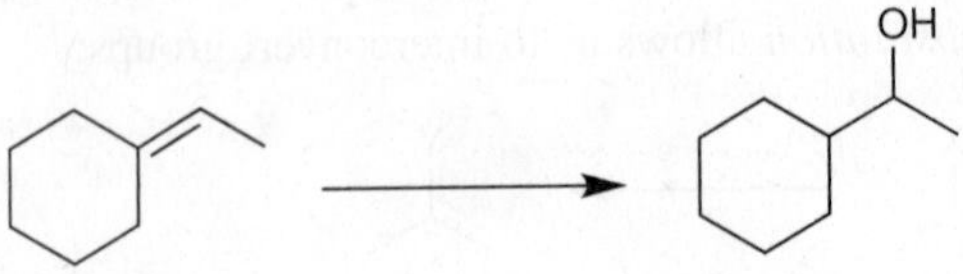

11.72

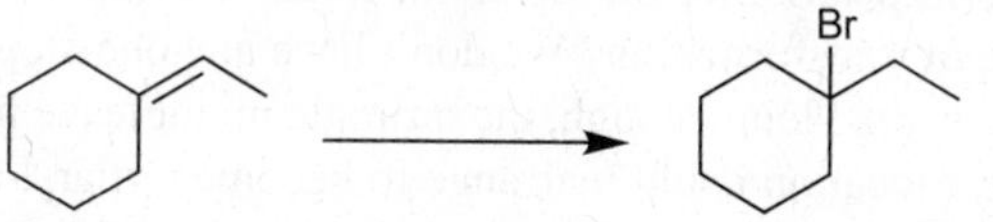

11.73

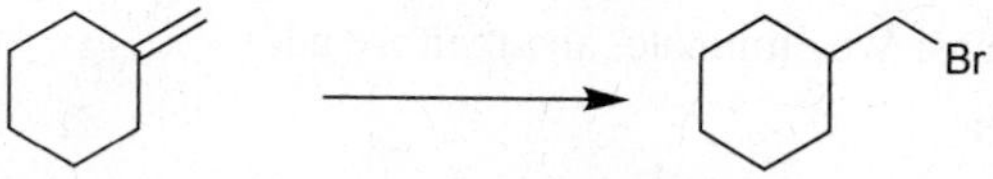

11.74

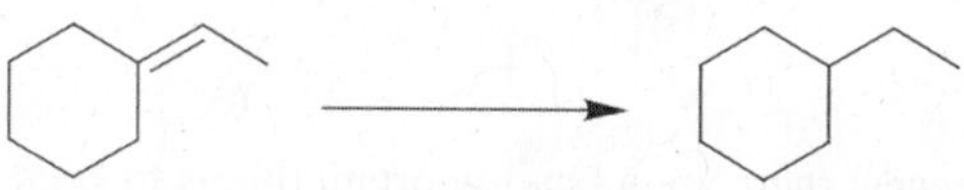

11.75

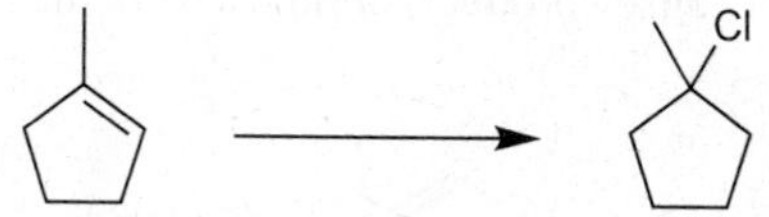

11.76 + En

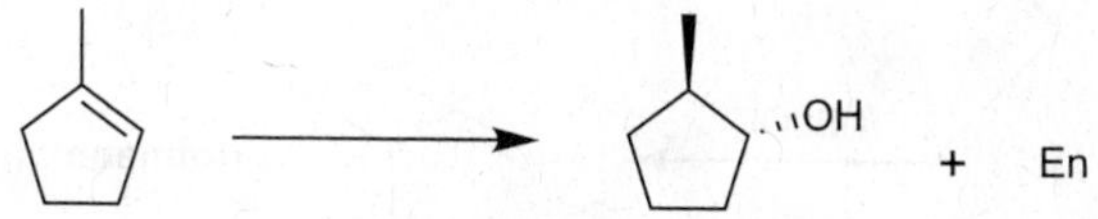

11.77

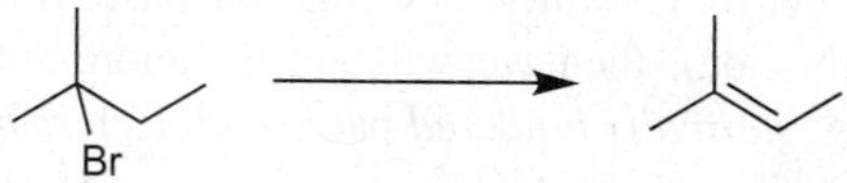

11.78

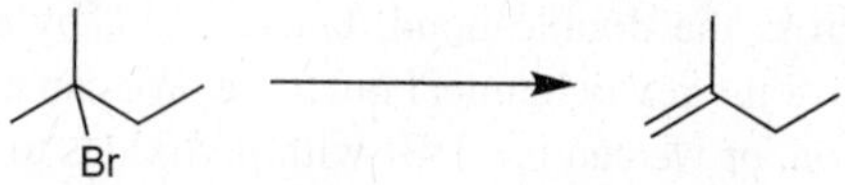

11.8B   Changing the Position of a Leaving Group

Now let's get some practice *combining* our reactions and proposing syntheses.
Consider the following transformation:

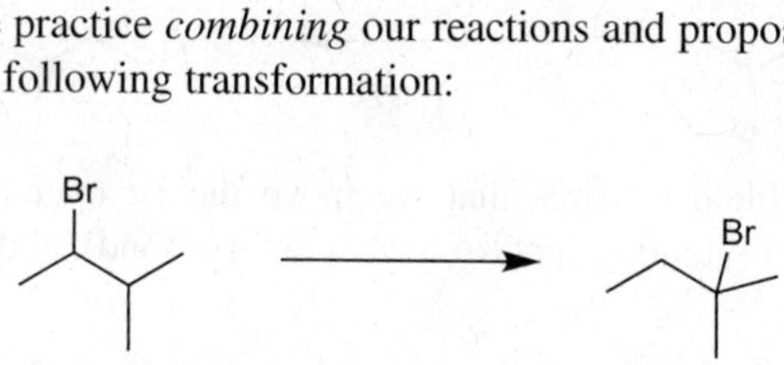

The net result is the change in position of the Br atom. It has "moved" over. How can we accomplish this type of transformation? We don't have any one-step method for doing this. Perhaps if we waited long enough, the bromide might leave in an S_N1 reaction, and the resulting carbocation could rearrange to become tertiary, and then bromide could re-attack. But that would take too long. Waiting for an S_N1 at a secondary substrate is not the best idea. We can do this much more quickly and efficiently if we do it in two steps: We eliminate, and then we add:

When doing this type of sequence, there are a few important things to keep in mind. In the first step (elimination), we have a choice regarding which way to eliminate: do we form the more substituted alkene (Zaitsev product) or do we form the less substituted alkene (Hofmann product):

We can control which product we get by carefully choosing our base. If we use a strong base (like methoxide or ethoxide), then we will get the more substituted alkene. However, if we use a strong, *sterically hindered* base, such as *tert*-butoxide, then we will get the less substituted alkene.

After forming the double bond, we must also carefully consider the regiochemistry of how we add HBr across the double bond. Once again, by carefully choosing our reagents, we can control the regiochemical outcome. We can either use HBr to force a Markovnikov addition, or we can use HBr with peroxides to force an *anti*-Markovnikov addition. Let's see an example:

EXERCISE 11.79 What reagents would you use to accomplish the following transformation?

Answer This problem requires that we move the Br over to the left. We can accomplish this by eliminating first (to form a double bond), and then adding across that double bond:

However, we must be careful to control the regiochemistry properly in each of these two steps. In the elimination step, we need to form the less substituted double bond (i.e., the Hofmann product), and therefore, we must use a sterically hindered base. Then, in the addition step, we need to place the Br on the less substituted carbon (*anti*-Markovnikov addition), so we must use HBr *with peroxides*. This gives us the following overall synthesis:

1) *t*-BuOK
2) HBr, ROOR

There is one more thing to keep in mind when using this type of technique: *hydroxide is a terrible leaving group*. Let's see what to do when dealing with an OH group. As an example, suppose we wanted to do the following transformation:

Our technique would suggest the following steps:

Eliminate   Add H-OH

The first step of our technique requires an elimination reaction to give the Hofmann product, so we must use an E2 reaction, employing a sterically hindered base. In order to do this, we must first convert the OH group into a good leaving group. This can be accomplished by converting the OH group into a tosylate, which is a much better leaving group than OH:

TsCl, pyridine

Bad Leaving Group   Good Leaving Group

If you have not yet learned about tosylates in your lecture course, you might want to consult your textbook for more information on tosylates.

After we have converted the OH into a tosylate, then we can do our technique (using a strong, sterically hindered base to eliminate, followed by *anti*-Markovnikov addition of H and OH):

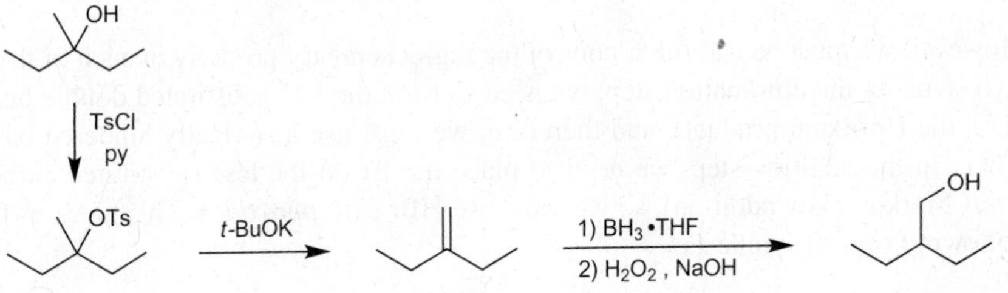

PROBLEMS What reagents would you use to accomplish each of the following transformations?

11.80

11.81

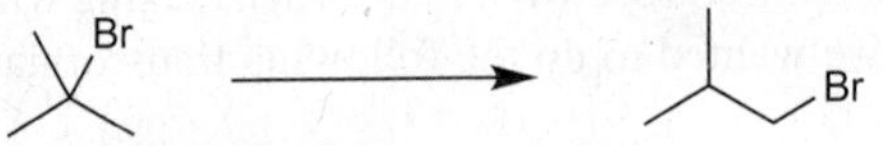

11.82

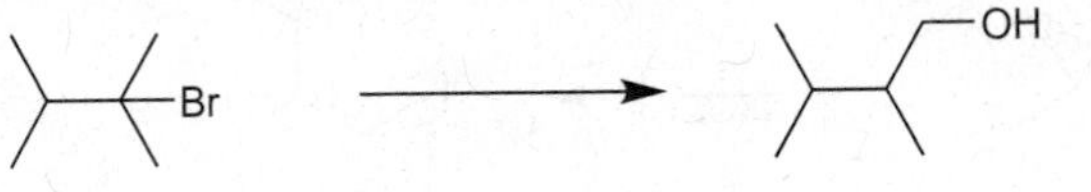

11.83

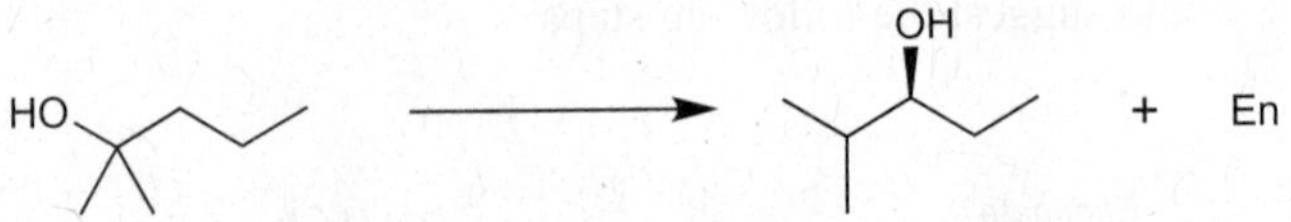

11.84

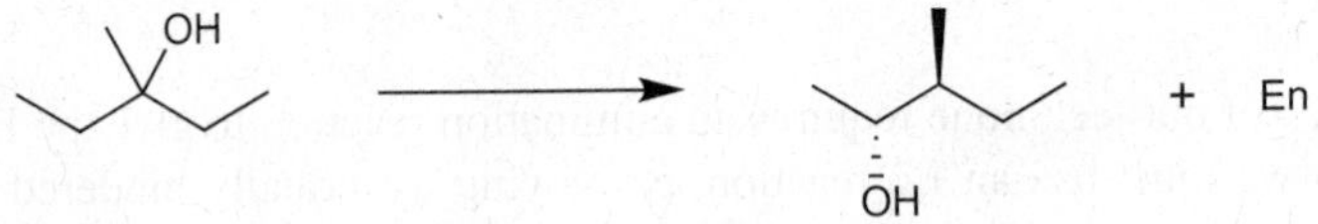

11.85

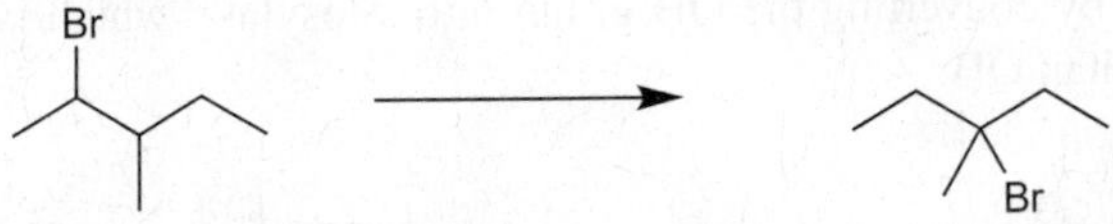

11.8C   Changing the Position of a π Bond

Until now, we have seen how to combine two reactions into one synthetic technique: eliminate and then add. Now, let's focus on another type of technique: add and then eliminate. Let's see an example:

EXERCISE 11.86 What reagents would you use to accomplish the following transformation?

Answer In this example, we are tasked with moving the position of a double bond. We have not seen a way to do the transformation above in one step. However, we can easily accomplish it with two steps—add and then eliminate:

In the first step above, we need a Markovnikov addition (Br needs to end up at the more substituted carbon), which we can easily accomplish by using HBr. The second step requires an elimination to give the Zaitsev product, which we can accomplish if we choose our base carefully (we will need to use a base that is NOT sterically hindered). Therefore, the overall synthesis is:

Take special notice of what we can accomplish when we use this technique: it gives us the power *to move the position of a double bond*. When using this technique, we must carefully consider the regiochemistry of each step. In the first step (addition), we must decide whether we want a Markovnikov addition (HBr), or an *anti*-Markovnikov addition (HBr with peroxides). Also, in the second step (elimination), we must decide whether we want to form the Zaitsev product or the Hofmann product (which we can control by carefully choosing our base, ethoxide or *tert*-butoxide). Get some practice with this technique in the following problems.

PROBLEMS What reagents would you use to accomplish each of the following transformations?

11.87

11.88

11.89

11.90

11.8D Introducing Functionality

In the techniques we have seen so far, the starting compound always had some functional group that we could manipulate. Either the starting material had a leaving group or it had a double bond. But what about situations in which the starting material has no functional groups—no leaving group, and no double bond? What can be done? In situations like this, there is only one possible course of action to take: *radical bromination*. Let's see an example:

EXERCISE 11.91 What reagents would you use to accomplish the following transformation?

Answer In this problem, we are starting with an alk*ane*. There are no leaving groups, so we cannot do a substitution or an elimination reaction. There are also no double bonds, so we cannot do an addition. It seems that we are stuck, with nothing to do. Clearly, our only way out of this situation is to introduce a functional group into the compound, via radical bromination. Radical bromination will place a Br at the most substituted position (the tertiary position), and then we can eliminate:

So, our overall synthesis is:

There are a few things to keep in mind when using this technique. First of all, radical bromination will selectively place a Br on the most substituted position. Therefore, you should always look for the tertiary position to see where the Br will go. Then, when doing the elimination, make sure to choose the base carefully, in order to achieve the desired regiochemistry. Let's get some practice with this.

PROBLEMS What reagents would you use to accomplish each of the following transformations?

11.92

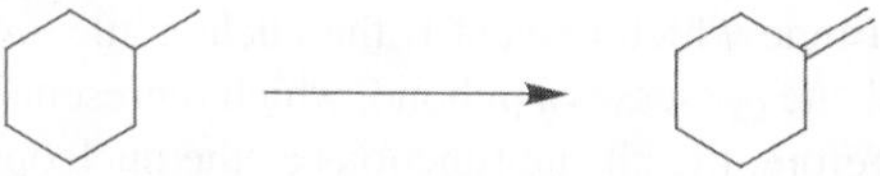

11.93

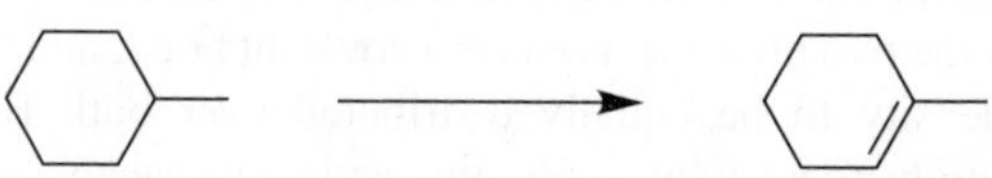

11.94

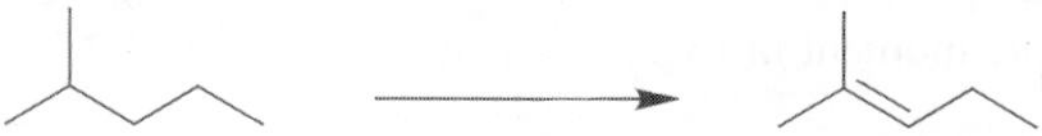

11.95

11.96

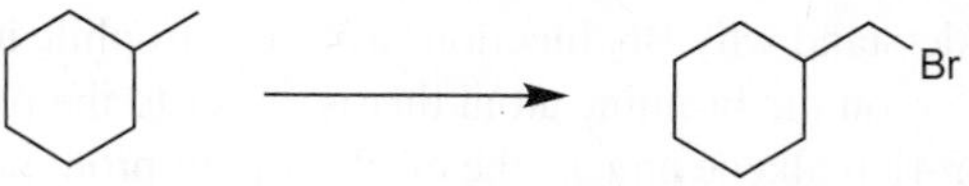

11.97

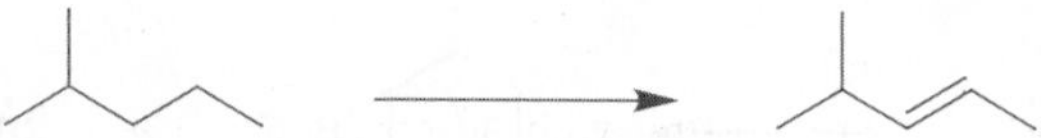

11.9 ADDING Br AND Br; ADDING Br AND OH

We now turn our attention to the next addition reaction: adding Br and Br across a double bond:

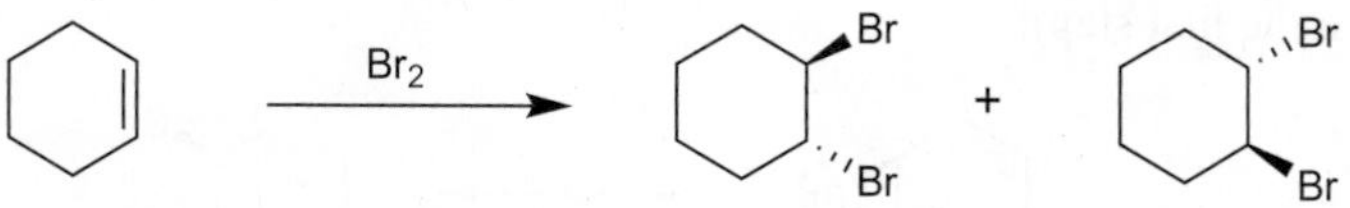

We will begin by analyzing the regiochemistry and stereochemistry of the above re-action. The regiochemistry will be irrelevant, regardless of what alkene we start with, because we are adding two of the same group (Br and Br). However, the stereo-chemistry *is* relevant, because we are creating two new stereocenters in the example above. If we carefully examine the products, we will notice that we have a pair of enantiomers. (*Caution:* Many students mistakenly believe that these two products are the same, but they are not—they are in fact enantiomers.) Our products represent the pair of enantiomers that we would get from an *anti* addition. So, we must try to understand *why* this reaction proceeds through an *anti* addition. To do this, we turn once again to the mechanism.

In the first step, we have an alkene reacting with Br_2. To understand this step of the mechanism, we must determine which reagent is the nucleophile, and which reagent is the electrophile. The alkene possesses a pi bond, which represents a region in space of electron density. Therefore, the alkene functions as the nucleophile.

This implies that Br_2 is the electrophile. But how does Br_2 function as an electrophile? The bond between the two bromine atoms is a covalent bond, and we therefore expect the electron density to be equally distributed over both Br atoms. However, an interesting thing happens when a Br_2 molecule approaches an alkene. The electron density of the pi bond *repels* the electron density in the Br_2 molecule, creating a temporary dipole moment in Br_2.

$$\delta+ \quad \delta-$$
$$Br-Br$$

As the Br_2 molecule gets closer to the alkene, this temporary effect becomes more pronounced. Now we can understand why Br_2 functions as an electrophile in this reaction: there is a temporary $\delta+$ on the bromine atom that is closer to the pi bond of the alkene. When the electron-rich alkene attacks the electron-poor bromine, we get the following first step of our mechanism:

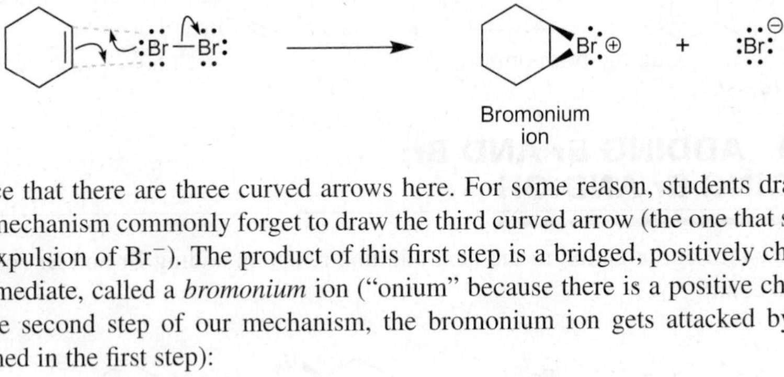

Bromonium
ion

Notice that there are three curved arrows here. For some reason, students drawing this mechanism commonly forget to draw the third curved arrow (the one that shows the expulsion of Br^-). The product of this first step is a bridged, positively charged intermediate, called a *bromonium* ion ("onium" because there is a positive charge). In the second step of our mechanism, the bromonium ion gets attacked by Br^- (formed in the first step):

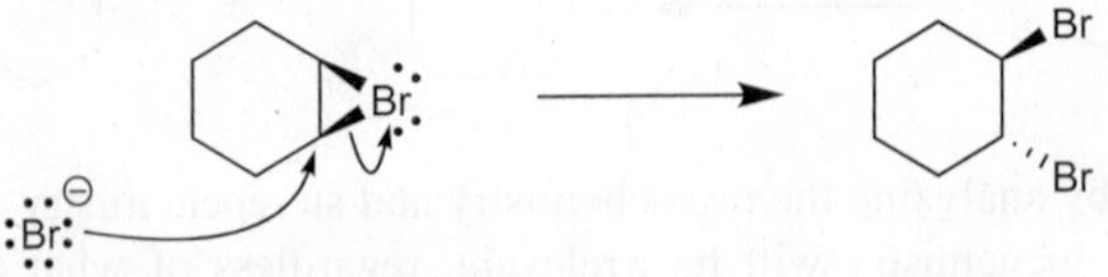

This step is simply an S_N2, and therefore, must be a back-side attack. In other words, the attacking bromide ion must come *from behind* (from behind the bridge), and therefore, we get an *anti* addition. There are some alkenes for which a *syn* addition predominates. Clearly, a different mechanism is operating in those cases. For the alkenes that you will encounter in this course, this reaction will always be an *anti* addition, proceeding via the mechanism that we showed.

Let's now summarize the profile for this reaction:

$\xrightarrow{Br_2}$	**Br** and **Br**
***Regio*chem**	not relevant
***Stereo*chem**	***anti***

This story gets more interesting when we use water as the solvent for the reaction. For example:

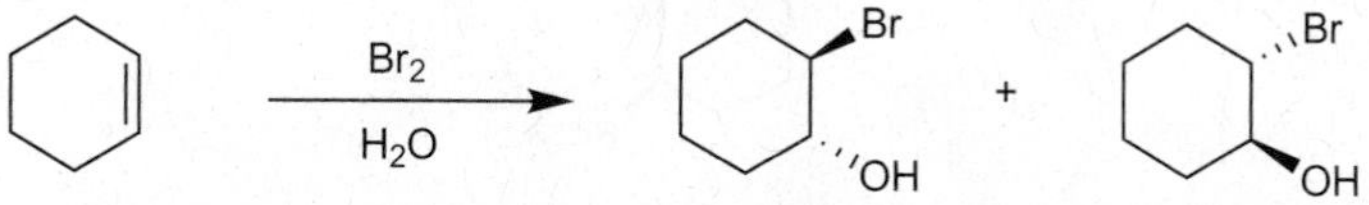

If we look at the products, we will see that we are now adding Br and OH, instead of Br and Br. In order to understand what is going on, we must look at the mechanism.

The first step is identical to what we saw just a few moments ago: the alkene attacks Br_2 to form a bridged, bromonium intermediate. But now we have a new possibility, because there are now two nucleophiles present: bromide and water. Rather than Br^- attacking the bromonium ion, a water molecule can attack instead, which will ultimately give the products shown above. The obvious question is: Why should H_2O attack instead of bromide? Isn't bromide a better nucleophile? Yes, it is true that bromide is a better nucleophile than H_2O. However, think about it from the point of view of a bromonium ion. It is a very high-energy intermediate (it has ring strain from the three-membered ring, and it has a positive charge on a bromine atom). Therefore, it is very eager to react with any nucleophile. It will not be very picky. It will react with the first nucleophile that it encounters. Since we are using water as the solvent here, the bromonium ion will most likely encounter a water molecule before it has a chance to get attacked by a bromide ion.

In the absence of water, we did not need to think about regiochemistry, because we were adding Br and Br. But now, in the presence of water, we are adding Br and OH (two different groups). As a result, the regiochemistry will be relevant if we start with an unsymmetrical alkene—for example,

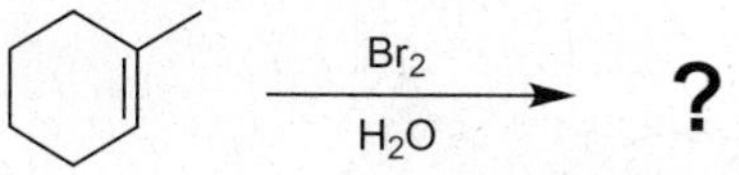

Which group will end up on the more substituted carbon? Br or OH? In other words, does water attack the more substituted carbon or the less substituted carbon? To answer this question, we need to look at the structure of the bromonium ion more

carefully than we did earlier. When we drew the bromonium ion earlier, the positively charged bromine atom was drawn to form a perfect three-membered ring (a nice triangle). But the bromine does not have to be in the center; it can lean to one side or the other:

Imagine if the bromine atom were leaning *all the way* to one side. That would just give us a carbocation:

It does not actually lean completely to one side, but it does lean a little toward one side—which gives more "carbocationic character" to the tertiary carbon:

The more substituted carbon (the tertiary carbon) can best handle this carbocationic character. Therefore, the more substituted carbon will possess more δ+ than the less substituted carbon. This means that the tertiary carbon atom will be close in character to an sp^2 hybridized carbon atom. It is not a full-fledged carbocation, so it is not fully sp^2 hybridized. But it is also not a regular sp^3 hybridized carbon. Rather, it is somewhere in between these two extremes (somewhere in between sp^2 and sp^3). Therefore, the geometry will not really be trigonal planar, nor will it be tetrahedral either. The geometry of the tertiary carbon atom above will be somewhere in between trigonal planar and tetrahedral. This helps explain how we can have an S_N2 attack at a tertiary center in the second step of the mechanism. Normally, we can't have an S_N2 at a tertiary center. But here it's OK, because the geometry is closer to being trigonal planar, which allows water to attack:

As a final step, deprotonation gives our product:

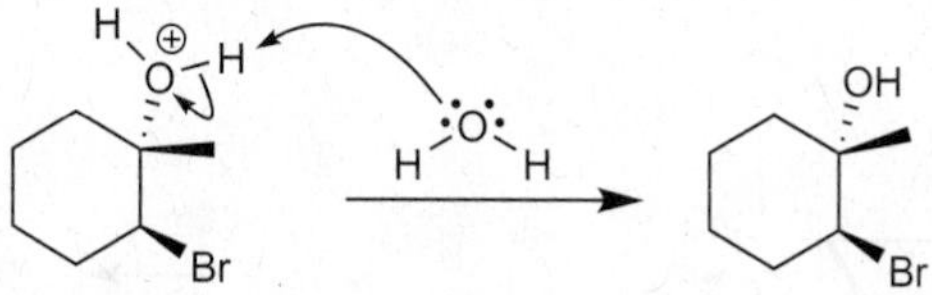

Notice that we use water to pull off the proton—we do not use hydroxide (HO^-), because there isn't much hydroxide around. It is always important to stay consistent with the conditions that are stated. In this case, the reagents are Br_2 and H_2O (not hydroxide). Therefore, we must use H_2O as a base to pull off the proton.

The final product is called a *halohydrin* (indicating that we have a halogen—Br—and an OH in the same compound). This reaction is commonly called *halohydrin formation*.

The profile of halohydrin formation can be summarized in the following chart:

$\xrightarrow[\text{H}_2\text{O}]{\text{Br}_2}$	**Br** and **OH**
*Regio*chem	OH goes on more substituted C
*Stereo*chem	*anti*

EXERCISE 11.98 Predict the products for each of the reactions below, and then propose a mechanism for formation of those products:

Answer **(a)** We are adding Br and Br, so the regiochemistry is irrelevant. But what about the stereochemistry? We look to see whether we are creating two new stereocenters. In this case, we are. So, the stereochemistry is relevant. We have explored the mechanism and justified why the reaction must be an *anti* addition. So, we must draw the pair of enantiomers that we would get from an *anti* addition. To do this properly, it will be helpful to redraw the alkene, as we have done many times before:

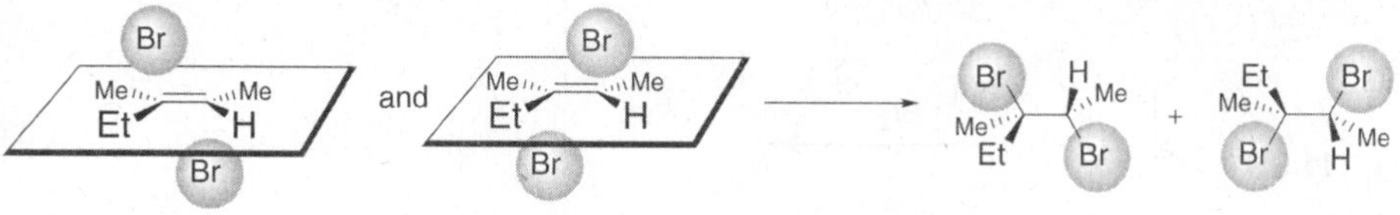

The mechanism of this reaction involves formation of a bromonium ion, followed by attack to open it up:

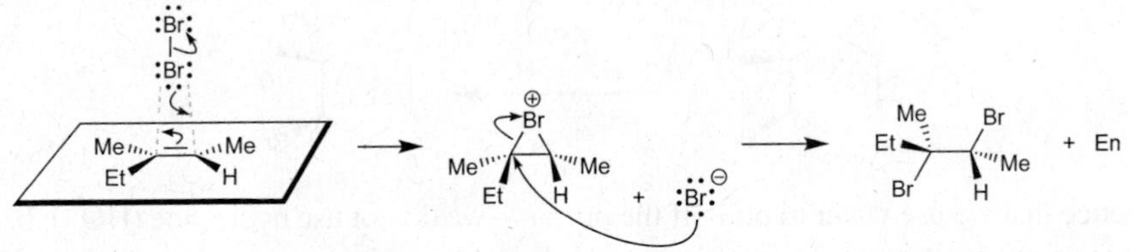

(b) In this problem, we are adding two different groups (Br and OH). So, the regiochemistry *is* relevant. We will need to draw our products with the OH on the more substituted carbon. What about the stereochemistry? We look to see whether we are creating two new stereocenters. In this case, we are. So, the stereochemistry *is* relevant. We have already explained why this reaction must be an *anti* addition. So, we must draw the pair of enantiomers that we would get from an *anti* addition. To do this properly, it will be helpful to redraw the alkene, like this:

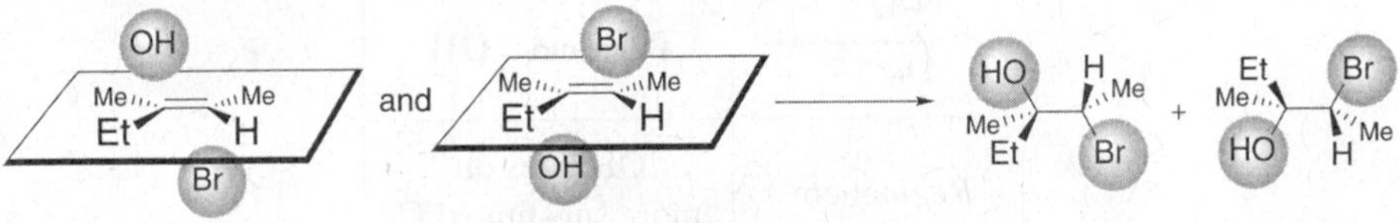

The mechanism of this reaction will have three steps: (1) formation of bromonium ion, (2) attack by water, and (3) deprotonation:

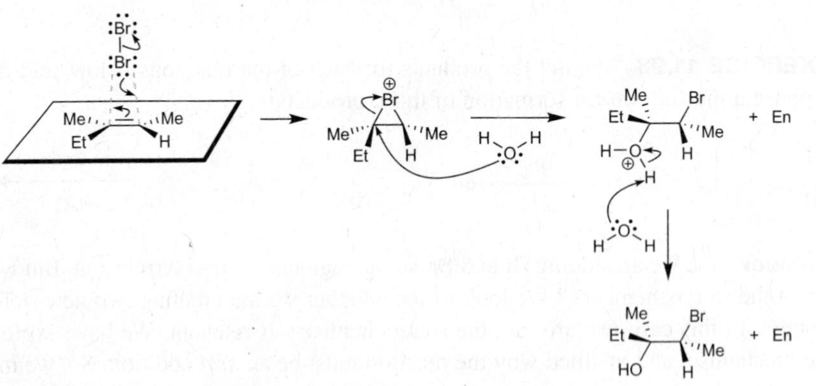

PROBLEMS Predict the products for each of the following reactions:

11.99

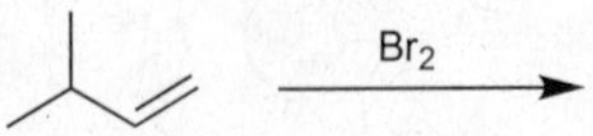

11.100

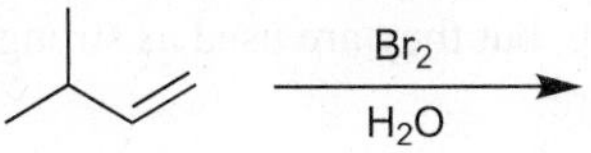

Br$_2$
H$_2$O

11.101

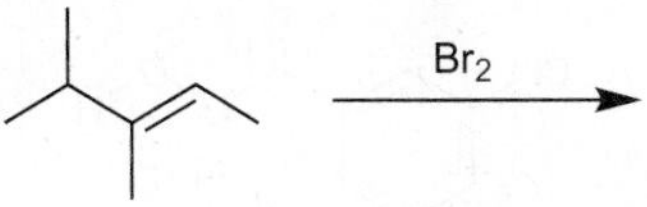

Br$_2$

11.102

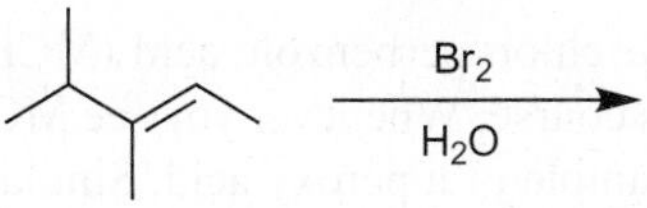

Br$_2$
H$_2$O

11.103

Br$_2$

11.104

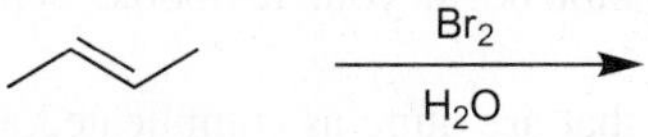

Br$_2$
H$_2$O

11.10   ADDING OH AND OH, *ANTI*

Now we will see how to add two OH groups across an alkene. It is possible to control whether the OH groups add via a *syn* addition, or via an *anti* addition (by choosing which reagents we will use to add the OH groups). In this section, we will learn how to add two OH groups in an *anti* addition. In the next section, we will learn how to add two OH groups in a *syn* addition.

To add two OH groups in an *anti* addition, we will employ a two-step process: we will first make an epoxide, and then we will open the epoxide with water under conditions of acid-catalysis:

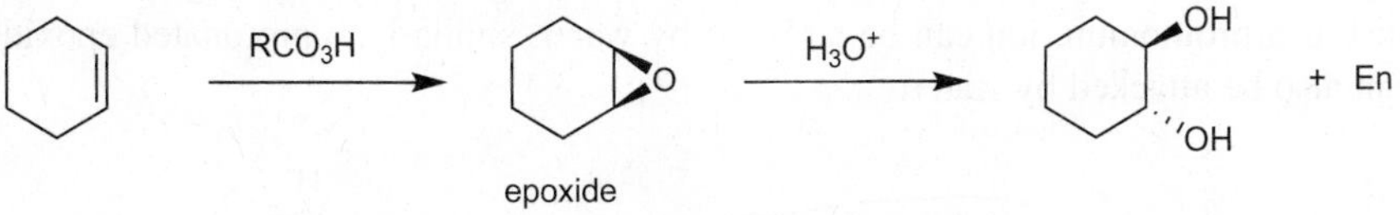

epoxide

We will now explore each aspect of this two-step process. In the first step, a peroxyacid (RCO$_3$H), sometimes called a per-acid, reacts with the alkene. Compare the structure of a carboxylic acid with the structure of a peroxy acid:

carboxylic acid   peroxy acid

Peroxy acids resemble carboxylic acids in structure, possessing just one additional oxygen atom. Peroxy acids are not very acidic, but they are used as strong oxidizing agents. Common examples of peroxy acids are:

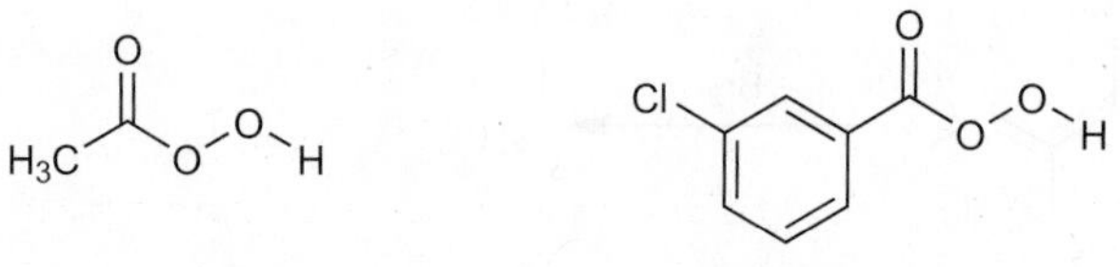

The second example above is called *meta*-chloroperbenzoic acid (MCPBA). It is perhaps the most common example in this course. Whenever you see MCPBA, you should immediately recognize it as an example of a peroxy acid. Similarly, whenever you see RCO$_3$H, you should also see it as the generic formula for a peroxy acid.

Peroxy acids will react with an alkene to form an epoxide. The mechanism is somewhat complicated, and may or may not be in your textbook, depending on which textbook you are using.

You will not see many mechanisms that are quite as complicated as this one, so we will not spend time on this mechanism. Let's turn our attention to the product of this reaction. We call the product an *epoxide*, which is the fancy term for a three-membered, cyclic ether.

We then take this epoxide and open it with water under conditions of acid-catalysis. Let's explore the mechanism of how this occurs. First, the epoxide is protonated:

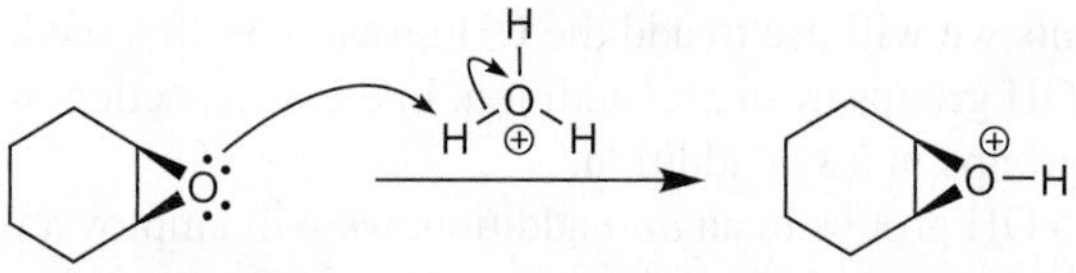

This proton transfer step produces an intermediate that is very similar to a bromonium ion (a three-membered ring with a positive charge on an electronegative atom). Just as a bromonium ion can be attacked by water, similarly, a protonated epoxide can also be attacked by water:

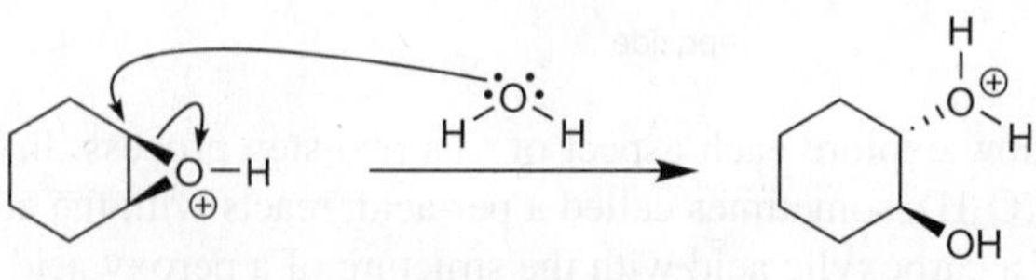

Once again (just like the attack of the bromonium ion in the previous section), water must attack from the back side, which explains the observed stereochemical preference for *anti* addition.

As a final step, we just need to deprotonate:

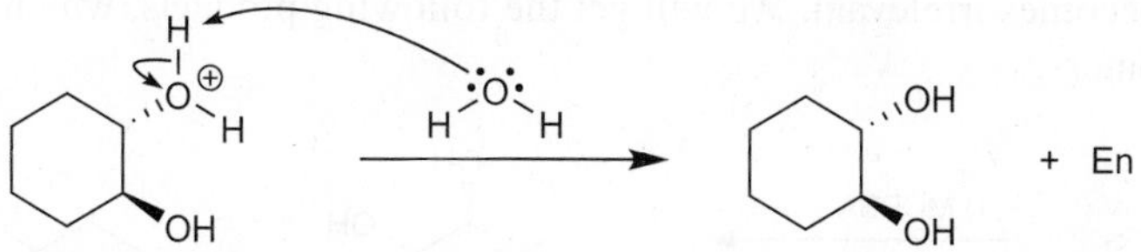

Once again, notice that we use water to deprotonate (rather than using hydroxide) in order to stay consistent with the conditions. In acidic conditions, we most certainly *cannot* use hydroxide ions in our mechanism.

Now we can summarize the profile for the two-step synthesis that we saw in this section:

1) MCPBA 2) H₃O⁺ →	**OH** and **OH**
*Regio*chem	not relevant
*Stereo*chem	*anti*

EXERCISE 11.105 Predict the products for each of the reactions below:

(a) 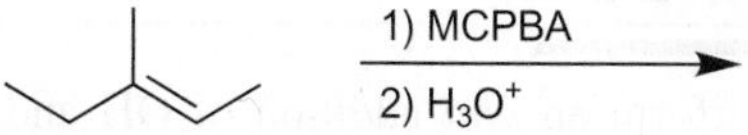 1) MCPBA 2) H₃O⁺

(b) 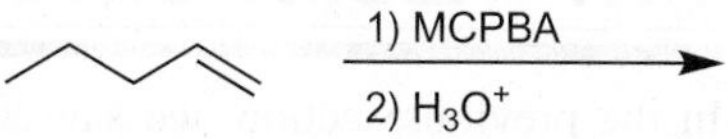 1) MCPBA 2) H₃O⁺

Answer **(a)** We are adding OH and OH, and therefore, regiochemistry will be irrelevant. What about stereochemistry? In this case, we are creating two new stereocenters, so we must carefully consider the stereochemistry of this reaction in order to draw the correct pair of enantiomers. This two-step synthesis gives an *anti* addition of OH and OH. Therefore,

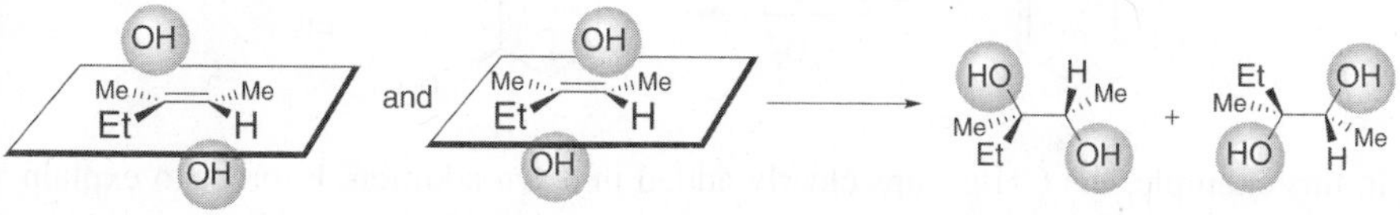

(b) In this example, we are *not* creating two new stereocenters. We are only creating one new stereocenter. So, the stereochemistry in this case becomes irrelevant. It is true that the reaction proceeds through an *anti* addition of OH and OH. However, with only one stereocenter in the product, the preference for *anti* addition becomes irrelevant. We will get the following products, which are a pair of enantiomers:

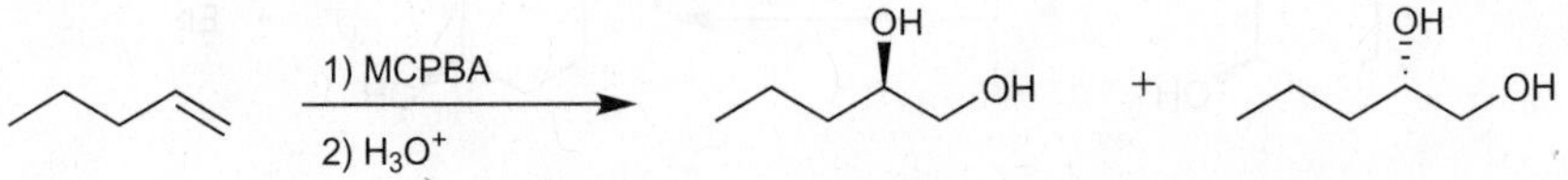
1) MCPBA
2) H_3O^+
OH
OH
+
OH
OH

PROBLEMS Predict the products for each of the following reactions:

11.106 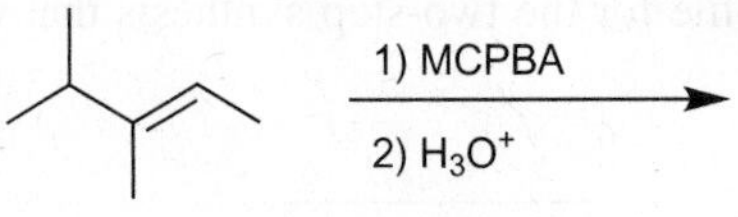
1) MCPBA
2) H_3O^+

11.107 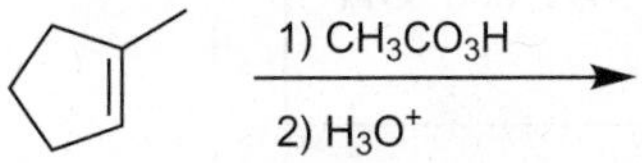
1) CH_3CO_3H
2) H_3O^+

11.108 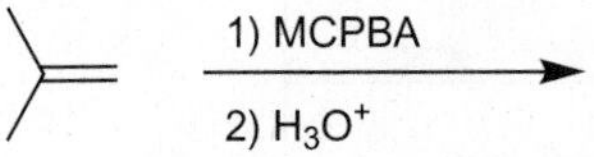
1) MCPBA
2) H_3O^+

11.109 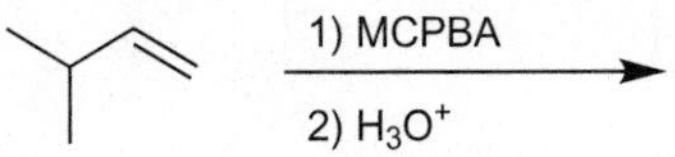
1) MCPBA
2) H_3O^+

11.11 ADDING OH AND OH, *SYN*

In the previous section, we saw how to perform an *anti* addition of OH and OH across a double bond. In this section, we will explore the conditions that will allow us to perform a *syn* addition of OH and OH across a double bond. This reaction is often called a *syn hydroxylation*.

Consider the following example:

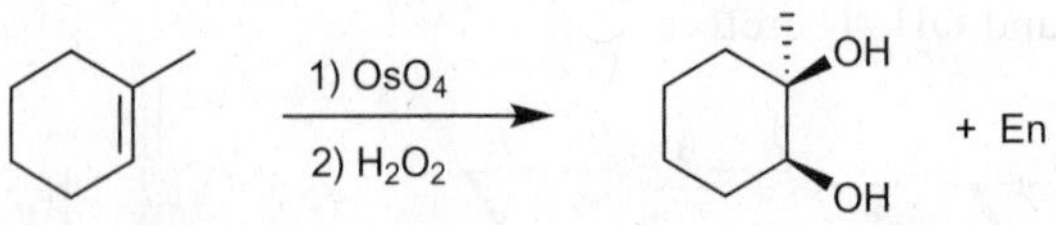
1) OsO_4
2) H_2O_2
OH
OH
+ En

In this example, the OH groups clearly added in a *syn* addition. In order to explain why this reaction proceeds via a *syn* addition, we must look at the first step of the mechanism:

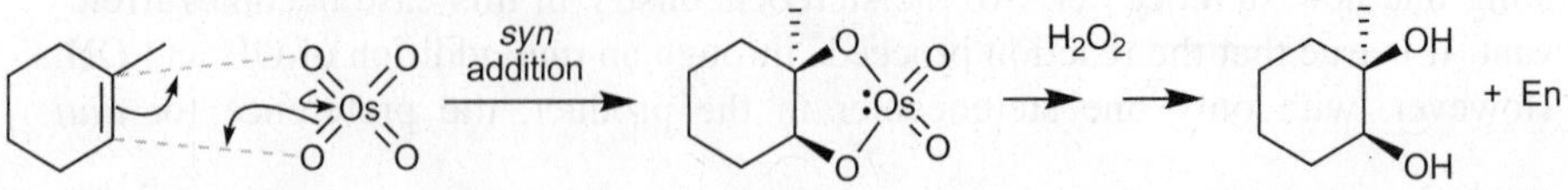
syn
addition
H_2O_2
OH
OH
+ En

It is this first step that allows us to understand why the reaction follows a *syn* addition. In this step, osmium tetroxide (OsO_4) adds across the alkene *in a concerted process*. In other words, both oxygen atoms attach to the alkene simultaneously. This effectively adds two groups *across the same face* of the alkene.

The pertinent details of this reaction can be summarized in the following chart:

$\dfrac{OsO_4}{H_2O_2}$	**OH and OH**
*Regio*chem	not relevant
*Stereo*chem	*syn*

The same transformation (*syn* addition of OH and OH) can also be accomplished with cold $KMnO_4$ and hydroxide. Again, we will only look at the first step of the mechanism:

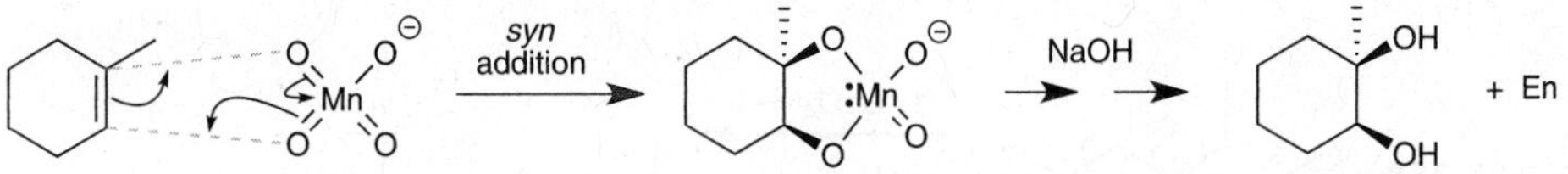

Once again, we have a concerted process that adds both oxygen atoms simultaneously across the double bond. Notice the similarity between the mechanisms of these two methods (OsO_4 vs. $KMnO_4$).

EXERCISE 11.110 Predict the products for each of the reactions below:

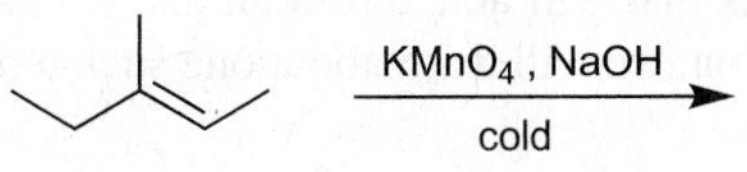

Answer In this reaction, we are adding two OH groups, so we don't need to think about regiochemistry. As always, stereochemistry will only be relevant if we are forming two new stereocenters. In this example, we are creating two new stereocenters, and therefore, we must be careful to draw only the pair of enantiomers that represent the products of a *syn* addition:

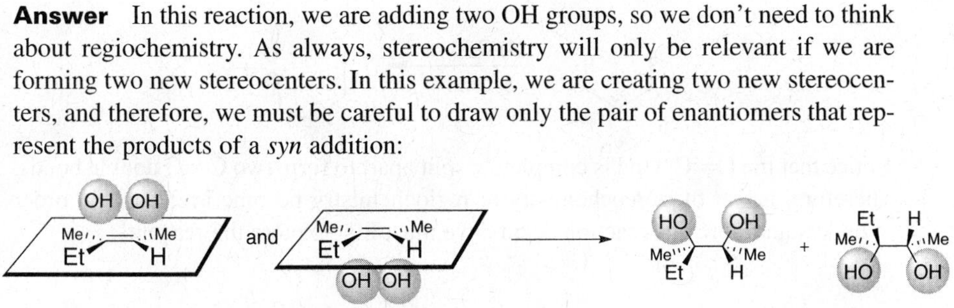

PROBLEMS Predict the products for each of the following reactions:

11.111

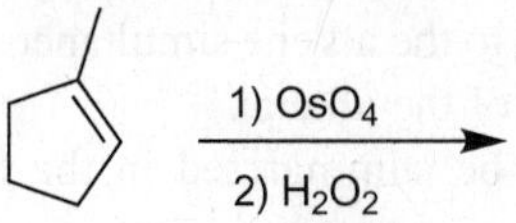

1) OsO$_4$

2) H$_2$O$_2$

11.112

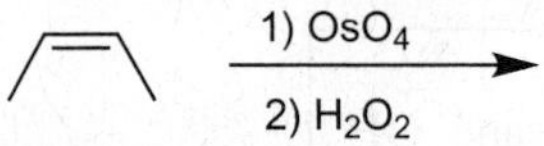

1) OsO$_4$

2) H$_2$O$_2$

11.113

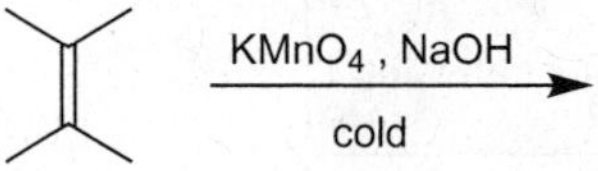

KMnO$_4$, NaOH

cold

11.114

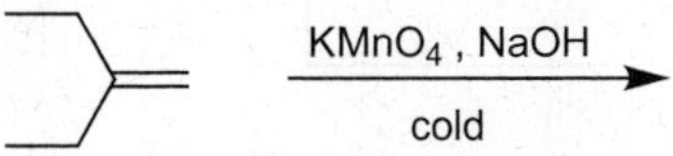

KMnO$_4$, NaOH

cold

11.115

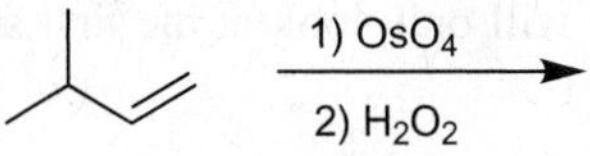

1) OsO$_4$

2) H$_2$O$_2$

11.116

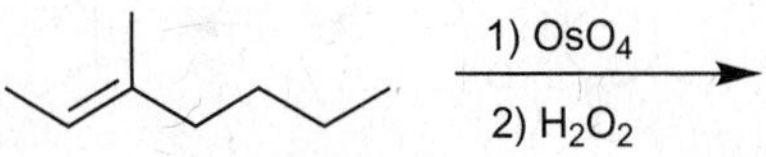

1) OsO$_4$

2) H$_2$O$_2$

11.12 OXIDATIVE CLEAVAGE OF AN ALKENE

There are many reagents that will add across an alkene and completely cleave the C=C bond. In this section, we will learn about one such reaction, called *ozonolysis*. Consider the following example:

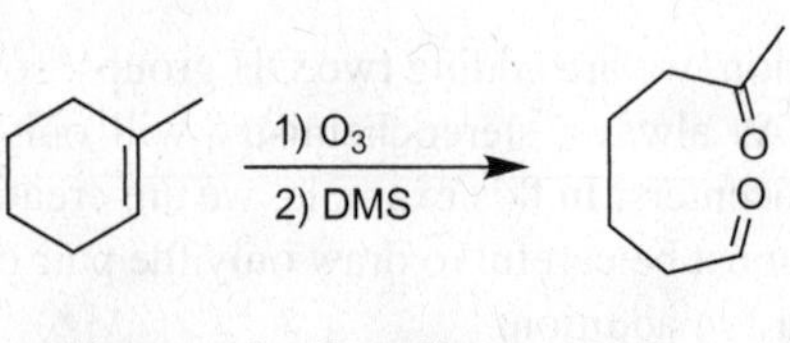

1) O$_3$

2) DMS

Notice that the C=C bond is completely split apart to form two C=O double bonds. Therefore, issues of stereochemistry or regiochemistry become irrelevant. In order to understand how this reaction occurs, we must first explore the reagents.

Ozone is a compound with the following resonance structures:

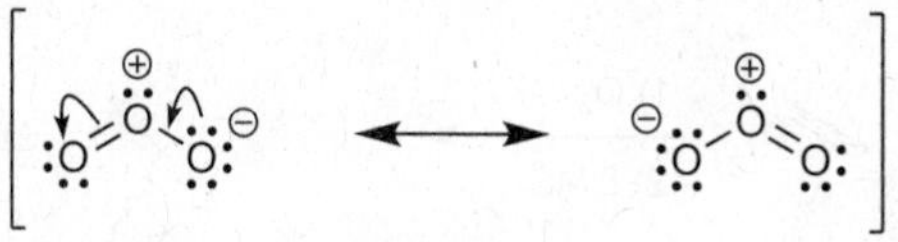

Ozone is formed primarily in the upper atmosphere, where oxygen gas (O_2) is bombarded with ultraviolet light.

As we did in the previous sections, we will only explore the first step of the mechanism:

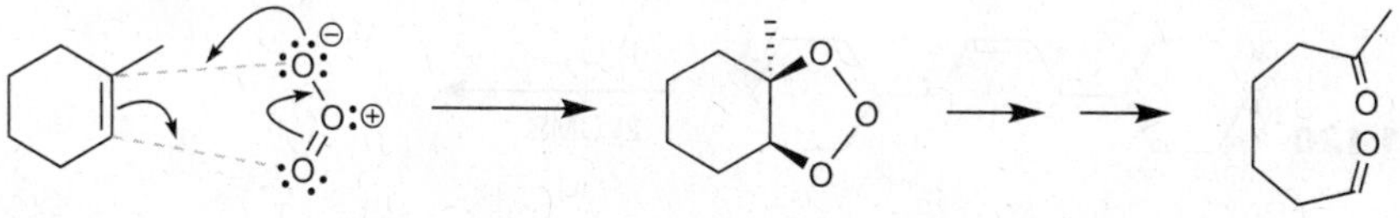

Compare this step to the mechanisms we saw in the previous section (*syn* dihydroxylation), and you should see striking similarities. The initial product (shown above) is called a molozonide, and it subsequently undergoes further rearrangements, before ultimately giving the product upon treatment with dimethyl sulfide (DMS). The structure of DMS is:

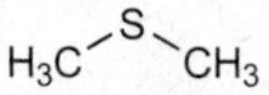

$$H_3C \diagdown S \diagup CH_3$$

DMS is a mild reducing agent. There are many other reducing agents that can be used in the final step of an ozonolysis, but DMS is common.

There is a simple technique for drawing the products of an ozonolysis: just split up each C=C bond into two C=O bonds. Let's see an example of how this works:

EXERCISE 11.117 Predict the products of the following reaction:

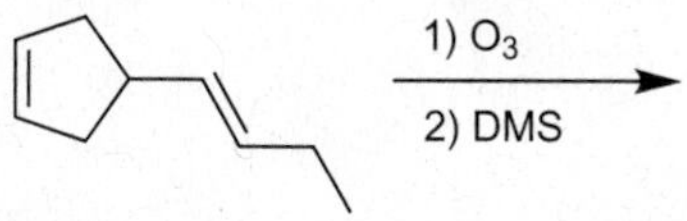

Answer There are two C=C double bonds in this compound. For each C=C double bond, we simply erase the C=C double bond, and place two C=O bonds in its place. Our answer looks like this:

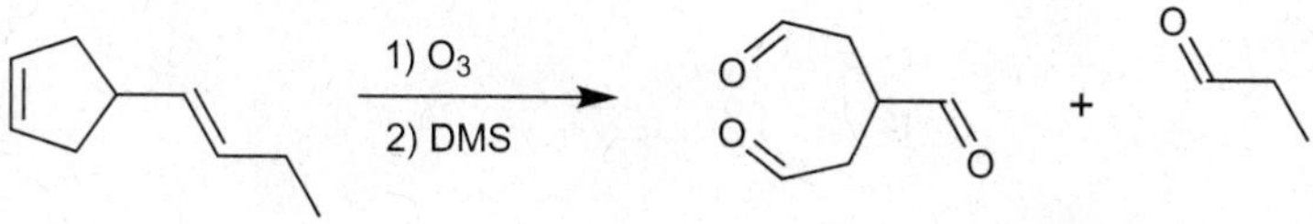

PROBLEMS Predict the products for each of the following reactions:

11.118

$$\text{(skeletal structure)} \quad \xrightarrow[\text{2) DMS}]{\text{1) O}_3}$$

11.119

$$\text{(skeletal structure)} \quad \xrightarrow[\text{2) DMS}]{\text{1) O}_3}$$

11.120

$$\text{(skeletal structure)} \quad \xrightarrow[\text{2) DMS}]{\text{1) O}_3}$$

11.121

$$\text{(skeletal structure)} \quad \xrightarrow[\text{2) DMS}]{\text{1) O}_3}$$

11.122

$$\text{(skeletal structure)} \quad \xrightarrow[\text{2) DMS}]{\text{1) O}_3}$$

11.123

$$\text{(skeletal structure)} \quad \xrightarrow[\text{2) DMS}]{\text{1) O}_3}$$

SUMMARY OF REACTIONS

Below is a diagram that shows the key reactions in this chapter. If you
pleted all of the problems in this chapter up to this point, you should be able to fill
in the reagents necessary for every transformation shown. Try doing that now. If you
have trouble remembering the reagents for a particular reaction, then just flip back
to the appropriate section, and find the reagents:

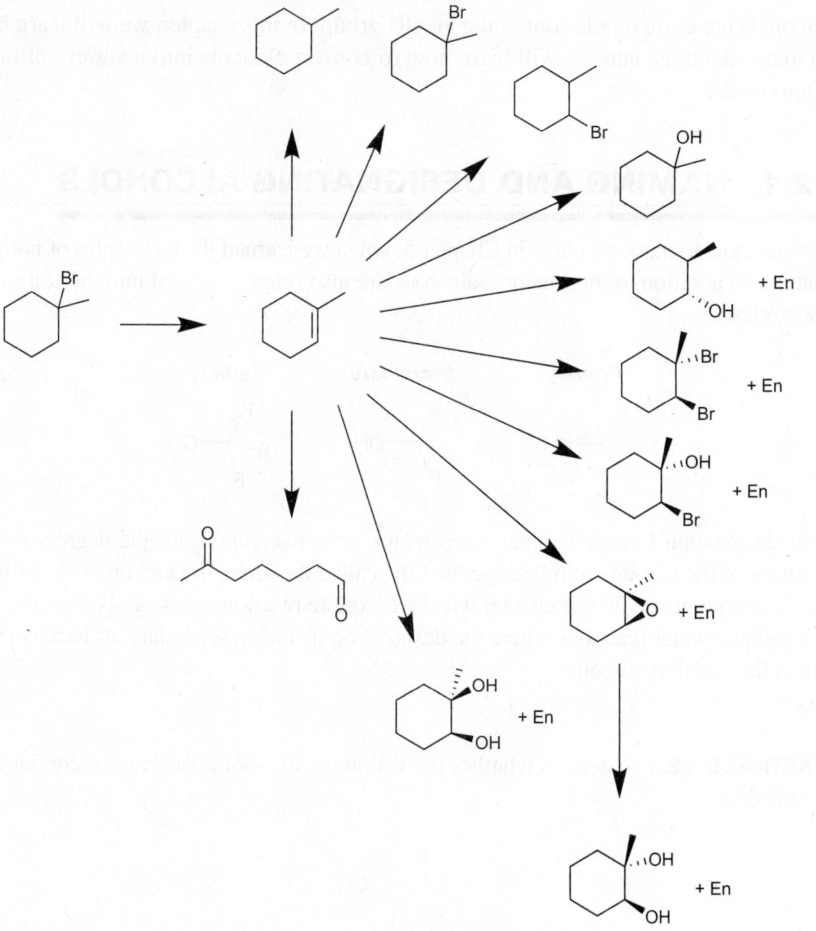

After you have filled in the necessary reagents above, you should study this diagram
carefully. Look it over, and make sure that every part of it makes sense to you. Go
over it ten times in your head. Review it until you are able to reconstruct the whole
diagram on a blank piece of paper (with nothing else in front of you).

CHAPTER 12

ALCOHOLS

Alcohols are compounds containing an OH group. In this chapter, we will learn how to make alcohols, and we will learn how to convert alcohols into a variety of other compounds.

12.1 NAMING AND DESIGNATING ALCOHOLS

We saw how to name alcohols in Chapter 5, when we learned the basic rules of nomenclature. In addition to their names, alcohols are also often classified into the following categories:

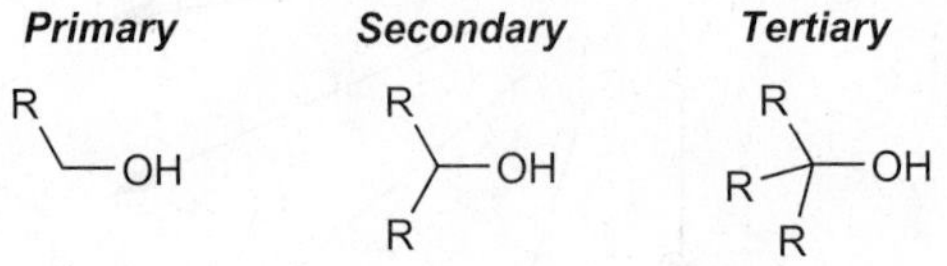

This designation scheme (primary, secondary, or tertiary) indicates the degree of substitution of the carbon atom bearing the OH, called the alpha (α) carbon. This will become important later in this chapter when we explore reactions of alcohols. Specifically, we will encounter reactions where the designation (primary, secondary, or tertiary) will affect the reaction outcome.

EXERCISE 12.1   Identify whether the following alcohol is primary, secondary, or tertiary:

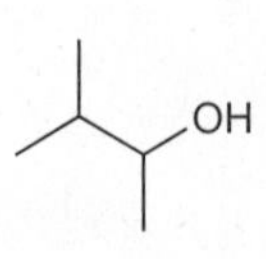

ANSWER   We begin by identifying the carbon atom connected directly to the OH group (the α carbon):

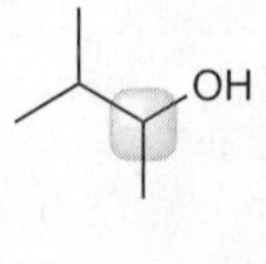

Now, we count the number of alkyl groups attached directly to the α carbon. In this case, there are two alkyl groups:

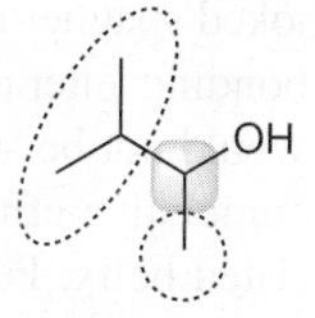

Therefore, this compound is a secondary alcohol.

PROBLEMS Identify whether each of the following alcohols is primary, secondary, or tertiary:

12.2 _____

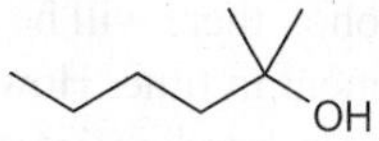

12.3 _____

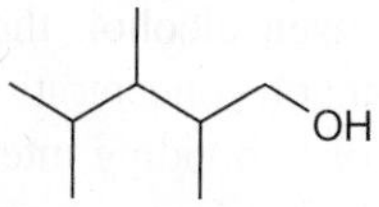

12.4 _____

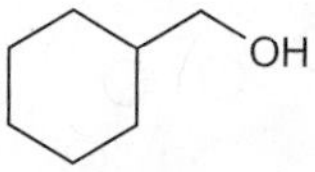

12.5 _____

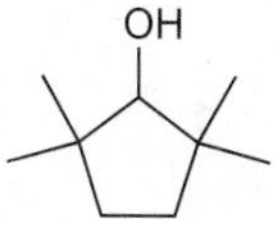

12.2 PREDICTING SOLUBILITY OF ALCOHOLS

An alcohol, _by definition_, will contain an OH group in its structure. Therefore, we expect hydrogen bonding to occur:

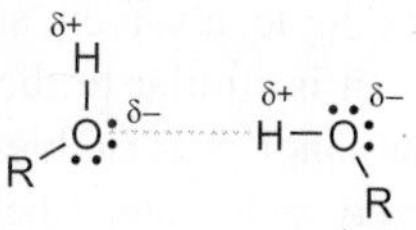

Hydrogen bonding does NOT refer to a type of bond (even though it is called hydrogen _bonding_). This terminology is somewhat misleading. H-bonding actually refers to a force of attraction between molecules (an intermolecular force). This attraction is temporary, or fleeting, in the liquid and gaseous states. As two alcohol molecules approach each other, they are momentarily attracted to one another. At low enough temperatures (if the alcohol is a solid), then these forces will actually hold together the alcohol molecules in a crystalline state. But in the liquid or gaseous state, the molecules are bouncing off each other, and they are NOT permanently held together.

Why the misleading terminology (H-*bonding*)? At some point during your high school studies, you were likely exposed to a model of the DNA helix, which looks like a twisted ladder. That twisted ladder is actually two very large molecules (each of which is like a piece of cooked spaghetti), twisted around each other. Each rung of the ladder is a hydrogen bonding interaction between the two molecules. Each individual H-bond, by itself, would not be strong enough to hold the two large molecules together. However, the cumulative effect (of millions of H-bonds) holds the two molecules together in a twisted helix. Perhaps this can help us understand why we call these interactions (the rungs of the ladder) hydrogen *bonding*. This also explains why it is so easy to "unzip" the helix.

You might learn about the structure of DNA at the end of your organic chemistry course. For right now, we will be focused on problems that deal primarily with small molecules; and therefore, for our purposes, we should think of H-bonding as an interaction; a type of intermolecular force.

For any given alcohol, the strength of the H-bonding interactions will be greatly dependent on concentration. For a concentrated alcohol, there will be a fairly decent amount of H-bonding interactions at any given moment in time. However, if we dilute the alcohol in a solvent that cannot form hydrogen bonding interactions with the alcohol, then the H-bonding effect will be minimal:

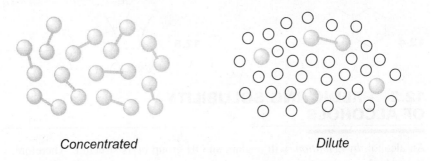

Concentrated   *Dilute*

Now consider diluting an alcohol in a solvent that *can* form hydrogen bonding interactions with the alcohol (for example, a solvent such as water). In such a case, the interaction is indeed very strong (it is similar to the effect of having a concentrated alcohol). This explains why methanol is miscible with water. The term *miscible* means that methanol can be mixed with water (they will dissolve in each other) *in all proportions*. However, not all alcohols are miscible with water. In fact, just the very small alcohols (methanol, ethanol, propanol, and *tert*-butanol) are miscible with water. To understand why, we must realize that every alcohol has two regions. The *hydrophobic* region, which does *not* interact well with water, and the *hydrophilic* region, which *does* interact well with water:

$$
\begin{array}{c}
\text{Hydrophobic} \\
\text{Region}
\end{array}
\quad
\begin{array}{c}
\text{H} \quad \text{H} \\
\text{H—C—O} \\
\text{H}
\end{array}
\quad
\begin{array}{c}
\text{Hydrophilic} \\
\text{Region}
\end{array}
$$

In the case of methanol, the hydrophobic end of the molecule is fairly small. This is true even of ethanol and propanol. But an interesting thing happens when we look at 1-butanol. The hydrophobic end of the molecule is now large enough to prevent miscibility:

Hydrophobic Region — OH Hydrophilic Region

Water and 1-butanol will mix, but _not in all proportions_. That is, 1-butanol is considered to be soluble in water, rather than miscible. The term *soluble* means that only a certain volume of 1-butanol will dissolve in a specified amount of water at room temperature.

As we consider an even larger hydrophobic region, solubility decreases. For example, 1-octanol has a very low solubility in water at room temperature:

Hydrophobic Region — OH Hydrophilic Region

Here is a rule of thumb that can help us predict solubility: in order to have a high water-solubility, there should be no more than 5 carbon atoms *per OH group*. A compound with two OH groups and 7 carbon atoms will be very soluble in water. A compound with one OH group and 7 carbon atoms will NOT be very soluble in water. There are, of course, many exceptions to this simplified rule of thumb, but the rule can be helpful whenever we need to make a quick prediction about the solubility of an alcohol.

EXERCISE 12.6 Predict whether the following alcohol will have a high or low solubility in water:

OH

ANSWER This compound has eight carbon atoms, and only one OH group. The hydrophobic region of the molecule is too large, and we expect the molecule to exhibit very low water solubility.

PROBLEMS Predict whether each of the following alcohols will have a high or low solubility in water:

12.7 OH

12.8 OH

12.9 HO⌒⋀⌒OH

12.10 (structure of a cyclopentane ring with OH and two methyl groups)

12.3 PREDICTING RELATIVE ACIDITY OF ALCOHOLS

As we learned in Chapter 3, in order to determine the relative acidity of a compound, we must carefully assess the stability of its conjugate base. For example:

To determine how acidic this alcohol is...   ...remove the proton...   ...and assess the stability of the conjugate base.

When we draw the conjugate base of an alcohol (as we did above), we see that the negative charge is on an oxygen atom. This is much more stable than a negative charge on a nitrogen atom, but it is not as stable as a negative charge on a halogen:

Increasing stability

Least stable   Most stable

Therefore, in terms of acidity, alcohols will lie somewhere between amines and hydrogen halides:

Increasing acidity

Least acidic   Most acidic

The pK_a for most alcohols will fall in the range of 15-18. Remember, a low pK_a value means that a compound is fairly acidic, and a high pK_a value means that it is not very acidic. Compare the following pK_a values:

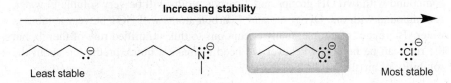

Increasing acidity

	R—H	R—NH₂	R—OH	X—H
pK_a values	Between 45 and 50	Between 35 and 40	Between 15 and 18	Between -10 and 3

In Chapter 3, we learned how to analyze the four factors that stabilize a negative charge (ARIO). The "I" in ARIO stood for "induction." We saw that neighboring halogens can withdraw electron density and stabilize the negative charge:

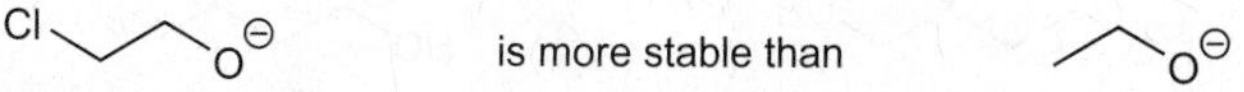

is more stable than

Therefore, the presence of halogens can lower the pK_a value of an alcohol to around 14. In other words, the presence of a halogen will make the compound more acidic.
The following two alcohols provide an interesting comparison:

Cyclohexanol
$pK_a = 18$

Phenol
$pK_a = 10$

The pK_a of cyclohexanol is 18, whereas the pK_a of phenol is 10. This means that phenol is eight orders of magnitude more acidic than cyclohexanol. In other words, phenol is 100 million times more acidic than cyclohexanol. Why is there such a huge difference? We can understand why cyclohexanol has a pK_a of 18, because that is within the expected range of an alcohol (15–18). So the real question is: why is the pK_a of phenol so low? Why is phenol so much more acidic than a regular alcohol? To answer this question, we will use the "R" of ARIO (*Resonance*) to explain the acidity of phenol:

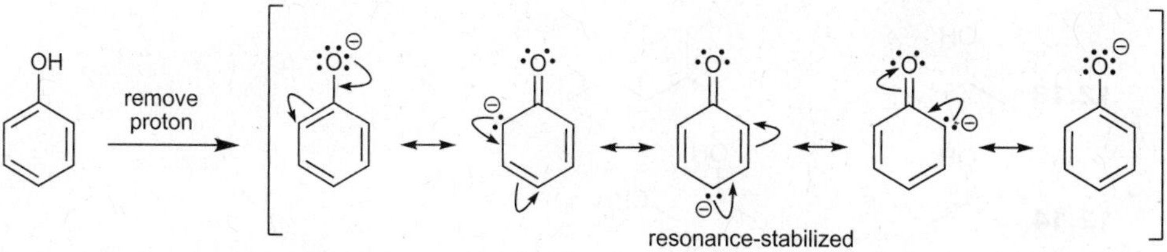

resonance-stabilized

The conjugate base of phenol is stabilized by resonance. This explains why phenolic protons are more acidic than typical alcoholic protons.

EXERCISE 12.11 Identify the most acidic proton in the following compound:

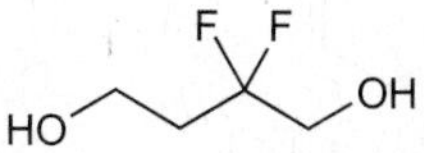

ANSWER The two OH groups will certainly be the two acidic protons in the compound, but we must choose which proton is more acidic. We must begin by drawing the conjugate base in each case, and then we will compare those conjugate bases, using our four factors (ARIO):

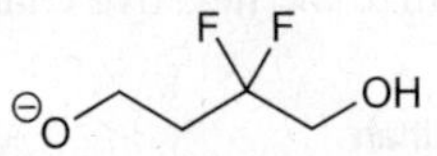

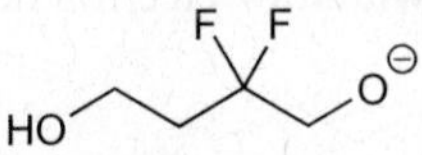

In both cases, the negative charge is on an oxygen atom, so our first factor (**A**RIO) does not help. Neither structure is stabilized by resonance, so our second factor (A**R**IO) also does not help. In this case, it is the third factor (induction) that indicates which structure is more stable. Both conjugate bases are stabilized by the electron-withdrawing effects of the two fluorine atoms. But the more stable conjugate base is the one where the two fluorine atoms are closer to the negatively charged oxygen atom:

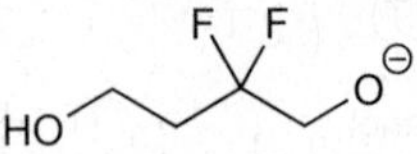

Therefore, we expect the following highlighted proton to be the most acidic proton:

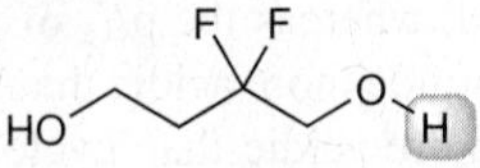

PROBLEMS   For each pair of compounds, identify the compound that is more acidic. Make sure that you can explain your choice in each case.

12.12

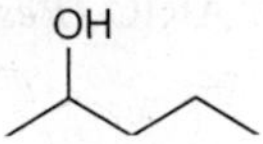

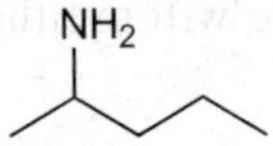

12.13

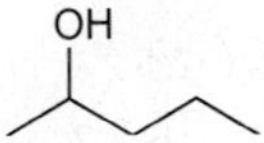

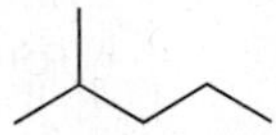

12.14

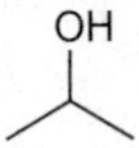

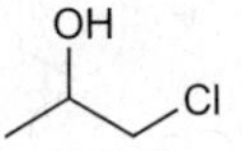

12.15

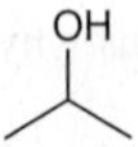

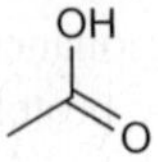

12.16

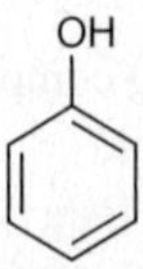

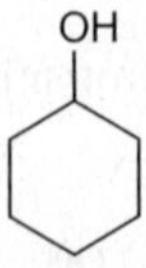

12.17

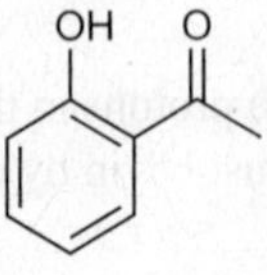

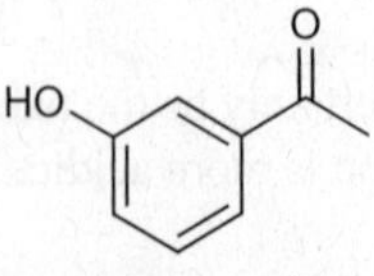

12.4 PREPARING ALCOHOLS: A REVIEW

We will learn many ways of making alcohols in this chapter. First, let's review the various reactions that we have already learned in previous chapters. All of these reactions can be used to produce an alcohol:

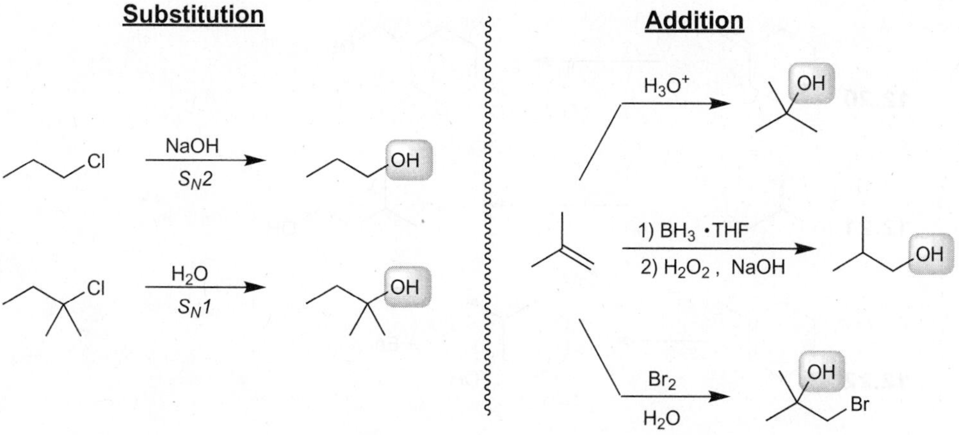

Substitution

Addition

Before we can learn *new* methods for making alcohols, you must make sure that you remember the methods that you have already learned (above). Let's get some practice:

EXERCISE 12.18 What reagents would you use to accomplish the following transformation:

ANSWER In this case, H and OH are added across the pi bond in an *anti*-Markovnikov addition. No stereocenters were formed, so the stereochemical outcome is not relevant. We have only seen one way to achieve an *anti*-Markovnikov addition of water across a pi bond: hydroboration-oxidation. Therefore, our answer is:

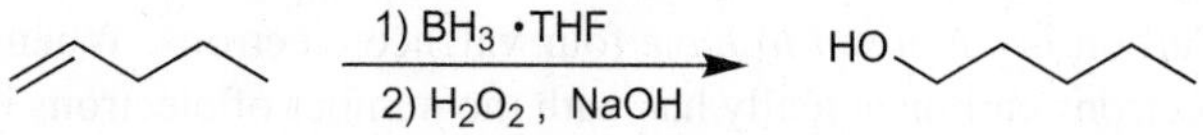

PROBLEMS What reagents would you use to accomplish each of the following transformations:

12.19

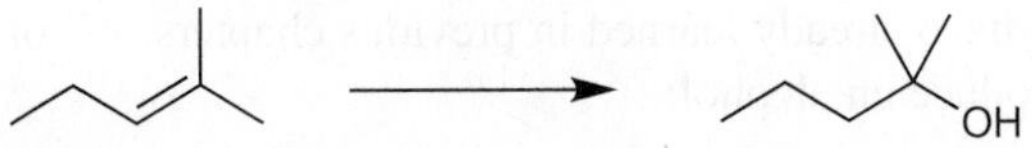

12.20

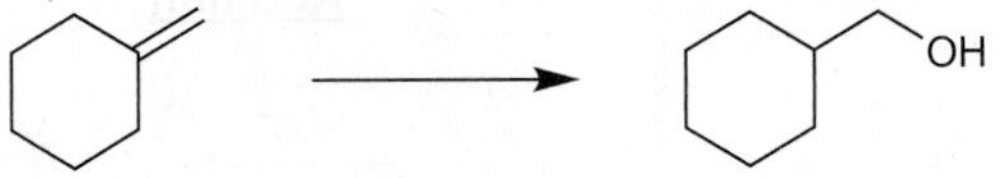

12.21

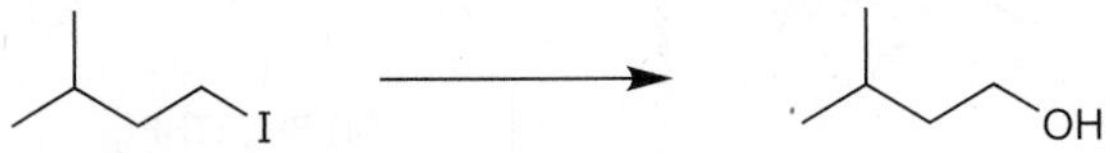

12.22

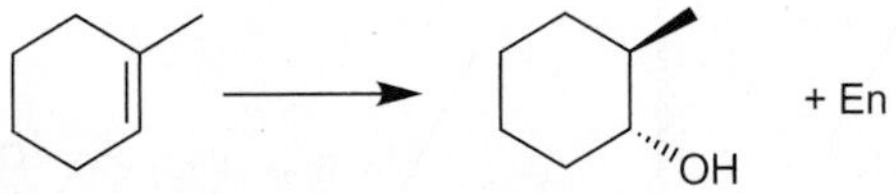

+ En

12.5 PREPARING ALCOHOLS VIA REDUCTION

In this section, we will learn how to prepare alcohols through a *reduction* process. In order to understand what the word *reduction* means, we must go back and review what oxidation states are. We will do that now:

There are two methods for counting electrons: *formal charge* and *oxidation state*. These two counting methods actually represent two flipsides of the same coin. To calculate formal charge, we treat all bonds as covalent, regardless of whether they are or not:

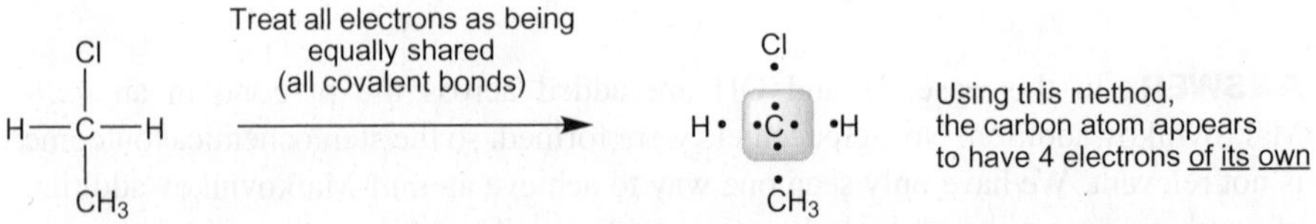

When we treat all bonds as covalent, the carbon atom appears to have four electrons of its own. Carbon is *supposed to have* four valence electrons. When we compare how many electrons carbon actually has with the number of electrons it is supposed to have, we see that everything is just right in this case. It is supposed to have four valence electrons, and it is clearly using four valence electrons. Therefore, there is no formal charge.

On the flipside, we will now calculate the *oxidation state* of the same carbon atom. To calculate the oxidation state of an atom, we treat all bonds as ionic, regardless of whether they are or not:

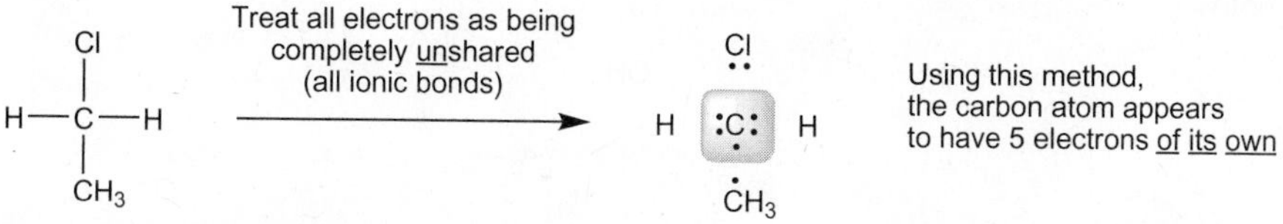

Treat all electrons as being completely unshared (all ionic bonds) → Using this method, the carbon atom appears to have 5 electrons of its own

For each bond, we give both electrons to the more electronegative atom. For the C—Cl bond, chlorine is more electronegative, so chlorine gets both electrons. For each of the C—H bonds, carbon is *very slightly* more electronegative than hydrogen, so carbon will get both electrons in each case. But for the C—C bond, neither atom is more electronegative than the other, so we must place one electron on each carbon atom. Now when we count the electrons, it appears as though the carbon atom is using five electrons of its own. But we know that carbon is only supposed to have four valence electrons. This carbon atom is using one extra electron, and therefore, this carbon atom will have an oxidation state of -1.

Formal charges and *oxidation states* represent two different ways of estimating electron density. Neither method is perfectly accurate. Each method assumes an extreme that is not true. Formal charges are calculated based on the assumption that all bonds are *covalent* (generally an incorrect assumption), and oxidation states are calculated based on the assumption that all bonds are *ionic* (generally an incorrect assumption). Earlier in this book, we focused our attention on formal charges exclusively. For purposes of this section, we will now focus our attention exclusively on oxidation states.

Carbon can range in oxidation state from -4 to $+4$:

- 4   **+ 4**

The oxidation state of the carbon atom in an alcohol will be dependent on the identities of the atoms that are attached to the carbon atom. Here are some examples (make sure that you can calculate and verify that the oxidation states shown here are correct):

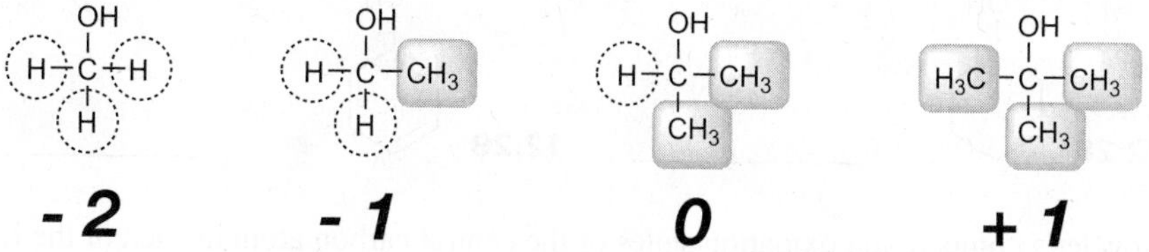

- 2   **- 1**   **0**   **+ 1**

Let's get practice:

EXERCISE 12.23 Calculate the oxidation state of the carbon atom highlighted below:

OH

ANSWER For each bond, we give both electrons to the more electronegative atom. For the C—O bond, oxygen is more electronegative, so oxygen gets both electrons. For each of the C—C bonds, we treat the electrons as being shared equally (one electron on one carbon atom, and the other electron on the other carbon atom). For the C—H bond, carbon is *very slightly* more electronegative than hydrogen, so carbon will get both electrons.

Treat all electrons as being completely unshared (all ionic bonds)

H_3C—C—CH_3 ⟶ $H_3C \bullet$ $\bullet C \bullet$ $\bullet CH_3$ Using this method, the carbon atom appears to have 4 electrons of its own

When we count the electrons that carbon is using now, it appears as though the carbon atom is using four electrons of its own. Next, we compare that number to the number of valence electrons that carbon is supposed to have (four valence electrons). This carbon atom is using exactly the right number of electrons, and therefore, this carbon atom will have an oxidation state of 0.

PROBLEMS Calculate the oxidation state of the carbon atom that is highlighted in each of the following compounds:

12.24 _____

12.25 _____

12.26 _____

12.27 _____

12.28 _____

12.29 _____

Now let's compare the oxidation states of the central carbon atom in each of the following compounds:

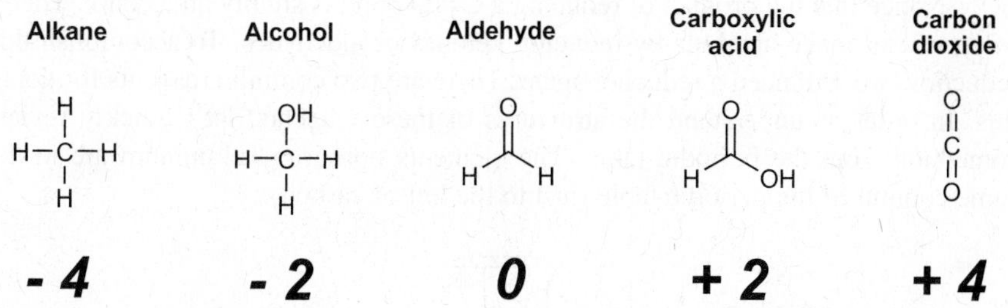

Alkane	Alcohol	Aldehyde	Carboxylic acid	Carbon dioxide
H−C−H (with H above, H below)	H−C−H (with OH above, H below)	(aldehyde structure, H and H)	(carboxylic acid structure, H and OH)	O=C=O
- 4	**- 2**	**0**	**+ 2**	**+ 4**

Notice the trend. Let's ignore the two extremes above (alkane and carbon dioxide), and let's focus on the middle three compounds: alcohols, aldehydes, and carboxylic acids. Carboxylic acids are at a higher oxidation state than aldehydes, which in turn, are at a higher oxidation state than alcohols. Now imagine that we are running a reaction that converts an alcohol into an aldehyde or a carboxylic acid. This reaction would constitute an increase in oxidation state. Whenever we run a reaction that increases the oxidation state, we say that an *oxidation* has occurred. Therefore, converting a primary alcohol into an aldehyde or a carboxylic acid is called an oxidation:

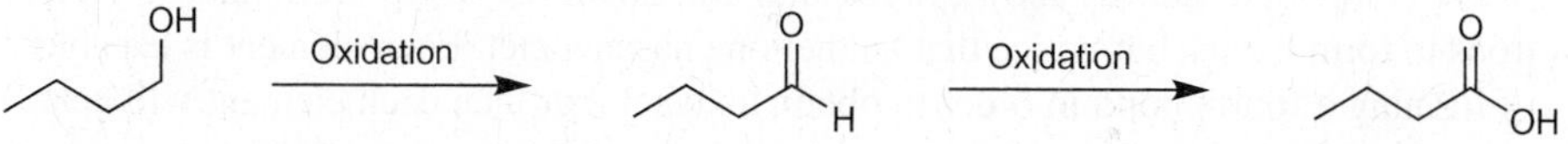

Similarly, it is also called an oxidation when we convert a secondary alcohol into a ketone:

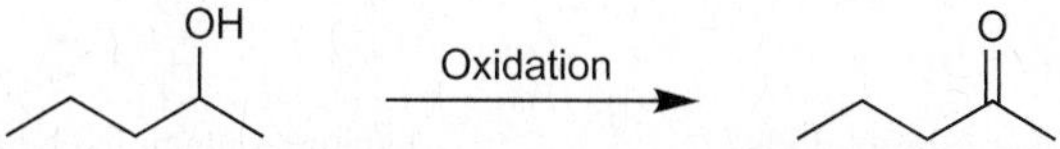

In order to accomplish an oxidation, there is always some other compound that must itself get reduced (that compound is called the oxidizing agent, because it causes the desired oxidation).

Whenever we run a reaction that involves a *decrease* in oxidation state, we say that a *reduction* has occurred. For example, converting a ketone or aldehyde into an alcohol:

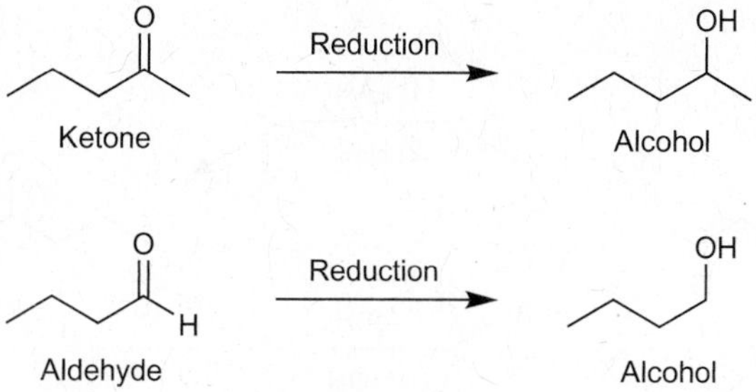

Ketone   Alcohol

Aldehyde   Alcohol

Notice that the product of reducing a C=O bond is simply an alcohol. Therefore, we can make alcohols by reducing ketones or aldehydes. To accomplish this reduction, we will need a reducing agent. There are two common reagents for doing this. In order to understand the structures of these reagents, let's quickly review something from the periodic table. The elements boron and aluminum are in the same column of the periodic table (just to the left of carbon):

Each of these elements (boron and aluminum) has three valence electrons. Therefore, each of these elements can comfortably form three bonds:

In the compounds shown above, boron and aluminum are using their valence electrons to form bonds, but notice that neither one has an octet. Each element is capable of forming a fourth bond in order to obtain an octet, but then each element will bear a formal charge of –1.

Sodium Borohydride
(NaBH₄)

Lithium Aluminum Hydride
(LAH)

In both of these compounds, the central atom (B or Al) has four bonds and a negative charge. In the first compound (NaBH₄), Na⁺ is used as the counter-ion. In the second compound (LAH), Li⁺ is used as the counter-ion. The choice of counter-ion is usually not relevant for our discussion, and we will choose to ignore it from now on.

Either one of these reagents (NaBH₄ or LAH) can attack a ketone or aldehyde to give an alcohol:

1) LAH

2) H₂O

NaBH₄

MeOH

The mechanisms are somewhat complex and beyond the scope of our course, but the following simplified mechanism will be useful:

As we carefully analyze the first step of this mechanism, we see that the reducing agent (LAH) is simply functioning as a source of H⁻. The mechanism of this first step is the same whether we use LAH or NaBH₄. Then, in the second step, a proton source is used to generate the alcohol.

We have now seen the way that LAH and NaBH₄ work. Essentially, they are both just sources of nucleophilic H⁻. You might wonder why we can't just use sodium hydride (NaH), like this:

This actually doesn't work well. Recall (Chapter 10) that nucleophilicity is dependent on polarizability (the size of the atom). Large atoms (like sulfur or iodine) are very polarizable, and therefore, they are excellent nucleophiles. Small atoms are not polarizable, and they make poor nucleophiles. H⁻ is as small as they come, and therefore H⁻ is not a great nucleophile. H⁻ is a good base (in fact, it is an excellent base), but it is not a good nucleophile. So, we use LAH or NaBH₄ as a "source" of nucleophilic H⁻. We can think of LAH and NaBH₄ as *delivery agents of nucleophilic H⁻*.

Although we can use either reagent to reduce a ketone or aldehyde, nevertheless, there are striking differences between LAH and NaBH₄. LAH is much more reactive than NaBH₄. If we look at the structure of LAH, we see that the central atom (bearing the negative charge) is aluminum. By contrast, in NaBH₄, the negative charge is on boron. Aluminum is much larger (more polarizable) than boron, and therefore, LAH is much more reactive. The higher reactivity of LAH is a topic that will be important in the second semester of organic chemistry. For now, the difference in reactivity is less apparent, because both LAH and NaBH₄ will react with ketones and aldehydes to give alcohols.

For purposes of our discussion here, there is one noticeable difference between LAH and NaBH₄. Whenever we use LAH, we must provide a proton source *after* the reaction (*after* LAH has had a chance to attack the ketone or aldehyde), because LAH is so reactive that it would violently react with water. However, when we use NaBH₄ (the milder reducing agent), we do not have to worry about

$NaBH_4$ reacting with the proton source. The proton source (methanol is often used) can be present in the reaction flask at the same time as $NaBH_4$. This is shown in the following way:

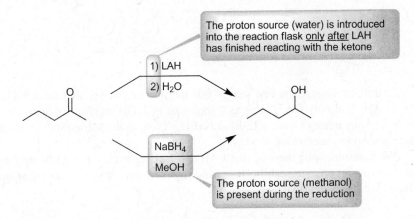

The proton source (water) is introduced into the reaction flask only after LAH has finished reacting with the ketone

1) LAH

2) H_2O

$NaBH_4$

MeOH

The proton source (methanol) is present during the reduction

Notice that with LAH, two separate steps are required.

So far, we have seen two sources of nucleophilic H^- (LAH and $NaBH_4$). There are many, many other hydride reagents that have been prepared (some that are even more reactive than LAH, and others that are even milder than $NaBH_4$), for example:

$$R-\overset{\overset{\displaystyle R}{|}}{\underset{\underset{\displaystyle R}{|}}{Al}}\overset{\ominus}{-}H$$

where R can be almost anything

By carefully choosing the R groups above (to be either electron-donating groups or electron-withdrawing groups), we can very carefully control the reactivity of the hydride reagent. For now, we will simply focus our attention on the two commonly used hydride reagents: LAH and $NaBH_4$.

Let's do some problems where we use LAH or $NaBH_4$ to reduce ketones or aldehydes into alcohols.

EXERCISE 12.30   Identify a ketone and any other reagents you would use to prepare the following alcohol:

OH

ANSWER   This alcohol could be prepared from the following ketone:

We can convert this ketone into our product (the desired alcohol) using either LAH followed by water, or NaBH$_4$ together with methanol:

PROBLEMS Identify the starting ketone or aldehyde you would use to prepare each of the following alcohols through reduction reactions:

12.31

12.32

12.33

12.34

12.35

12.36

12.6   PREPARING ALCOHOLS VIA GRIGNARD REACTIONS

In the previous section, we saw that ketones and aldehydes can be attacked by a suitable source of H$^-$. In a similar way, ketones and aldehydes can also be attacked by a suitable source of R$^-$. Compare these two reactions:

Both H^- and R^- can attack a ketone or aldehyde to give an alcohol. The main difference is the effect on the carbon skeleton. With H^-, the carbon skeleton does not change at all. But with R^-, the carbon skeleton gets larger. We are forming a C—C bond. We will soon see that this is very important for synthesis problems. For now, let's focus on how we can make R^- in the first place. After all, a negative charge on a carbon atom is not very stable (and therefore not trivial to make).

There are many ways to get a negative charge on a carbon atom. Later in this course, you will spend a lot of time learning about special C^- compounds. For now, we will just learn about one such compound, called a Grignard reagent:

$$R-MgX$$

where R is an alkyl group. In a Grignard reagent, there is a carbon atom that is directly connected to a magnesium atom. If we compare the electronegativity values of C and Mg, we will see that C is much more electronegative than Mg. Therefore, carbon pulls more strongly on the electron density and will develop a negative charge.

Grignard reagents are formed by inserting magnesium in between a C—X bond (where X is a halogen):

$$R-X \xrightarrow{\text{Mg}} R-Mg-X$$

Grignard reagent

The mechanism for this insertion of magnesium is beyond the scope of this course, and therefore, we will not go into it. For now, we should just know that we can insert Mg into a C—X bond (where X is Cl, Br, or I). Here are some examples:

Once a magnesium atom is inserted between a C—X bond, the newly formed C—Mg bond is somewhat ionic in character (because carbon is so much more electronegative than magnesium). Therefore, there are two acceptable ways to draw a Grignard reagent:

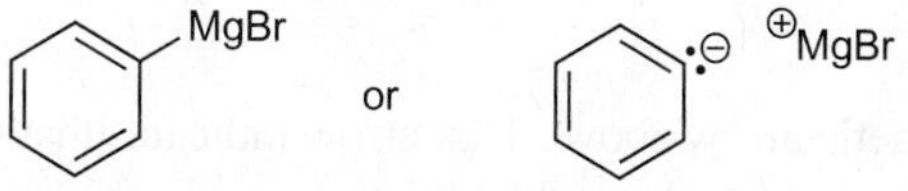

or

Neither drawing is absolutely correct. The first drawing assumes a perfectly *covalent* bond, which is certainly not accurate. The second drawing assumes an *ionic* bond. The reality actually lies somewhere in between these two extremes, although it is a lot closer to being ionic.

When a Grignard reagent attacks a ketone or aldehyde, it forms an alcohol, with a newly installed R group:

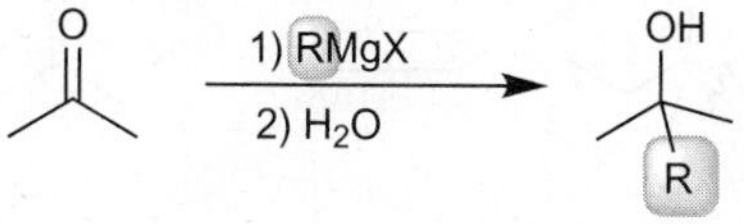

Here are two specific examples:

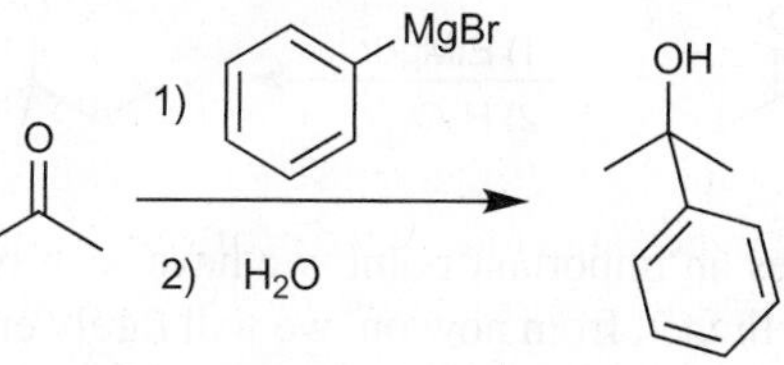

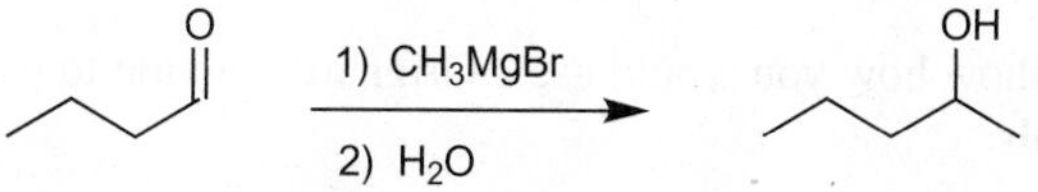

Grignard reagents also attack other compounds possessing a C=O bond (such as esters) to produce alcohols. For now, let's just focus on Grignard reagents attacking ketones and aldehydes.

EXERCISE 12.37 Show how you would use a Grignard reaction to prepare the following alcohol:

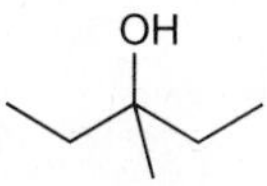

ANSWER   Working backwards, we could have formed the following bond (see the squiggly line), using a Grignard reaction:

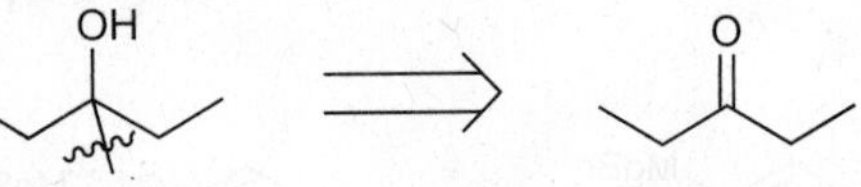

Notice the retrosynthetic arrow above. That arrow indicates that we can make the desired alcohol from the ketone above, like this:

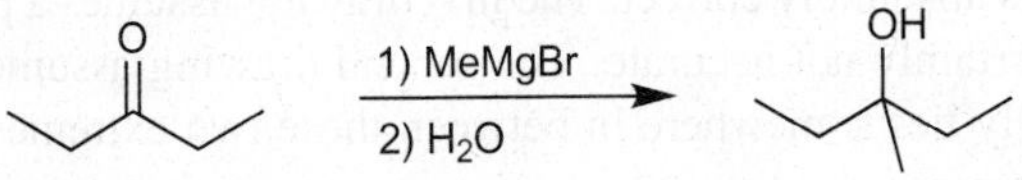

Alternatively, we could have formed this bond:

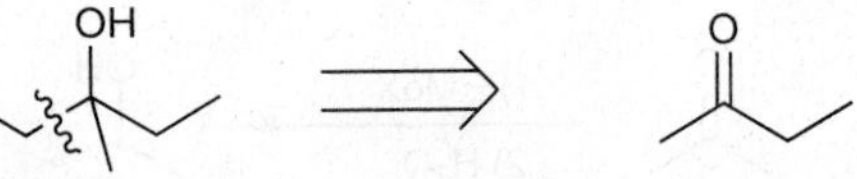

Once again, notice the retrosynthetic arrow above, which indicates that we can make the desired alcohol from the ketone above, like this:

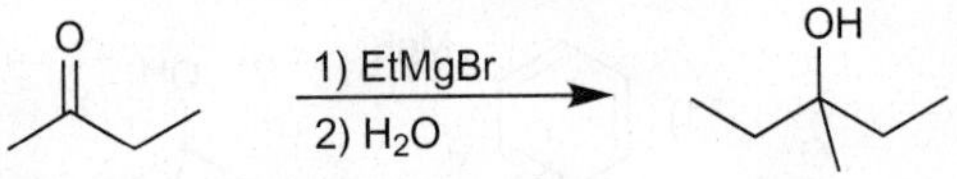

This problem illustrates an important point: we have seen two perfectly correct answers to this problem. In fact, from now on, we will rarely encounter synthesis problems that have only one solution. More often, we will find synthesis problems that have more than one acceptable answer.

PROBLEMS   Show how you would use a Grignard reaction to prepare each of the following alcohols:

12.38

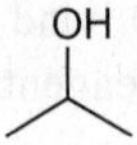

12.39

12.40

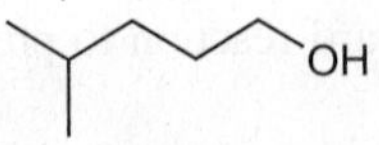

12.41

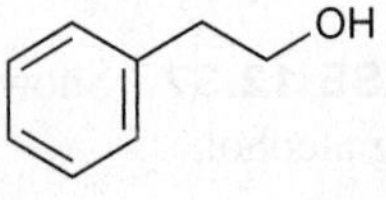

12.42

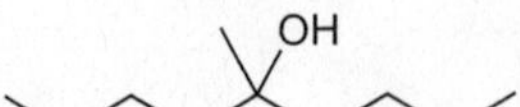

12.43

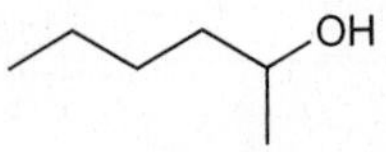

In this section, we learned that a Grignard reagent can attack a ketone or aldehyde to give an alcohol. This reaction is so incredibly important because we are not only converting a C=O bond into an alcohol, but we are also introducing an R group into the compound:

This is a *C—C bond-forming reaction*. Until now, we have only seen one kind of C—C bond-forming reaction (the alkylation of terminal alkynes). Now we have a second approach for creating a C—C bond. We must put this reaction into our synthetic "tool-bag." Whenever we have a synthesis problem, we must always ask two questions: is there a change in the carbon skeleton, and is there change in the functional group (as we will see in the next chapter). If we have a synthesis problem where the carbon skeleton is getting larger, then we know we will have to create a C—C bond. And so far in this book, we have seen two ways to do that. We can alkylate a terminal alkyne, or we can use a Grignard reagent to attack a ketone or aldehyde.

You must make sure that you have this reaction at your fingertips, since it will appear several times throughout the rest of your organic chemistry course. Let's practice a bit. We will begin with a few problems that are just one-step syntheses (Grignard) to make sure you got it. Then, the last two problems in this problem set will be multi-step syntheses (using a Grignard reaction *and* other reactions we have seen).

PROBLEMS Propose a plausible synthesis for each of the following transformations:

12.44

12.45

12.46

12.47

12.48

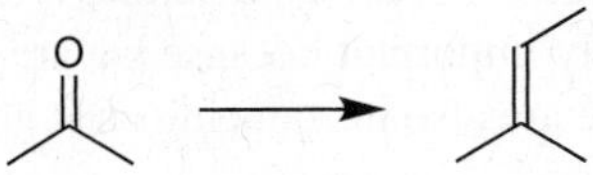

12.49

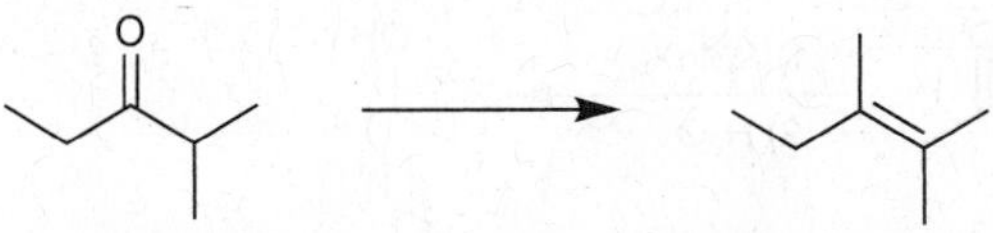

12.7 SUMMARY OF METHODS FOR PREPARING ALCOHOLS

So far, we have seen many ways to make alcohols. We can make primary, secondary, or tertiary alcohols through a variety of methods. As a review, identify the reagents necessary to accomplish each of the following transformations:

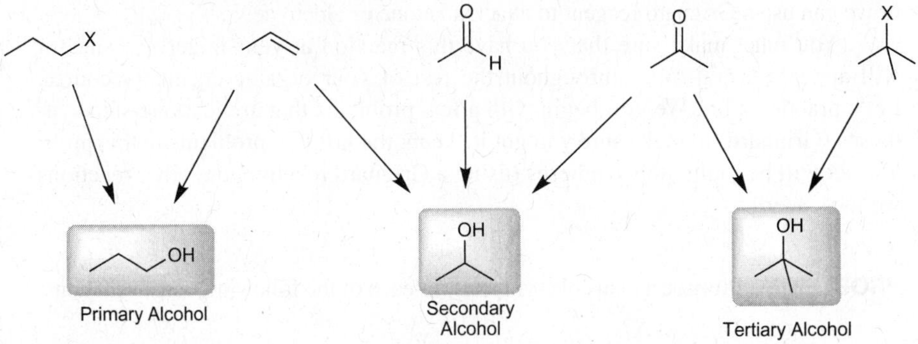

Primary Alcohol

Secondary Alcohol

Tertiary Alcohol

EXERCISE 12.50 Using any reagents of your choice, show three different ways of preparing the following alcohol:

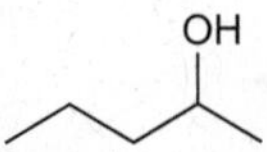

ANSWER This is a secondary alcohol. We have seen many ways to make secondary alcohols. We can start with a ketone, and reduce with LAH or NaBH₄:

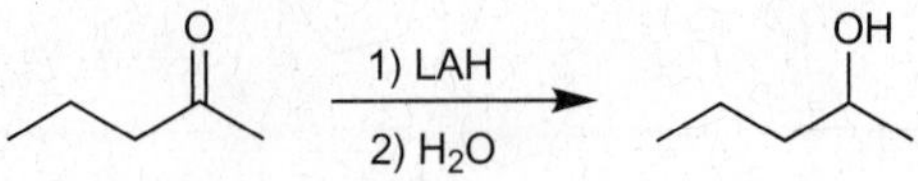

1) LAH

2) H₂O

Or, we can start with an aldehyde, and perform a Grignard reaction:

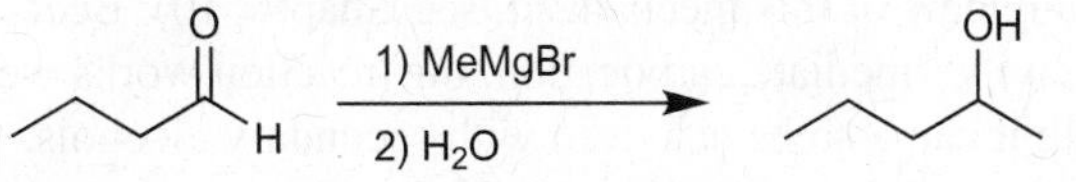

Or we can start with an alkene and perform an addition (acid-catalyzed hydration):

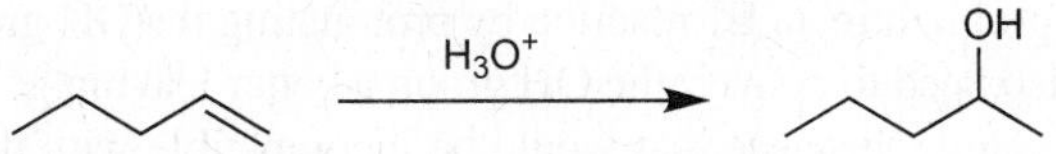

PROBLEMS   Using any reagents of your choice, show at least two ways to make each of the following alcohols:

12.51

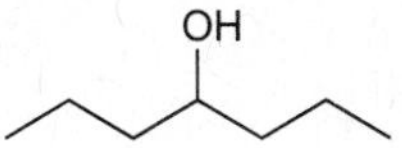

12.52

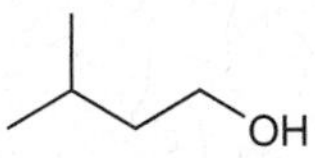

12.53

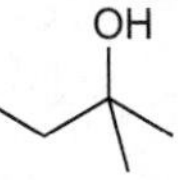

12.54

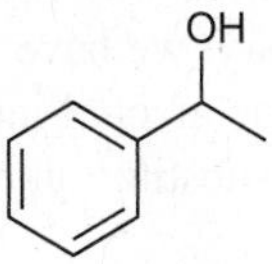

12.55

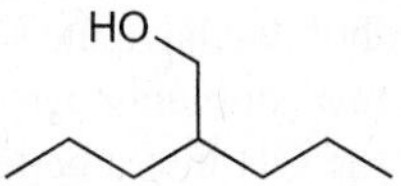

12.56

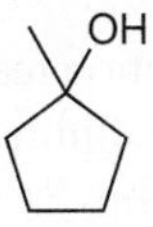

12.8   REACTIONS OF ALCOHOLS: SUBSTITUTION AND ELIMINATION

Now that we have seen how to make alcohols, we will focus our attention on reactions of alcohols. Let's start by reviewing reactions that we have already seen: substitution and elimination. Let's begin our review with elimination reactions first. We have seen two types of elimination reactions: E1 an E2.

To eliminate an OH group under E1 conditions, we need acidic conditions:

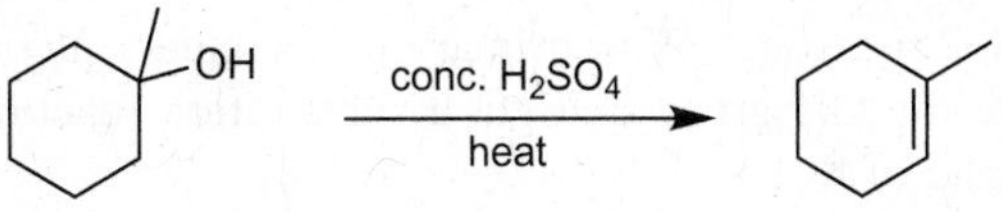

In this reaction, acidic conditions are used in order to protonate the OH group, effectively converting the OH group into an excellent leaving group. Then, the leaving group leaves to give an intermediate carbocation, which then loses a proton to give the alkene (for a review of this mechanism, see Chapter 10). Because the E1 mechanism involves an intermediate carbocation, this reaction works well with a tertiary alcohol, although it can also be achieved with secondary alcohols.

For primary alcohols, we will *not* be able to form a carbocation. Therefore, we generally cannot use an E1 reaction on primary alcohols. Instead, we use an E2 reaction. But we still have the same problem—OH is still a terrible leaving group. We took care of that problem in an E1 reaction by protonating the OH group. In an E2 reaction, we will also need to convert the OH group a better leaving group. But we cannot use a proton source, because that would be incompatible with the strongly basic conditions that are required in order to perform an E2 process. Therefore, if we want to use E2 conditions, we will first need to convert the OH group into a tosylate group:

In the last step, *tert*-butoxide was used to favor elimination over substitution (see Section 10.10). In summary, we have seen that we can use either an E1 process or an E2 process to convert an alcohol into an alkene.

Now let's consider substitution reactions involving alcohols:

$$R-OH \longrightarrow R-X$$

If we want to perform a substitution reaction with an alcohol, we have the same issue that we had when we explored elimination reactions a few moments ago—the OH group is not a good leaving group. So, we must convert the OH into a better leaving group. There are several ways to do that for substitution reactions. We will look at four different ways:

1) *Via an SN1 process.*   With a tertiary alcohol, we can use an SN1 process. We simply use an acid to protonate the OH group, converting it into an excellent leaving group. The first two steps of this reaction are identical to the E1 process that we just saw:

2) *Via an SN2 process.*   With primary or secondary alcohols, we can still protonate the OH group and get a substitution reaction (it will just be SN2 instead of SN1):

In this reaction, a carbocation would be too unstable to form. After protonating the OH group (converting it into a better leaving group), the leaving group is expelled when the nucleophile attacks, in an S_N2 process.

This reaction works well with HBr, but it does not work so well with HCl. Chloride is smaller than bromide, and therefore, chloride is less polarizable than bromide. This means that chloride is not as nucleophilic as bromide. Although chloride is still fairly nucleophilic, the reaction is slow, and we can help it along with a catalyst, such as $ZnCl_2$:

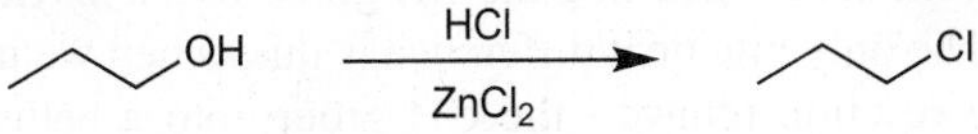

The effect of $ZnCl_2$ is similar to the effect of protonating an OH group, but $ZnCl_2$ just does a better job than H^+:

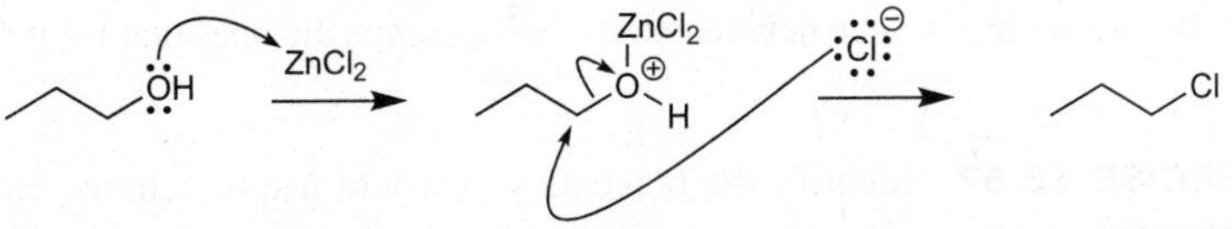

3) A third way to perform a substitution reaction is to convert the OH group into a tosylate, and then do an S_N2 reaction:

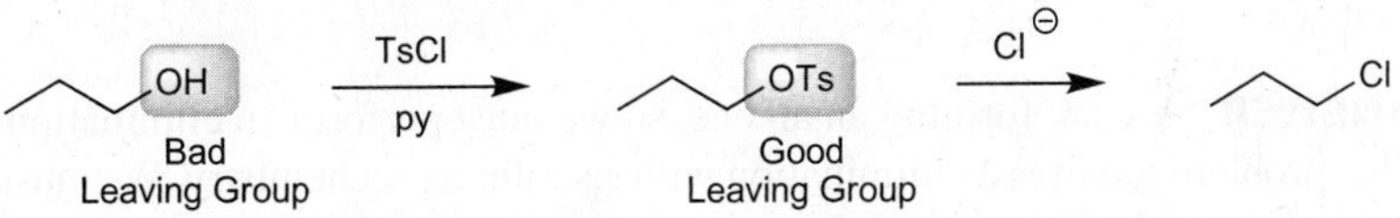

4) There is one other way to convert an OH group into a better leaving group for the purpose of doing a substitution reaction. We can use thionyl chloride ($SOCl_2$) to convert an alcohol into an alkyl chloride:

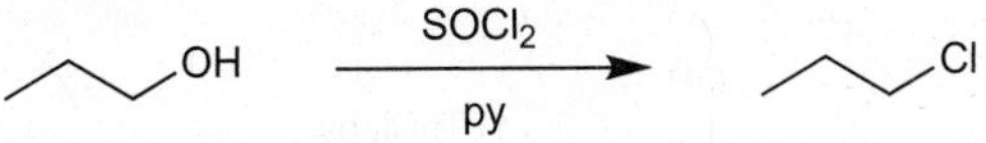

A mechanism for this process is shown below:

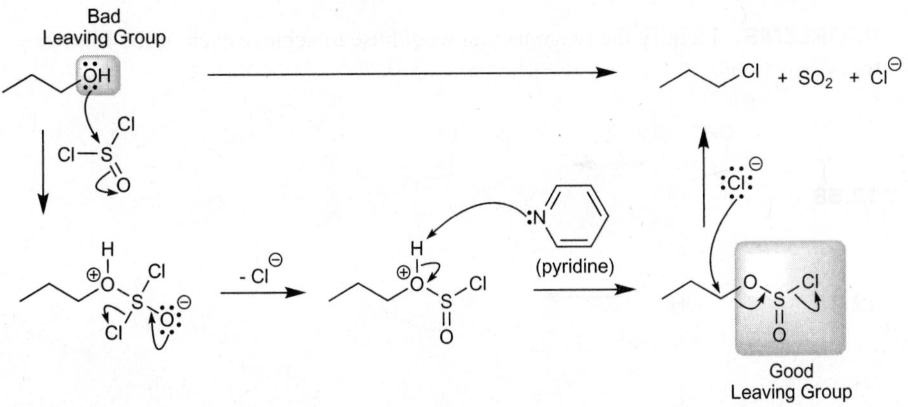

Notice that the first three steps simply convert a bad leaving group into a good leaving group. Also notice that SO_2 (a gas) is produced as a side product. This gas can escape from the reaction flask, thereby pushing the equilibrium toward formation of products. In fact, if the gas is free to leave as it is formed, the reaction will be pushed to completion.

This last reaction might seem like a new reaction. But it really is not so new. It is just an S_N2 reaction where the OH group is first converted into a better leaving group, and then chloride functions as the nucleophile. This reaction is really not all that much different than converting the OH group into a tosylate and then attacking with chloride. The only practical difference is this: when we use $SOCl_2$, everything happens in one reaction (convert the OH group into a better leaving group, *and* chloride attacks).

In this section, we have started looking at reactions of alcohols. So far in this section, we have focused on the details of familiar reactions (substitution and elimination). Before we learn some new reactions, let's practice the ones that we just reviewed:

EXERCISE 12.57 Identify the reagents you would use to achieve the following transformation:

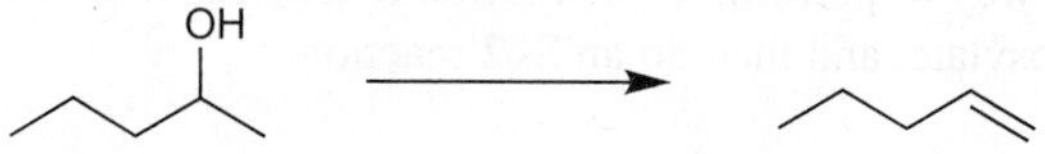

ANSWER We are forming an alkene, so we must perform an elimination reaction. This problem requires an elimination with specific regiochemistry. We must form the less substituted alkene (the Hofmann product). Therefore, we will need to use a sterically hindered base (*tert*-butoxide). The obstacle is that an OH group is a terrible leaving group. So, we must first convert the OH group into a better leaving group. We can do that by converting it into a tosylate, which will then undergo elimination when treated with *tert*-butoxide:

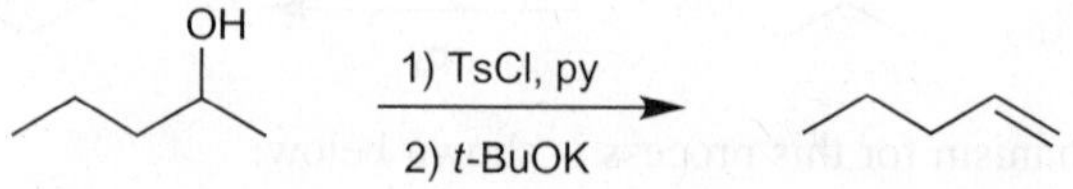

1) TsCl, py
2) *t*-BuOK

PROBLEMS Identify the reagents you would use to achieve each of the following transformations:

12.58

12.59

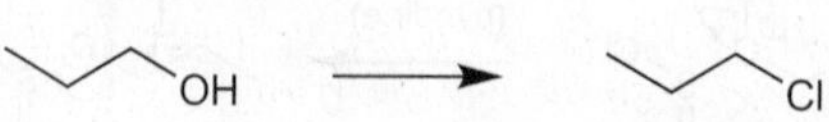

12.60

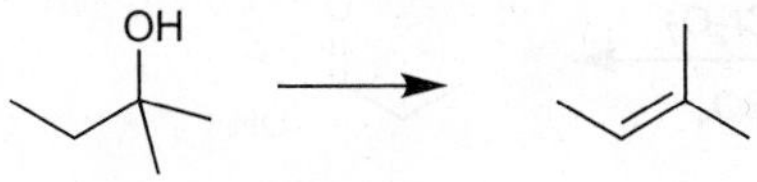

12.61

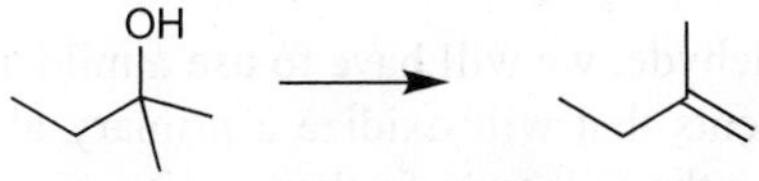

12.9 REACTIONS OF ALCOHOLS: OXIDATION

Earlier in this chapter, we learned definitions for the terms oxidation and reduction. We saw that oxidation involves an increase in oxidation state. For example, oxidation of a secondary alcohol will produce a ketone:

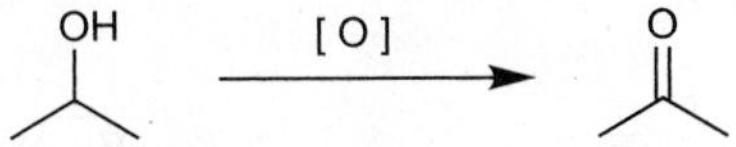

Notice the term used over the arrow, [O]. This term means that we are performing an oxidation. There are many oxidizing agents that can be used to accomplish this transformation. Chromic acid (H_2CrO_4) is a good example. Chromic acid can be formed from mixing sodium dichromate ($Na_2Cr_2O_7$) and sulfuric acid:

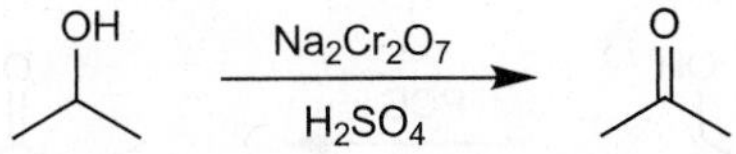

In the example above, a *secondary* alcohol was oxidized to give a ketone. However, when we start with a *primary* alcohol, we will have two choices: 1) we can oxidize to give an aldehyde, or 2) we can oxidize even further to the carboxylic acid:

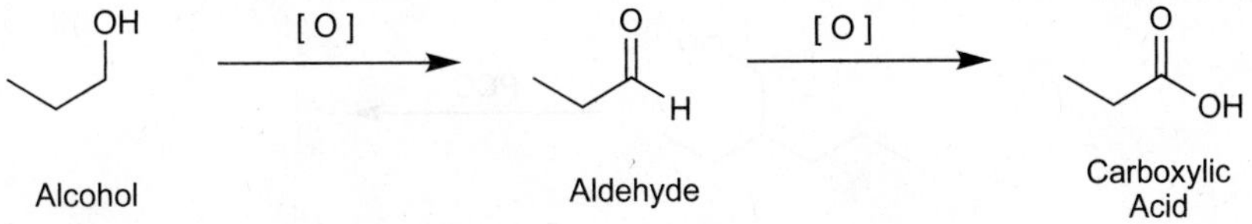

Alcohol   Aldehyde   Carboxylic Acid

We can control how far we oxidize by carefully choosing our reagents. If we want to go all the way up to a carboxylic acid, then we just use chromic acid:

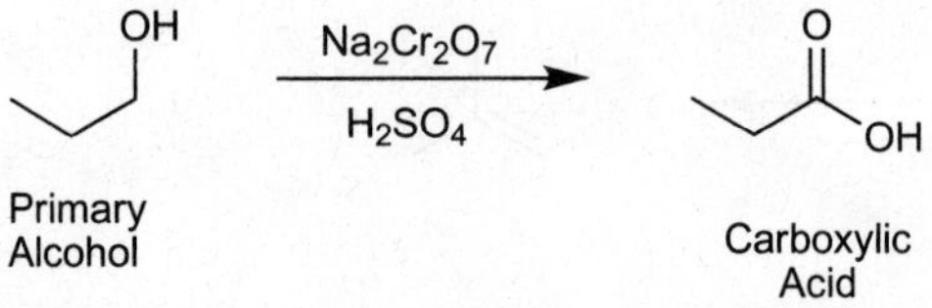

Primary Alcohol → (Na$_2$Cr$_2$O$_7$ / H$_2$SO$_4$) → Carboxylic Acid

However, if we want to stop at the aldehyde, we will have to use a milder oxidizing agent. There are many such reagents that will oxidize a primary alcohol to give an aldehyde (and will not oxidize the aldehyde further to give a carboxylic acid):

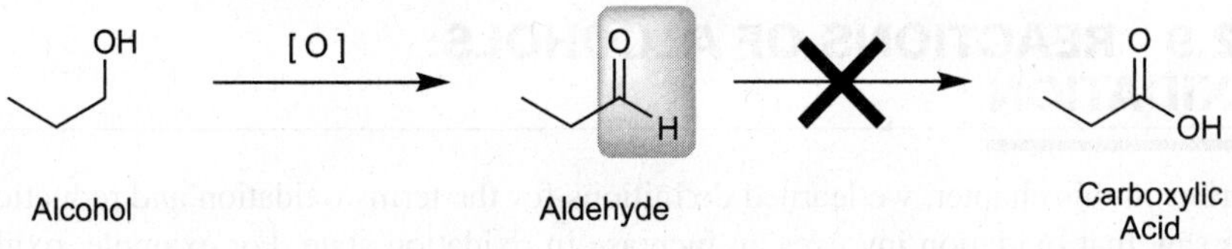

Alcohol → [O] → Aldehyde ✗ Carboxylic Acid

One such example is pyridinium chlorochromate, or PCC for short, which is prepared in the following way:

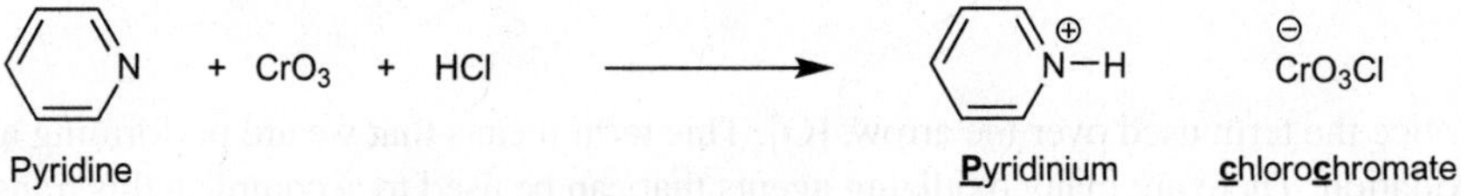

Pyridine + CrO$_3$ + HCl ⟶ Pyridinium chlorochromate

PCC is a mild oxidizing agent, and it will oxidize a primary alcohol to give an aldehyde. For example:

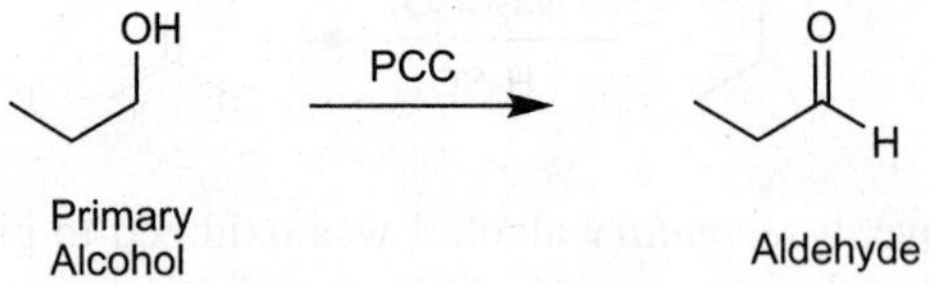

Primary Alcohol → PCC → Aldehyde

EXERCISE 12.62 Predict the major product of the following reaction:

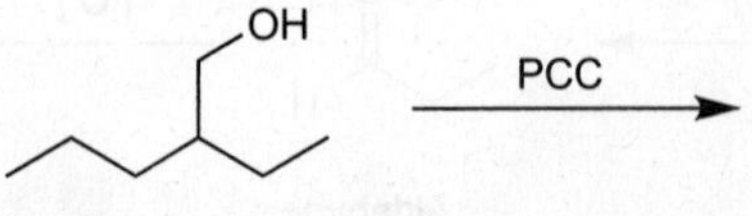

OH → PCC →

ANSWER We are starting with a primary alcohol, and we are oxidizing with PCC. This will give an aldehyde as a product (rather than oxidizing all the way up to a carboxylic acid):

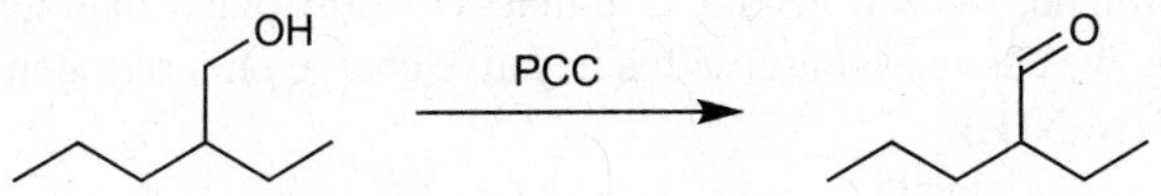

PROBLEMS Identify what reagents you would use to accomplish each of the following transformations:

12.63

12.64

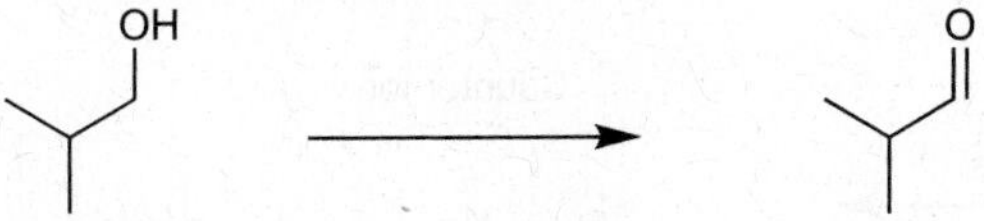

12.65

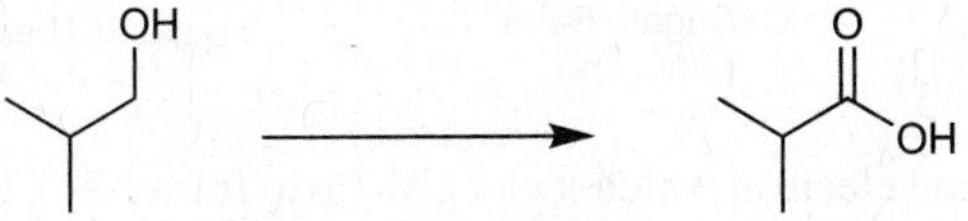

12.66

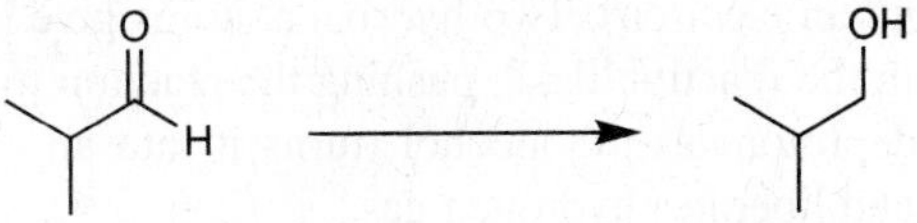

12.67

12.68

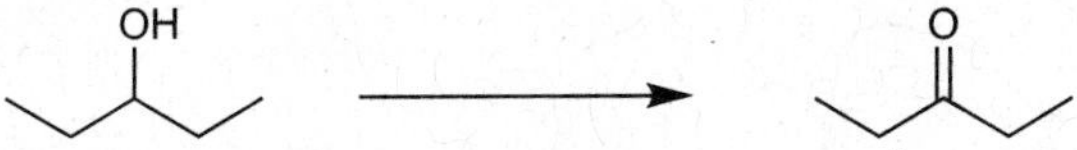

12.10 CONVERTING AN ALCOHOL INTO AN ETHER

Alcohols can be deprotonated with a strong base:

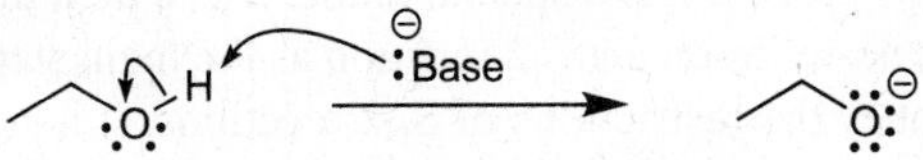

A fairly strong base is required to deprotonate an alcohol. By removing the proton, we are forming a negative charge on an oxygen atom (an alkoxide ion). Therefore, in order to deprotonate, we will need a base that is even stronger than an alkoxide ion. As an example, we can use a base with a negative charge on a nitrogen atom, such as sodium amide (NaNH$_2$):

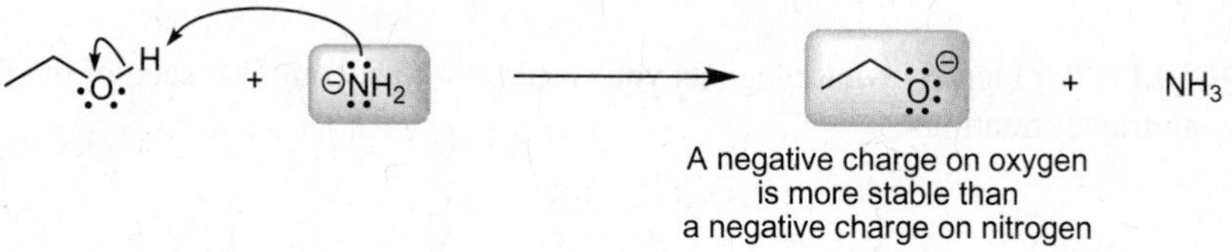

A negative charge on oxygen
is more stable than
a negative charge on nitrogen

However, there is a simpler way to deprotonate an alcohol. We can just use elemental sodium (Na), like this:

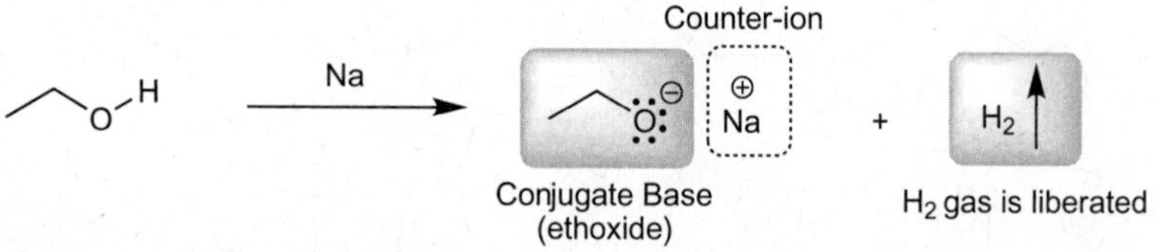

Conjugate Base
(ethoxide)

H$_2$ gas is liberated

Elemental sodium (Na) has one electron which it can give up to form Na$^+$. This electron can combine with the alcoholic proton to form a hydrogen atom (remember that a hydrogen atom is a proton *and* an electron). Two hydrogen atoms give hydrogen gas (H$_2$), which can escape from the reaction flask, pushing the reaction to completion. This process effectively deprotonates the alcohol (turns it into an alkoxide), forms Na$^+$ as the counter-ion, and liberates hydrogen gas.

Now that we have seen how to deprotonate an alcohol (to form an alkoxide), the obvious question is: what can we do with alkoxide ions? We have already seen that alkoxide ions can be used as strong bases. But alkoxides can also function as strong nucleophiles. For example, consider the following S$_N$2 reaction:

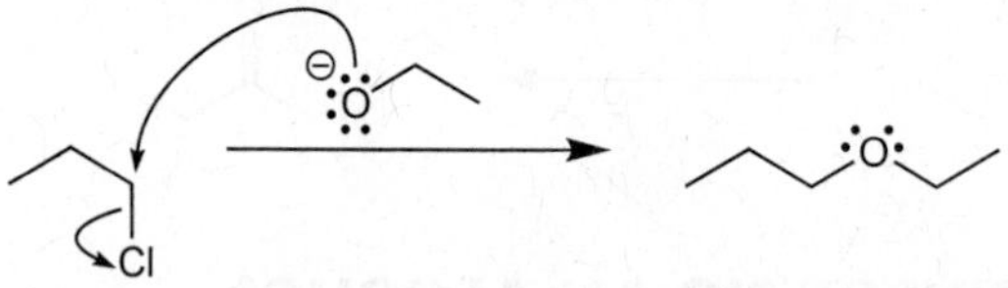

This is not a new reaction. This is just an S$_N$2 reaction. We are simply using the alkoxide ion (ethoxide in this case) to function as the attacking nucleophile. But notice the net result of this reaction: we have combined an alcohol and an alkyl halide to form an ether. This process has a special name. It is called the *Williamson Ether Synthesis*. This process relies on an S$_N$2 reaction as the main step, and therefore, we must be careful to obey the restrictions of S$_N$2 reactions. It is best to use a primary alkyl halide. Secondary alkyl halides cannot be used because elimination will predominate over substitution (as seen in Sections 10.9), and tertiary alkyl halides certainly cannot be used.

EXERCISE 12.69   Show how you could use a Williamson Ether synthesis to make the following compound:

ANSWER   We are using a Williamson Ether synthesis, so we will need to start with an alcohol and an alkyl halide to form the ether linkage. Working backwards (retrosynthetic analysis), we get the following:

This retrosynthetic analysis tells us that we can make our product from sodium propoxide and propyl chloride. We can prepare sodium propoxide by treating propanol with Na. In summary, our answer is:

PROBLEMS   Starting with 1-propanol, and using any other reagents of your choice, show how you could use a Williamson Ether synthesis to make each of the following compounds:

12.70

12.71

CHAPTER 13

SYNTHESIS

Synthesis is really just the flipside of predicting products. In any reaction, there are three groups of chemicals involved: the starting material, the reagents, and the products:

$$\text{Starting material} \xrightarrow{\text{Reagents}} \text{Products}$$

When the products are not shown, then you have a "predict the product" problem:

$$\text{Starting material} \xrightarrow{\text{Reagents}} ?$$

When the reagents are not shown, then you have a synthesis problem:

$$\text{Starting material} \xrightarrow{?} \text{Products}$$

Synthesis problems can be easy (if they are only one step) or they can be difficult (if they are more than one step). When you begin learning reactions in your course, you will start to encounter synthesis problems in your textbook. At first, you will get one-step problems, and as the course progresses, you will see multistep syntheses. In a multistep synthesis, you can often end up with a product that looks very different from the starting material. For example, look at the following series of reactions below. Don't concentrate on how the changes were made. For now, just focus on the fact that each reaction changes the compound only slightly, but in the end, we end up with a product completely different from the starting material:

Starting Material

Product

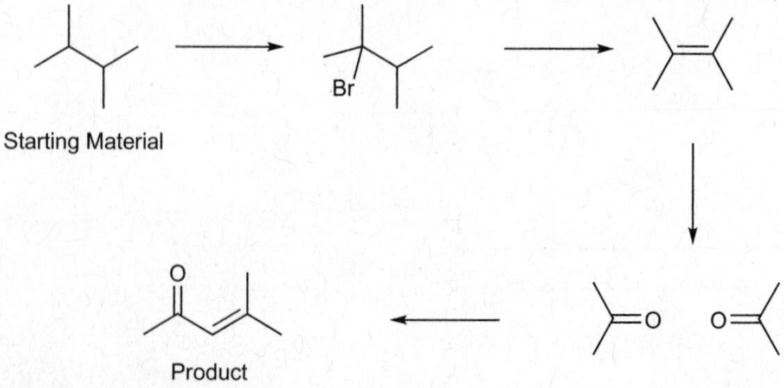

It can only take three or four steps before the problem can get quite difficult. If you convert the sequence above into a synthesis problem, it would look like this:

If you are having trouble with synthesis problems when you first encounter them, the worst thing you can do is to give up and say: "Oh, well, I'm not good at synthesis problems." As the course moves on, this attitude will slowly kill your grade in the course. To see why this is so, let's compare organic chemistry to a game of chess.

Imagine that you are learning how to play chess. You first learn about the pieces: how they are named, how to set up the board, and so on. Then you learn how each piece moves and how they capture each other. When you start playing your first game, you realize that there is quite a bit of strategy involved. Most strategies involve thinking more than just one move in advance. It is not good enough to know only how to move the pieces. You also need to think about how to plan out the next few moves so that you can coordinate an attack on your opponent's pieces. Imagine how silly it would be to take the time to learn how to move the pieces, but to then say to yourself that you are not good at strategy. Imagine thinking that you will keep playing chess, but you just won't be good at that one aspect of the game. That would be silly, because that one aspect of the game is the whole game itself. You either need to learn how to strategize, or just don't play chess. There is no in-between.

Organic chemistry is very much the same. Synthesis is all about strategizing. You need to think a few moves ahead, and you must learn how to do this. You cannot tell yourself that you are not good at synthesis problems, and therefore you will just focus on the other aspects of organic chemistry. Synthesis *is* organic chemistry. The second half of the course is all about learning reactions and applying them in syntheses. Everything that you have learned so far has prepared you for synthesis. The only way to become proficient at synthesis is to *practice*. Don't be lazy, and don't think that you can get through the course without learning how to propose syntheses. If you do, you will find that your performance in the course will spiral down to a point that will make you very unhappy.

There are a few techniques that will make you feel more comfortable with synthesis problems, and there are exercises that you can go through to increase your proficiency in doing synthesis problems. That's what this chapter is all about.

13.1 ONE-STEP SYNTHESES

As we mentioned earlier, one-step syntheses are the first synthesis problems you will encounter. They will never be more difficult than predicting products. Before you can move on to multistep syntheses, you first need to feel comfortable with one-step syntheses.

To do this, we need to make a list of reactions, but we will leave out the reagents, so that we can repeatedly photocopy the list and get practice filling in the reagents.

As you learn more and more reactions, this list will grow. With every five new reactions, you should photocopy all of the reactions that you have recorded here. Then, start filling in the reagents on the photocopy. Repeat this procedure whenever you have entered five new reactions.

If you keep up with this exercise as the course progresses, you will be in very good shape for solving one-step synthesis problems. The hardest challenge that you will face is keeping up with the work and not waiting until the night before the exam. If you wait (as most students do), you will find it very difficult to spend the time that it takes to master this material. Don't make that mistake. The secret to success in this course is to do a little bit every night (rather than cramming on the night before the exam). Cramming might work well for other courses, but it doesn't work well in organic chemistry.

Begin your list on the next page.

For now, skip forward a few pages. We have some techniques to go over that will help you solve synthesis problems.

Remember not to fill in the reagents or the mechanisms. For each reaction, just draw the starting material in front of the arrow and the products after the arrow. Leave the space above the arrow empty. You will fill in the reagents when you photocopy these pages:

Now photocopy this page, and try to fill in the reagents on your photocopied page.

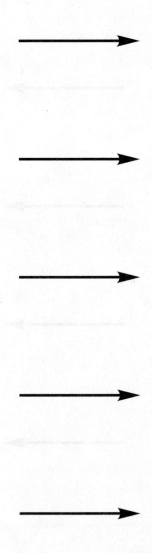

Now photocopy this page again, and fill in the reagents for every reaction on this page.

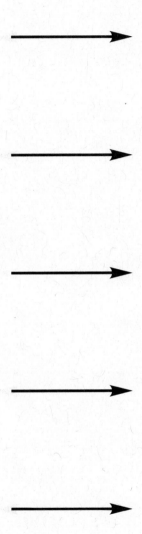

Now photocopy this page AND the previous pages, and fill in all of the reagents.

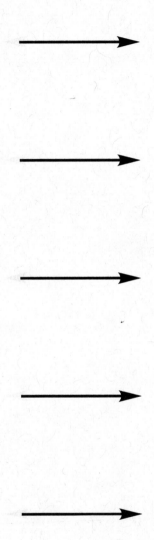

Now photocopy this page AND the previous pages, and fill in all of the reagents.

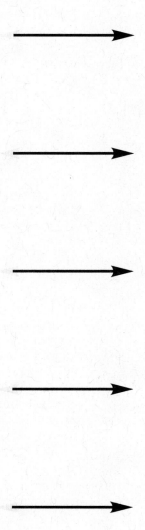

Now photocopy this page AND the previous pages, and fill in all of the reagents.

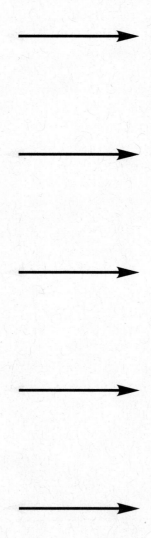

Now photocopy this page AND the previous pages, and fill in all of the reagents.

Now photocopy this page AND the previous pages, and fill in all of the reagents.

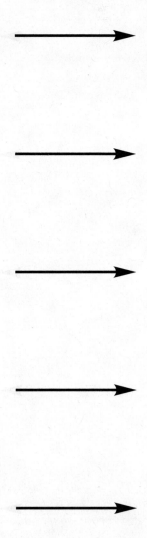

Now photocopy this page AND the previous pages, and fill in all of the reagents.

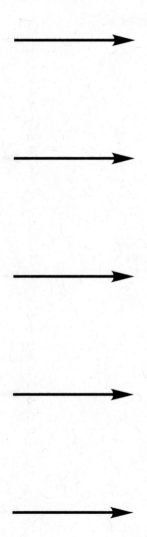

Now photocopy this page AND the previous pages, and fill in all of the reagents.

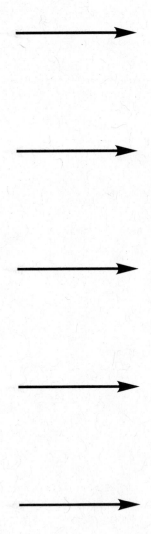

Now photocopy this page AND the previous pages, and fill in all of the reagents.

If you cover more than 30 reactions and need more space to continue, then you can just use a regular piece of paper to keep your list going.

13.2   MULTISTEP SYNTHESES

To prepare yourself for solving multistep syntheses, you need to learn how to think in more than one move. If you carefully review your list of reactions, you will find that the products of some reactions are the starting material for other reactions. For example, you will find that some reactions are used to form double bonds, and other reactions add reagents across double bonds. So if you pair up all of the possibilities, you will create a list of many two-step syntheses. By studying these two-step possibilities, you will begin to get familiar with seeing syntheses that are more than one step.

Let's see an example of what we mean. Below is one reaction that forms a double bond. It starts with an alkyne, and you will certainly learn this reaction at some point:

Now consider one of the reactions where reagents react with a double bond:

If we put these two reactions together into a two-step synthesis, we get the following:

You should now get some practice with this. You will probably learn around five methods for making double bonds and probably around 10 reactions that involve reagents reacting with double bonds. If you put together all of the possibilities, you will find that there are around 50 possibilities, depending on exactly how many reactions you learn. Clearly, you cannot keep a list like this as you go through the course. The list would be too long to study. And if you try to consider three-step syntheses, you will find that the number of permutations is too large to even compile such a list. It's just like our analogy to a game of chess.

In chess, you cannot possibly memorize every possible orientation of all of the pieces and then memorize the best move for each of those possibilities. There are too many permutations. Instead, you learn how to analyze each situation and as time goes on, you get better and better at it. By familiarizing yourself with certain permutations, you will get better at figuring things out as you go along. So let's start with the list that we talked about above—the approximately 50 possible two-step syntheses that involve forming a double bond and then doing something to that double bond.

Again, you should not try to make a list like this throughout your entire course. This task would be impractical. But if you make this first list of roughly 50 syntheses, you will learn how to start thinking in more than one step. It is important for you to get accustomed to thinking this way. Take a separate sheet of paper and try to create this list using the reactions in the beginning of your course. If you do not get a chance to write down all 50, that's OK. As long as you begin the process and draw at least 10 or 20 of them, then you will start to understand what it is like to think in more than one step.

After you have done this, we can start focusing on the main techniques for analyzing problems that display permutations that you have never seen. That is what the next section is all about.

13.3 RETROSYNTHETIC ANALYSIS

When you see a synthesis problem for the first time, you are not expected to immediately know the answer. I cannot stress this enough. It is so common for students to get overly anxious when they see synthesis problems that they cannot solve. Get used to it. This is the way it is supposed to be. Going back to our chess analogy, you don't need to make a move as soon as it is your turn. You are allowed to think about it first. In fact, you are supposed to think about it first. So, how do you begin thinking about a multistep synthesis problem where you do not immediately see the solution? The most powerful technique is called *retrosynthetic analysis*. This means that you analyze the problem backward. Let's see how this works with an example:

The synthesis problem above is a multistep synthesis problem, because we do not have a single reaction that allows us to do this transformation in just one step. So the best way to start is to first look at the product and work our way backward.

We see that the product is a dibromide. So we ask ourselves: Do we know any way of making a dibromide? You can see that to answer this question, you must have first mastered one-step syntheses. If you have not yet done this for all of the reactions that you have learned so far, you will need to go back to the beginning of this chapter and do that first (if you are a student in this situation, continue reading for now, so you can see where this is all going).

So we should be able to recognize that we know how to make dibromides from double bonds. We draw the alkene that would have been used to form the product:

Now we are one step closer to solving this problem. The next step is to ask if there is a way to turn the starting material into this double bond. And there is. We just do an elimination reaction to get the double bond. So now we have solved our synthesis by working backward:

Notice that the stereochemistry and regiochemistry needs to work out for every step. You cannot use a step that has the wrong stereochemistry or regiochemistry. I suppose you could have memorized all possible two-step syntheses from the reactions in your textbook, and then you would have gotten this problem right away (maybe . . .), but that is not a practical approach. What will you do for three-step or four-step syntheses? You need to get accustomed to thinking backward. The more practice you can get, the better off you will be.

Here is where we run into a big problem. There is no way for me to give you problems that are appropriate. Every course goes at its own pace, in its own order, and with exams at different points in the course. I cannot give problems that will be perfectly appropriate for every student everywhere. So, how are we going to get practice? Very simply. You are going to make your own problems, as described in the next section.

13.4 CREATING YOUR OWN PROBLEMS

Creating your own problems is easier than it sounds. You just choose any reaction from the lists that you have been making (in Chapters 8 and 13). Then look at the product of that reaction and choose another reaction that you have learned that will transform that compound into something else. We work backward to solve synthesis problems, but we work forward to create a synthesis problem. At each step, draw the product of that reaction and then move on to the next step. After you have gone two or three or four steps, erase everything in the middle. Just draw the very first compound and the final product. Draw an arrow between them, and you have a synthesis problem.

There is one catch. You will not find that problem to be very challenging, because you are the one who made it. So here is what you should do. Find a friend in the course, and each of you should make up 10 or 20 problems. Then you switch off with each other. You will find that this is a very effective method for studying. The larger your study group becomes, the more effective it will be. Don't be shy. You will need to work with a friend to get the practice that you need, not to mention the valuable peer support. If you are reading this book, then chances are that other students in your course have this book also. They will have the same need that you do. Team up with them.

Even if you cannot find a friend with whom you can swap problems, it will still be a useful exercise to create your own problems. The process of creating problems by itself is a worthwhile process. It will help you get accustomed to thinking in multiple steps for synthesis problems.

To summarize, these are the keys to becoming proficient at solving synthesis problems:

1. Master the one-step syntheses by constant review.

2. Train yourself to work backward when solving a problem.

3. And, finally, get lots of practice.

Chapter 1

1.2) 6
1.3) 6
1.4) 5
1.5) 7
1.6) 6
1.7) 8
1.8) 4
1.9) 10
1.10) 9
1.11) 10

1.13)

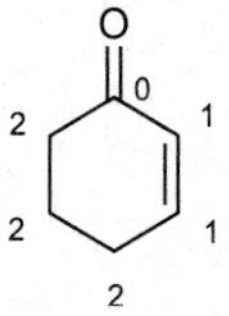

1.14)

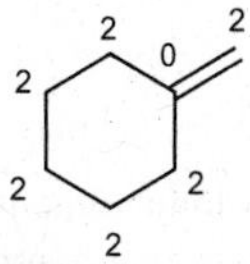

1.15)

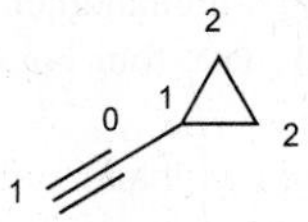

1.16)

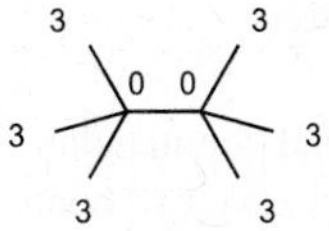

1.17)

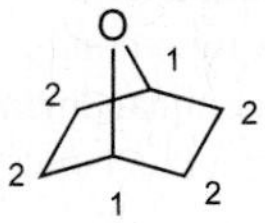

1.18)

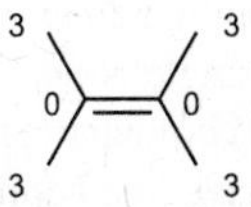

1.19)

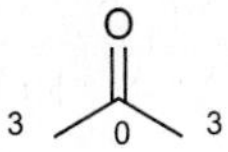

1.20)

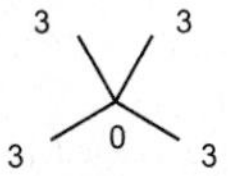

1.21)

1.22)

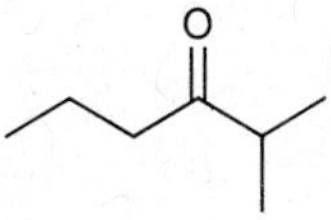

1.23)

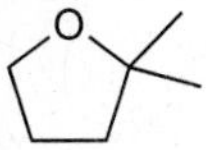

1.24)

1.25) Substituted a Cl with an OH
1.26) Added two OH groups across a double bond
1.27) Eliminated H and Cl to form a double bond
1.28) Added Br and Br across a double bond
1.29) Eliminated H and H to form a double bond
1.30) Substituted an I with an SH
1.31) Eliminated H and H to form a triple bond
1.32) Added H and H across a triple bond
1.34) No charge
1.35) Positive
1.36) Negative
1.37) No charge
1.38) Positive
1.39) Negative
1.40) Positive
1.41) Negative
1.42) No charge

349

1.43) Positive
1.44) No charge
1.45) No charge

1.47)

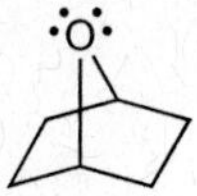

1.48)

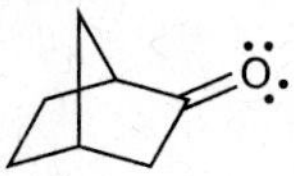

1.49)

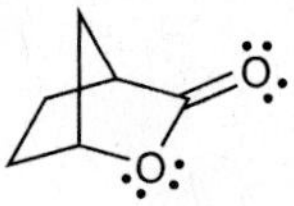

1.50)

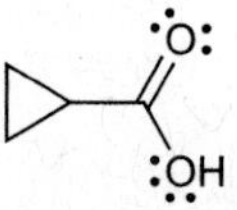

1.51)

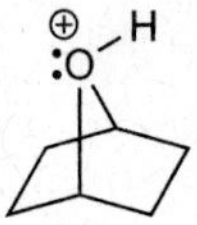

1.52)

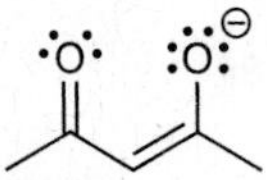

1.54)

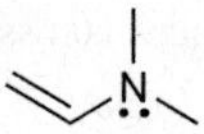

1.55)

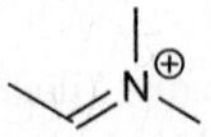

1.56)

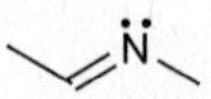

1.57)

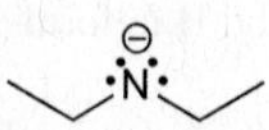

1.58)

1.59)

1.60)

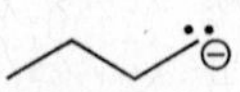

1.61)

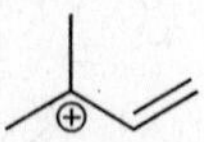

1.62)

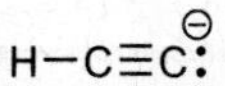

$$H-C\equiv C\colon^{\ominus}$$

1.63)

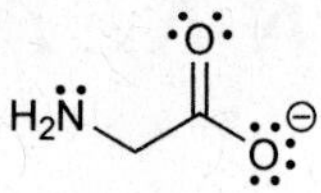

1.64)

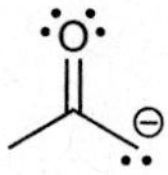

1.65)

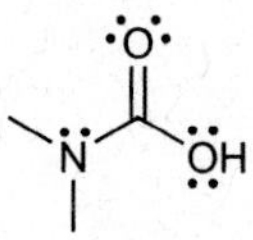

1.66)

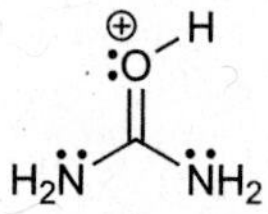

1.67)

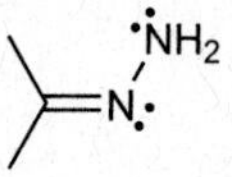

1.68)

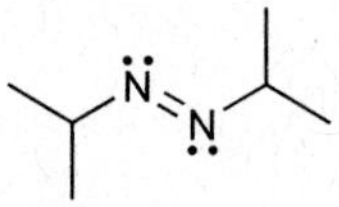

Chapter 2

2.2) Violates second commandment—nitrogen cannot have five bonds
2.3) Violates second commandment—nitrogen cannot have four bonds and one lone pair
2.4) Violates second commandment—oxygen cannot have three bonds and two lone pairs
2.5) No violation
2.6) Violates second commandment—carbon cannot have five bonds
2.7) Violates first commandment—cannot break a single bond
2.8) Violates second commandment—carbon cannot have five bonds
2.9) No violation
2.10) Violates first commandment—cannot break a single bond
2.11) Violates second commandment—carbon cannot have five bonds
2.12) No violation

2.14)

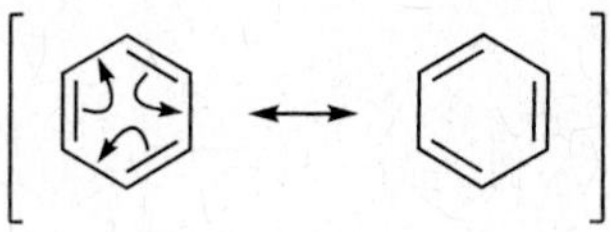

2.15)

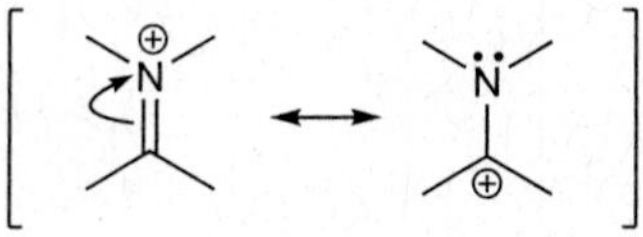

2.16)

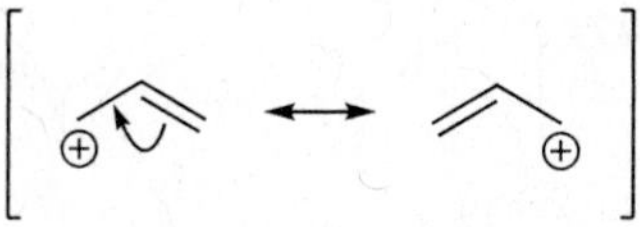

2.17)

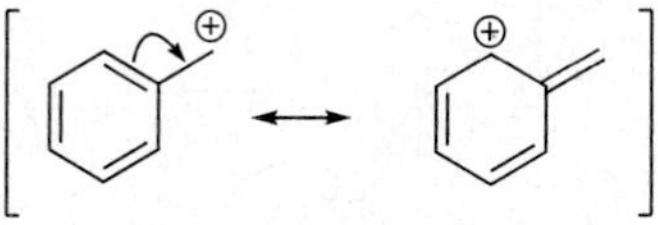

2.18)

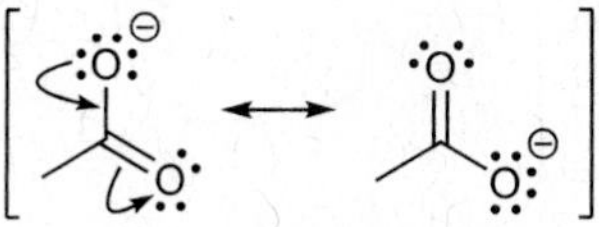

2.19)

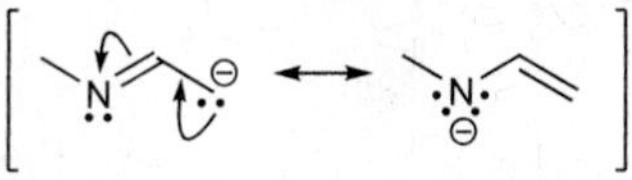

2.21)

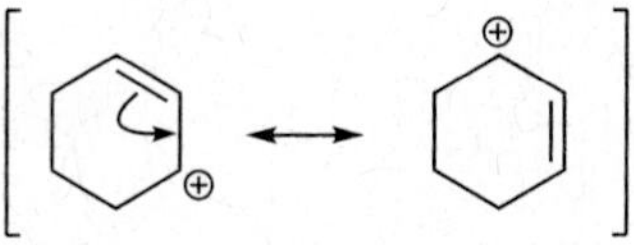

2.22)

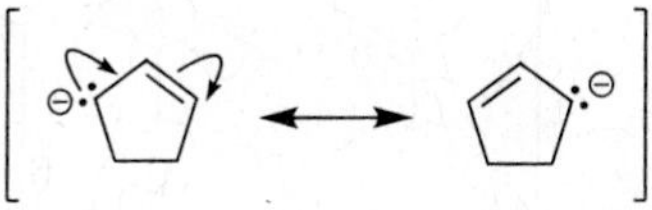

2.23)

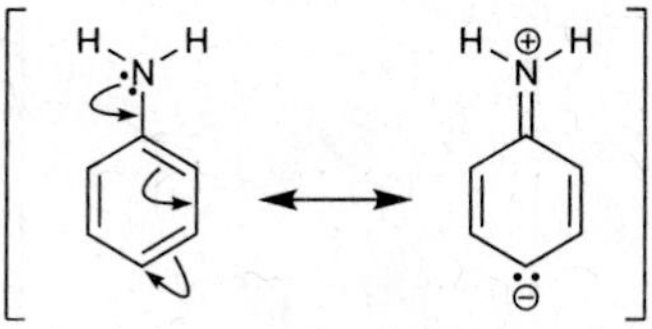

2.24)

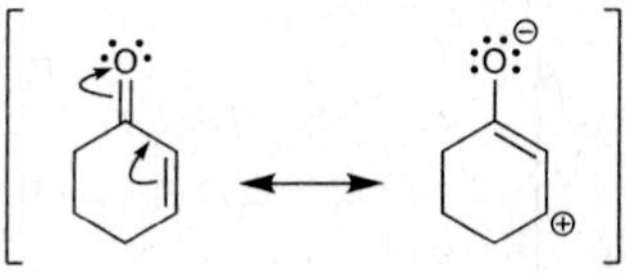

2.25)

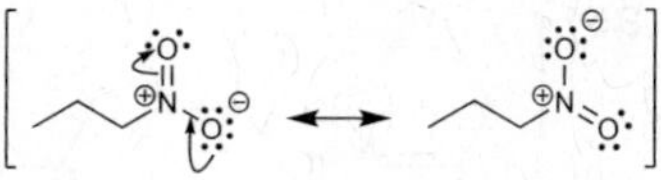

2.26)

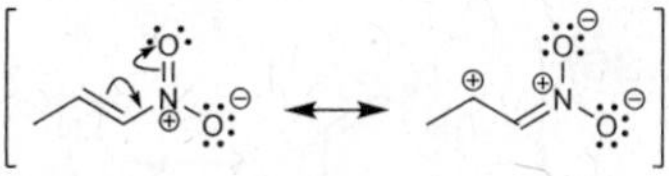

2.27)

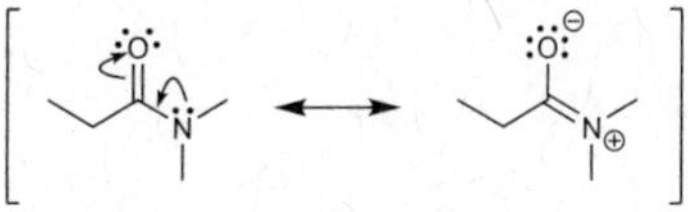

2.28)

2.30)

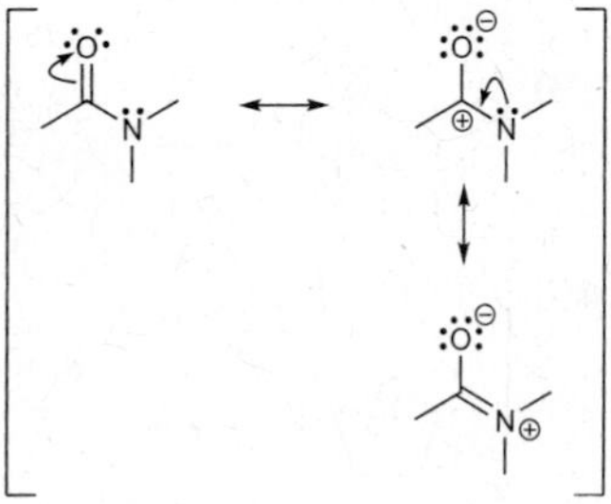

2.32)

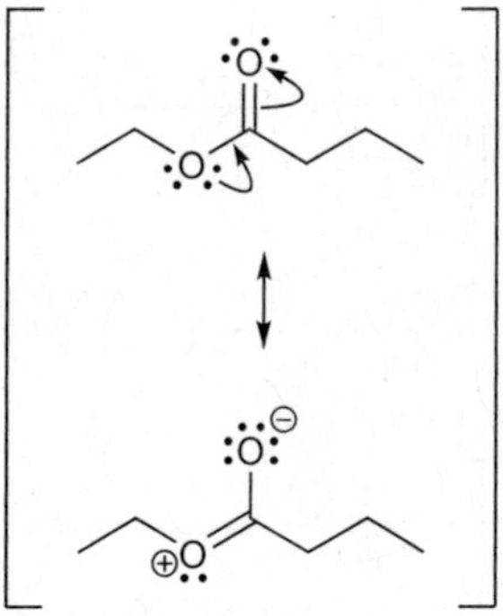

2.33)

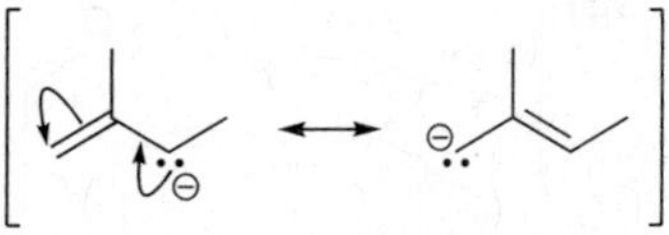

2.34)

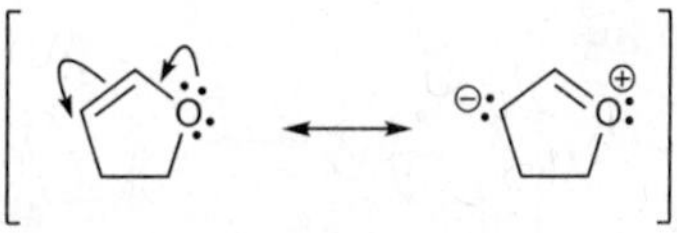

2.35)

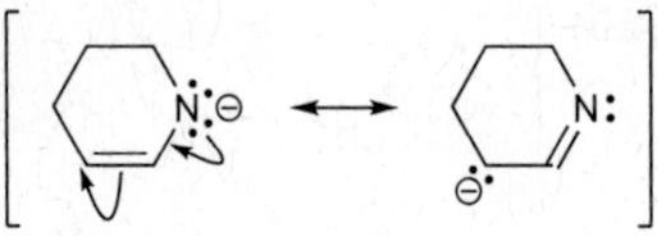

2.36)

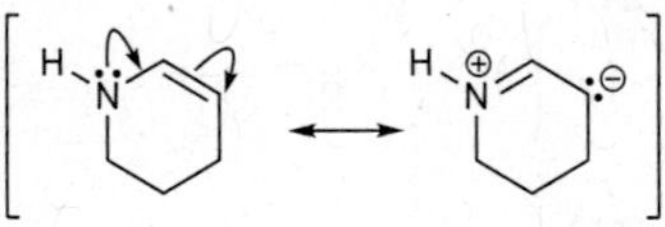

2.37)

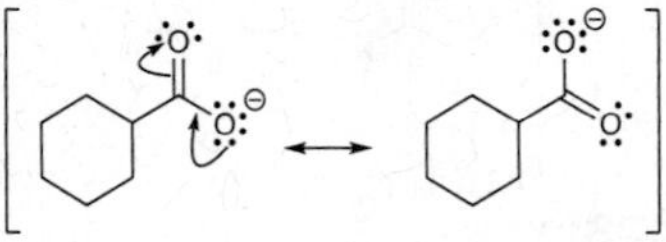

2.38)

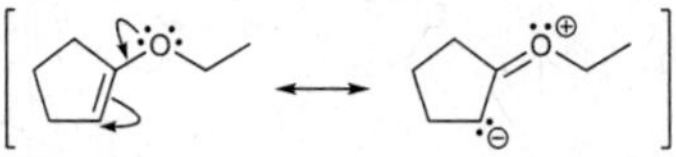

2.39)

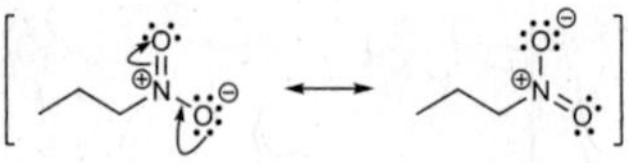

2.40)

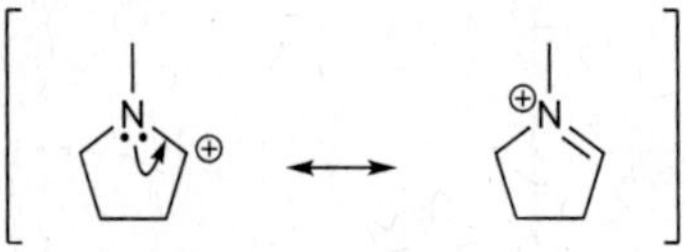

2.41)

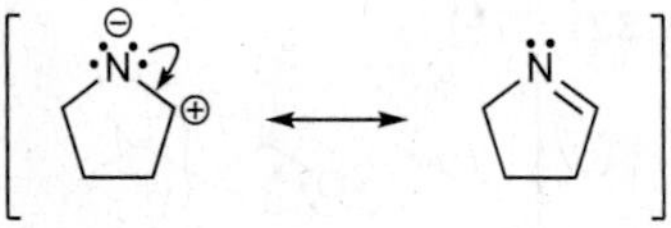

2.42)

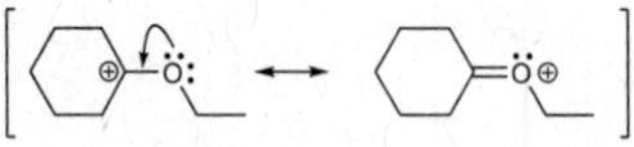

2.43)

2.44)

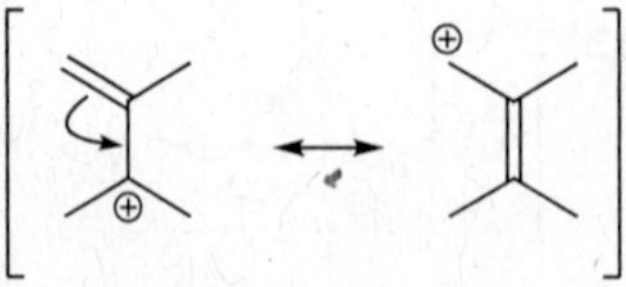

2.45)

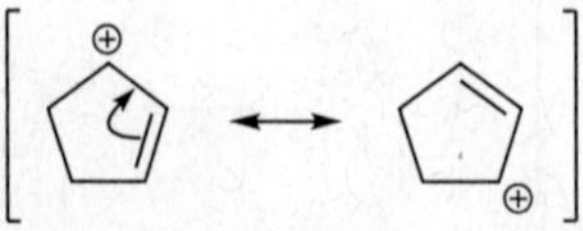

2.46

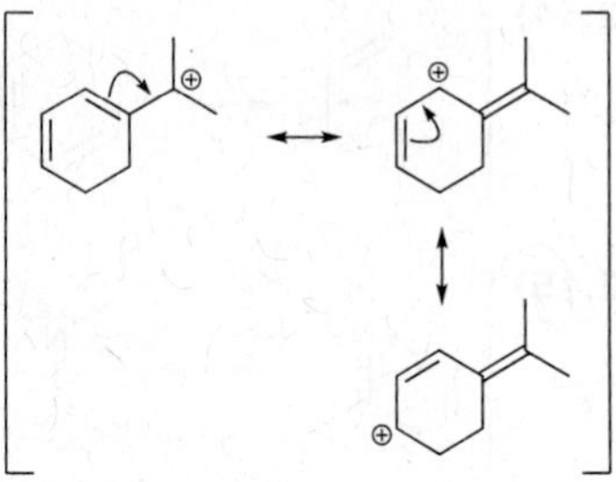

2.47)

2.48)

2.49)

2.50)

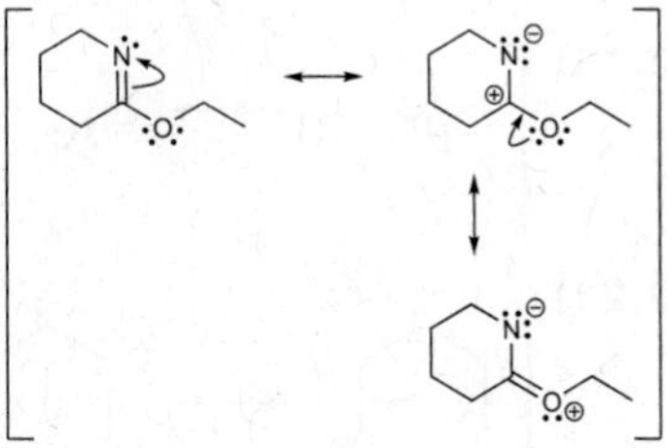

2.51)

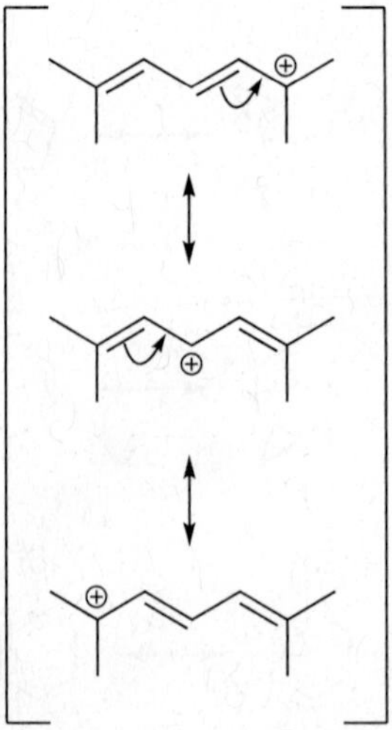

2.52)

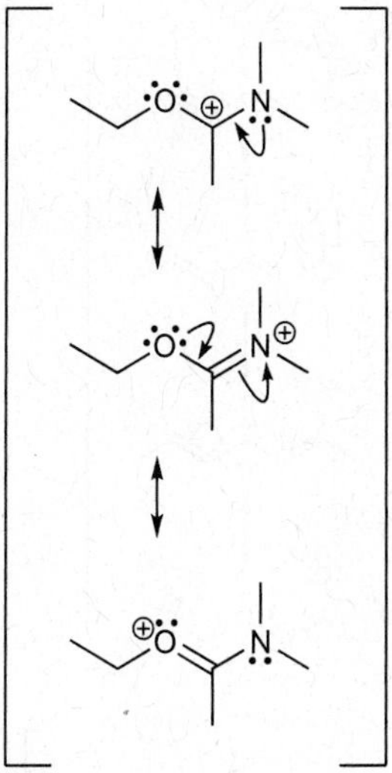

2.53)

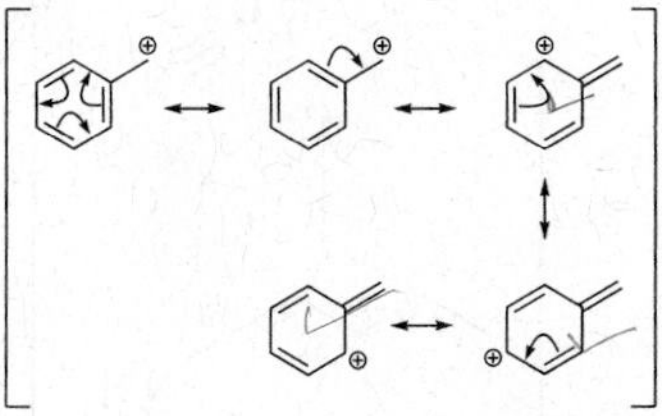

2.54)

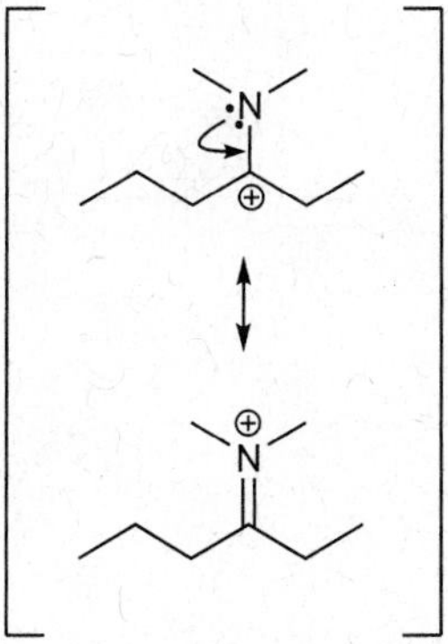

2.55)

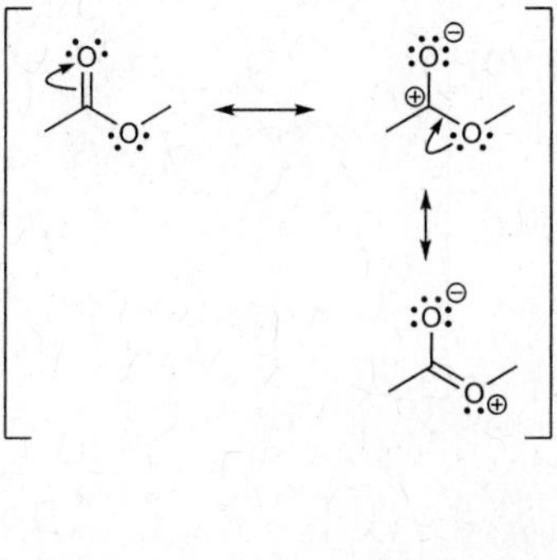

2.56)

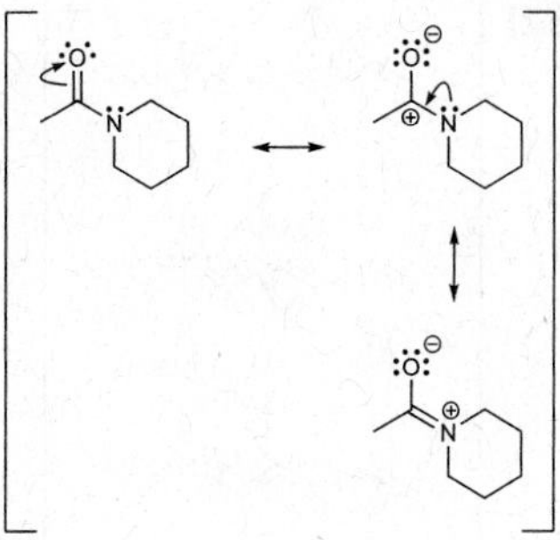

2.57)

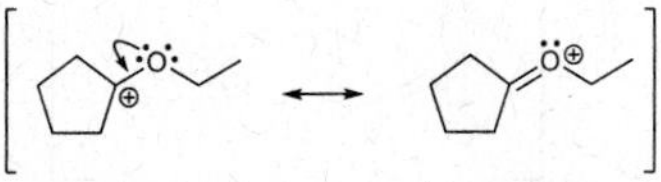

2.58)

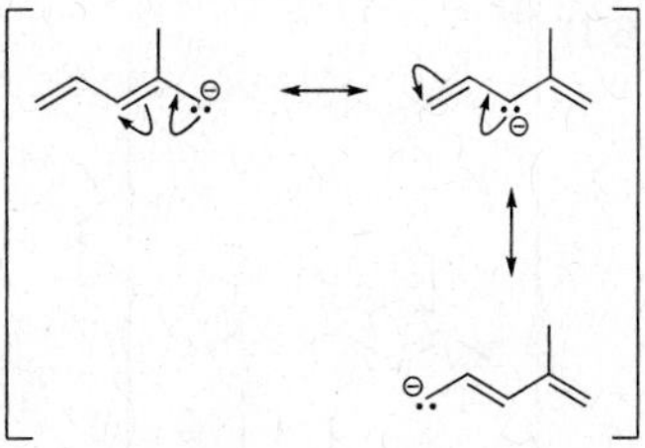

2.59)

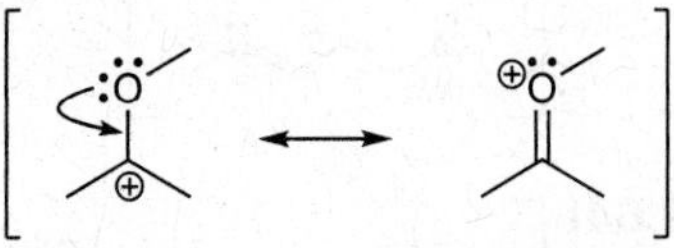

2.60)

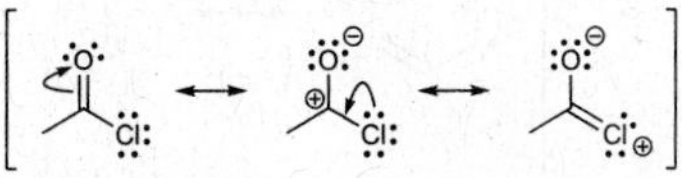

2.61)

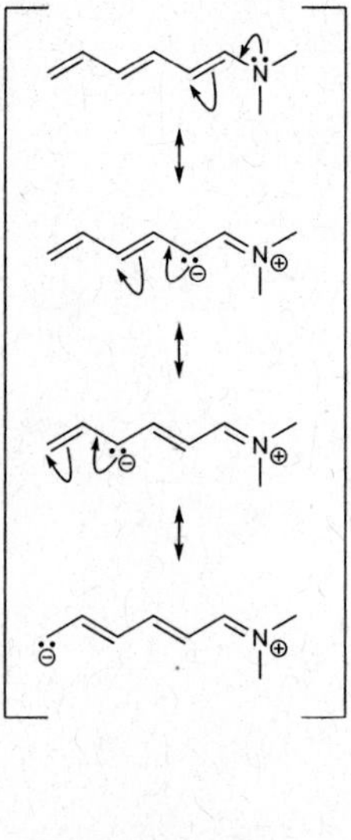

2.62)

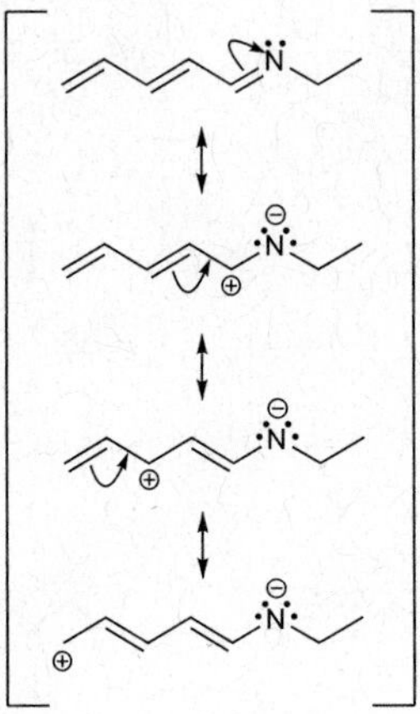

2.63)

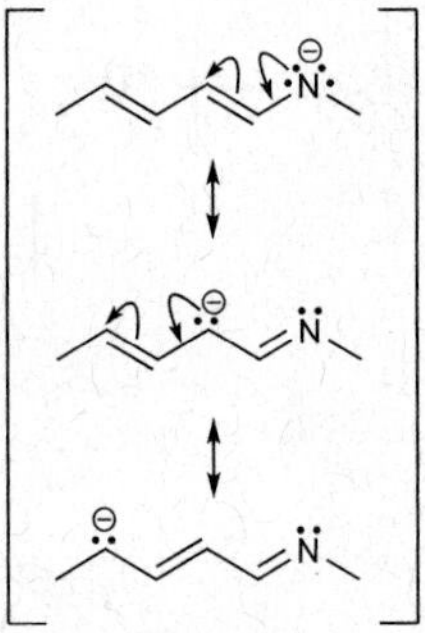

2.64)

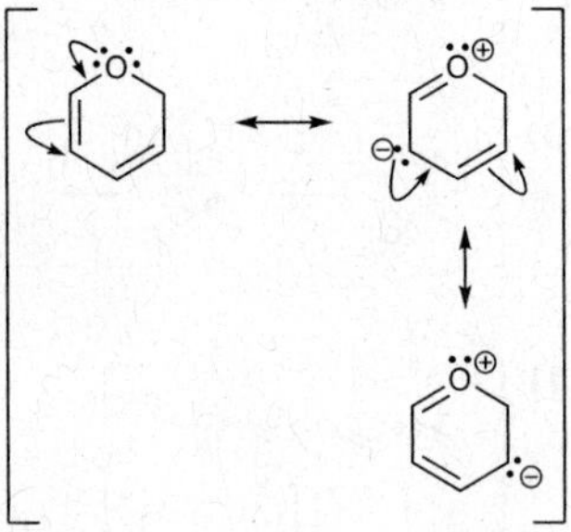

2.65)

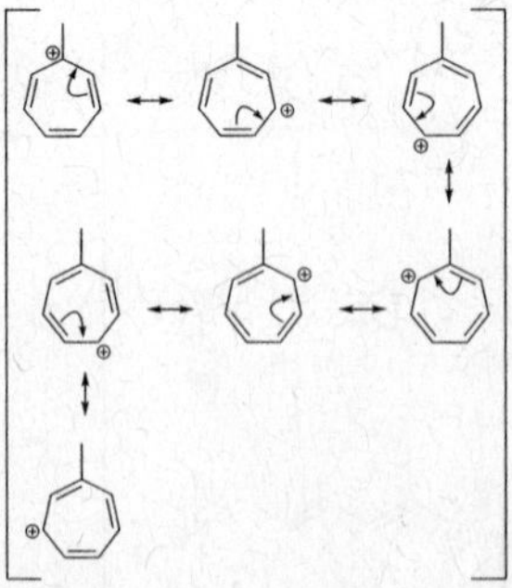

2.66)

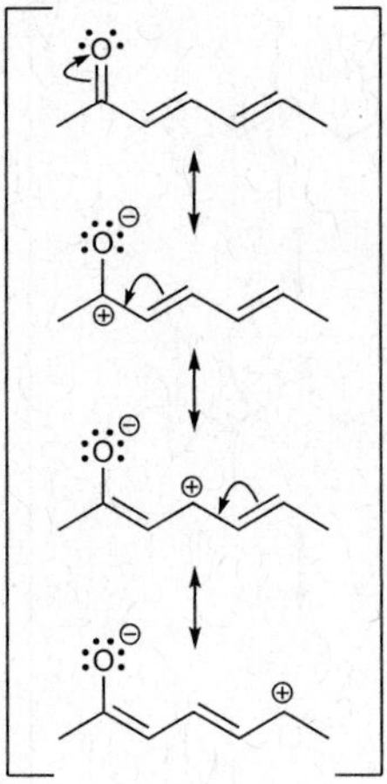

2.67)

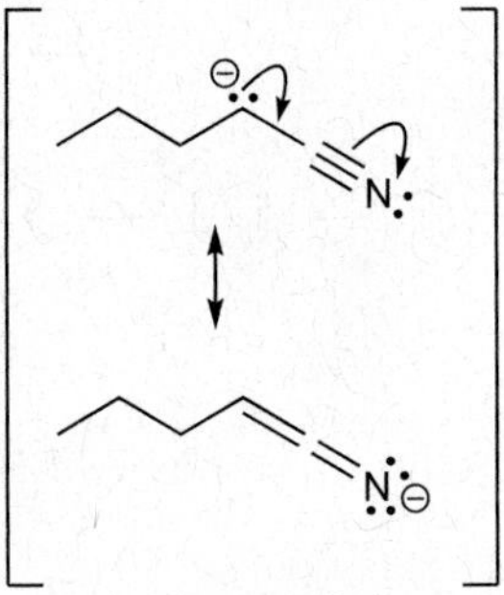

2.68)

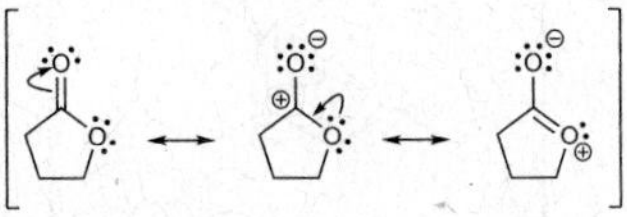

2.69)

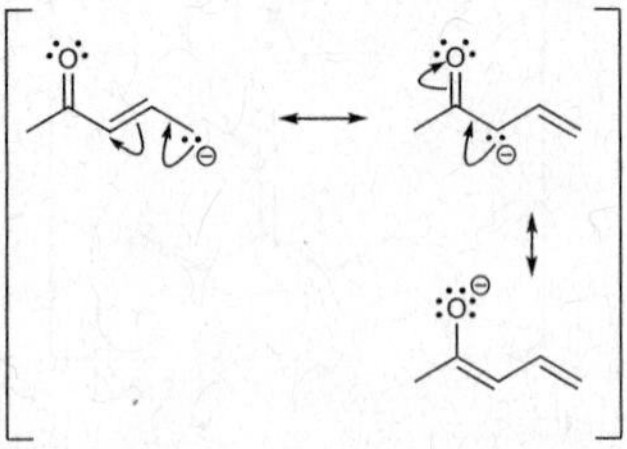

2.70)

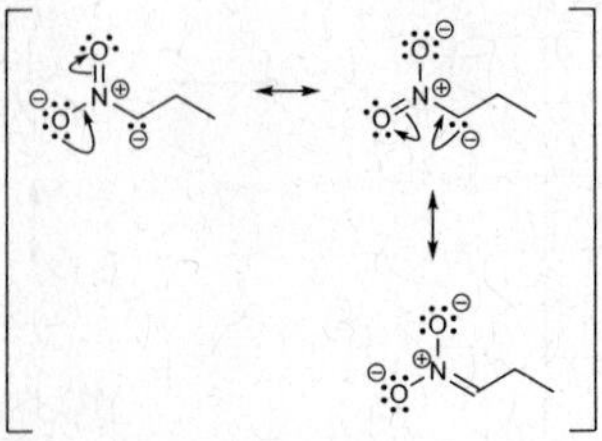

2.71)

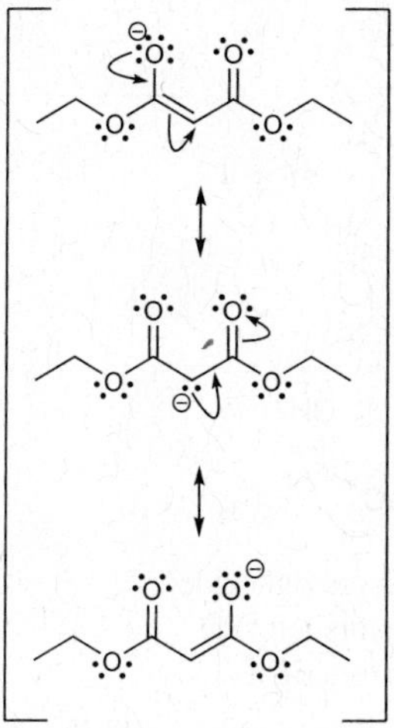

2.72)

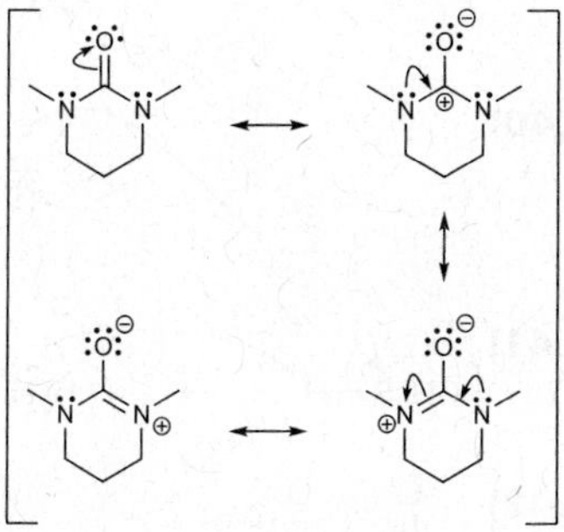

2.73)

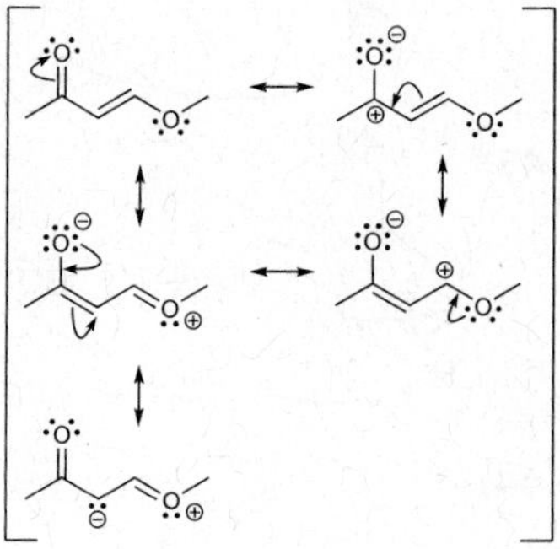

Chapter 3

3.2)

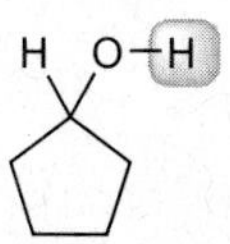

3.3)

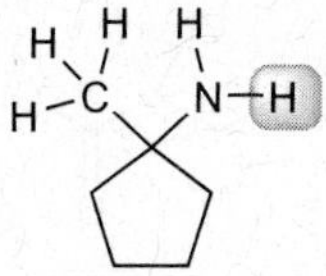

3.4)

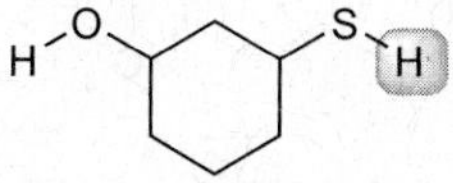

3.5)

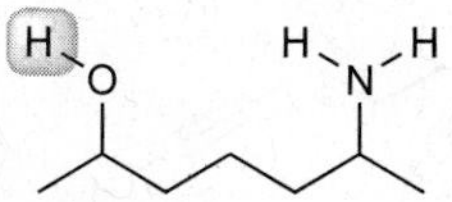

3.7)

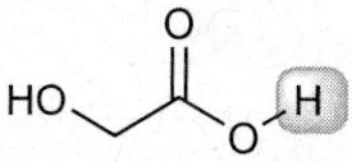

3.8)

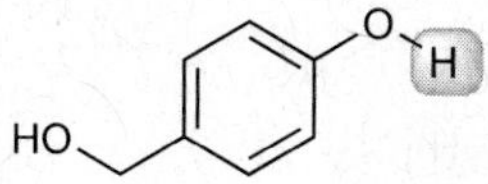

3.9)

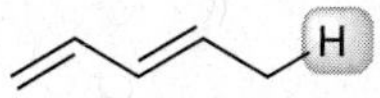

3.10)

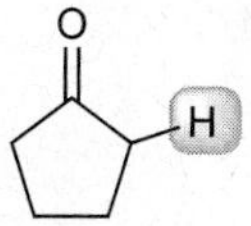

3.11)

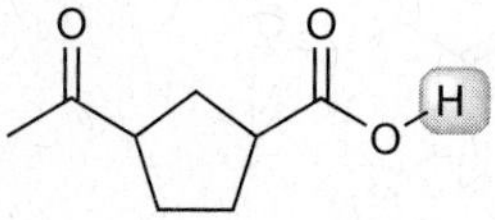

3.12)

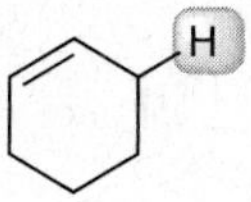

3.14)

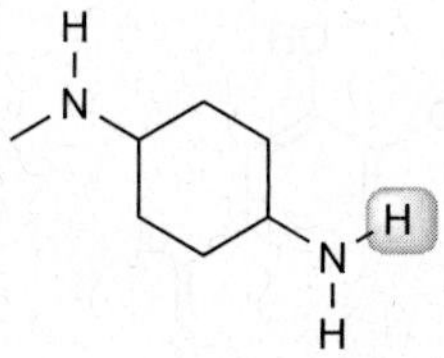

3.15)

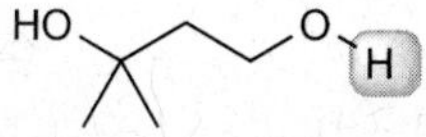

3.16)

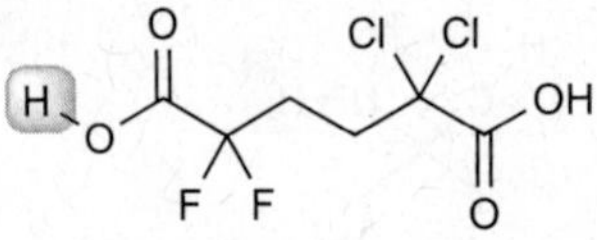

3.19)

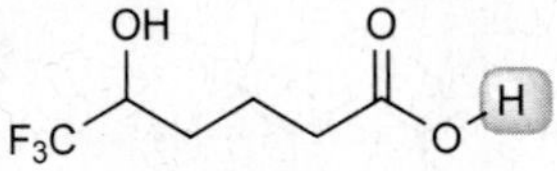

3.20)

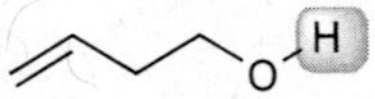

3.21)

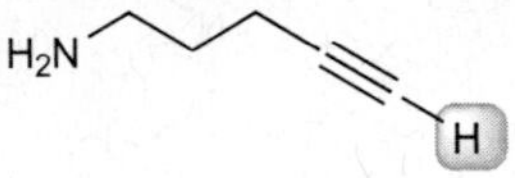

3.22)

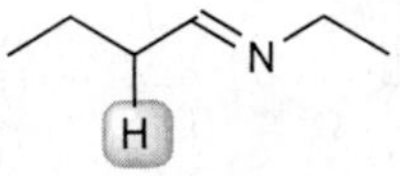

3.23)

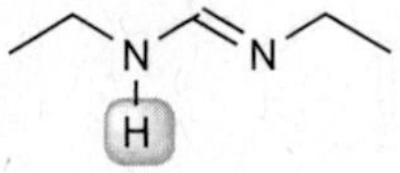

3.24)

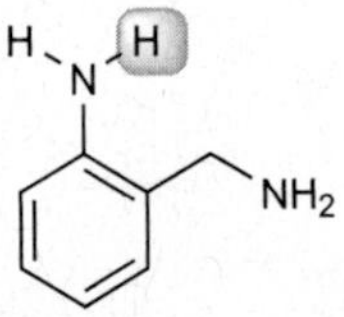

3.25)

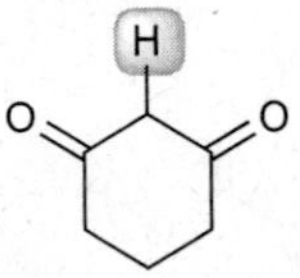

3.26)

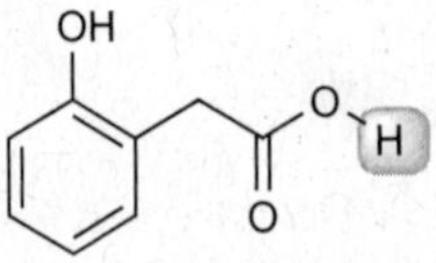

3.27)

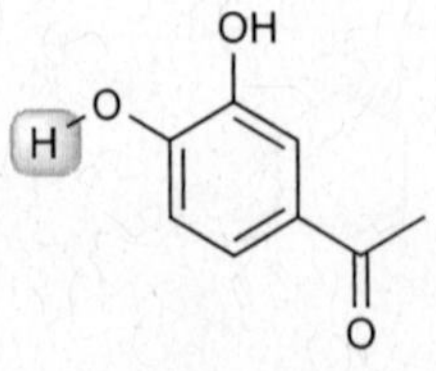

3.28) HI

3.29) CH$_3$SH

3.30) H$_2$O

3.31)

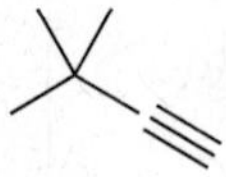

3.32)

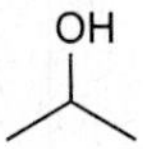

3.33)

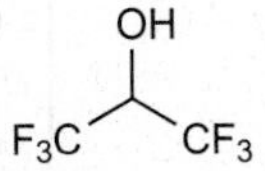

3.35) Favors right side
3.36) Favors left side
3.37) Favors right side

3.39)

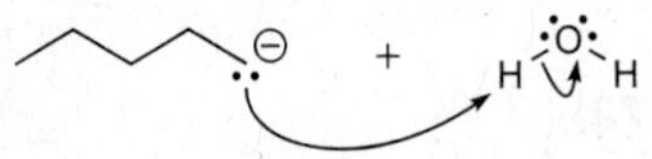

3.40)

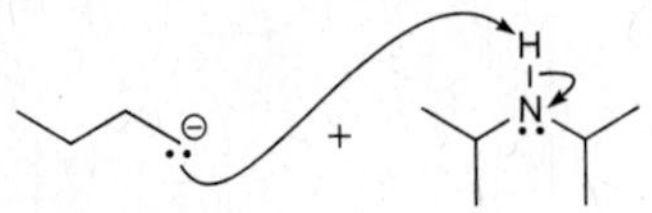

3.41)

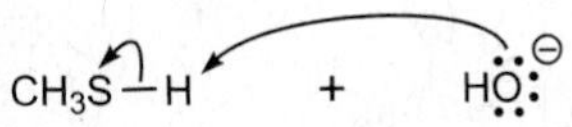

3.42)

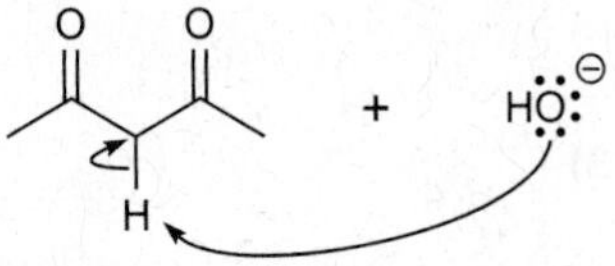

3.43)

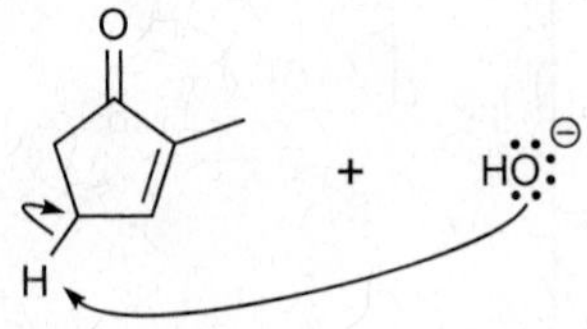

3.44)

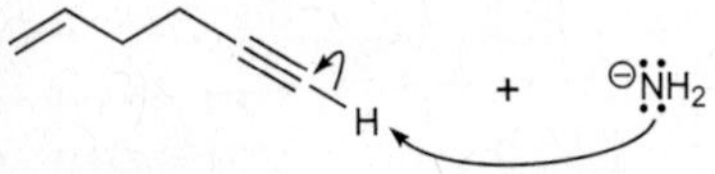

3.45)

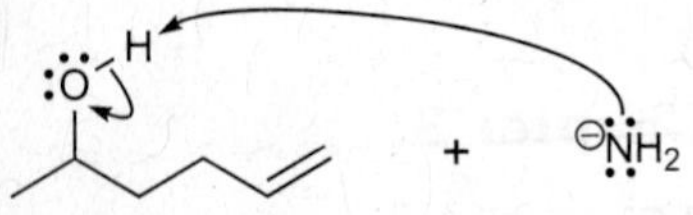

Chapter 4

4.2) sp^2
4.3) sp

4.4) sp^3

4.5) sp^2

4.6) sp^3

4.7) sp

4.8)

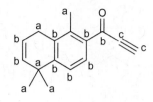

a = tetrahedral
b = trigonal planar
c = linear

4.10) All are sp^2 and trigonal planar

4.11)

a = tetrahedral, sp^3
b = trigonal planar, sp^2

4.12)

a = tetrahedral, sp^3
b = trigonal planar, sp^2
c = linear, sp

4.13)

a = tetrahedral, sp^3
b = trigonal planar, sp^2
c = trigonal pyramidal, sp^3

4.14)

a = tetrahedral, sp^3
b = trigonal pyramidal, sp^3
c = bent, sp^3

4.15)

a = tetrahedral, sp^3
b = trigonal planar, sp^2
c = linear, sp

4.16)

a = tetrahedral, sp^3
b = trigonal planar, sp^2

4.17)

a = tetrahedral, sp^3
b = trigonal planar, sp^2

4.18) The nitrogen atom is sp^2 hybridized and trigonal planar. The oxygen atom is sp^3 hybridized and bent.

4.19) The oxygen atom on the left is sp^2 hybridized and trigonal planar. The oxygen atom on the right is sp^3 hybridized and bent.

4.20) The oxygen atom is sp^2 hybridized and bent.

Chapter 5

5.2) –one
5.3) –oate
5.4) –al
5.5) –amine
5.6) –ol
5.7) –ol
5.8) –al
5.9) –one
5.10) –oic acid
5.12) –en-
5.13) –yn-
5.14) –dien-
5.15) –trien-
5.16) –trien-
5.17) –endiyn-
5.19) hex
5.20) hept
5.21) hex
5.22) non
5.23) oct
5.24) hex
5.25) hex
5.26) hex
5.27) pent
5.29) Two chloro groups
5.30) bromo, iodo
5.31) Five methyl groups
5.32) Six fluoro groups
5.33) Methyl
5.34) chloro, *tert*-butyl
5.35) amino, bromo, chloro, fluoro
5.36) iodo, fluoro, bromo
5.37) isopropyl
5.38) ethyl, hydoxy
5.40) *trans*
5.41) *trans*
5.42) *trans*
5.43) *cis*
5.44) *cis*
5.45) *trans*

5.47)

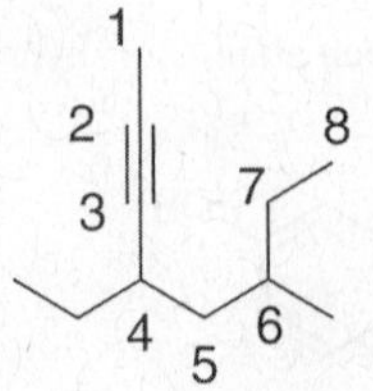

5.48)

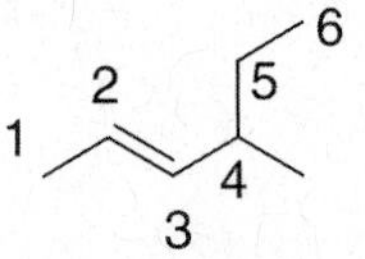

5.49)

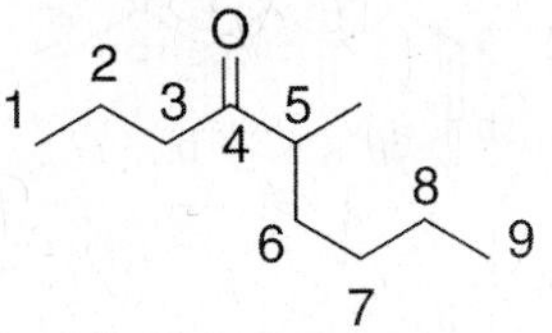

5.50)

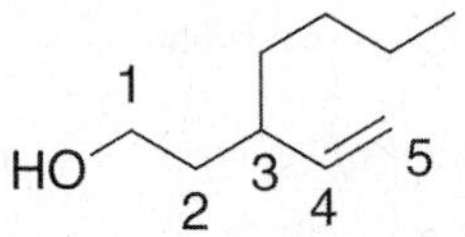

5.51)

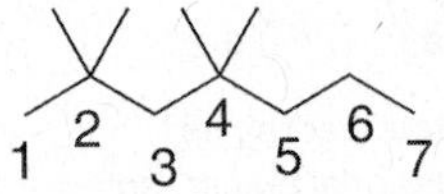

5.52)

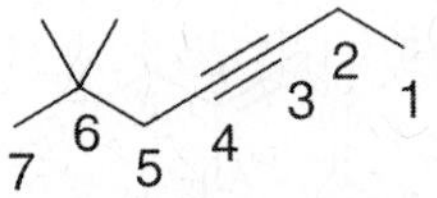

5.53)

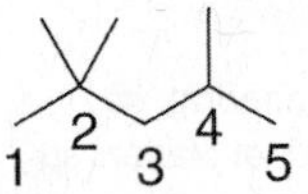

5.54)

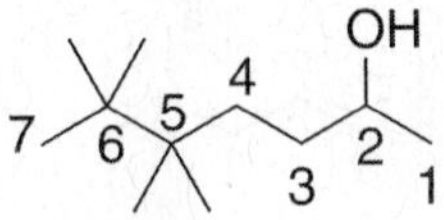

5.55)

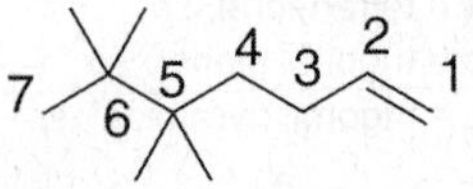

5.57) *trans*-4-ethyl-5-methyloct-2-ene
5.58) 4-ethylnonan-3-ol
5.59) 4,4-dimethylhex-2-yne
5.60) 4,4-dimethylcyclohexanone
5.61) 2-chloro-4-fluoro-3,3-dimethylhexane

5.62) *cis*-3-methylhex-2-ene
5.63) 2-ethylpentanamine
5.64) 2-propylpentanoic acid
5.65) *trans*-oct-2-en-4-ol
5.66) *trans*-5-chloro-6-fluoro-5,6-
dimethyloct-2-ene

Chapter 6

6.2)

6.3)

6.4)

6.5)

6.6)

6.7)

6.9)

6.10)

6.11)

6.12)

6.13)

6.14)

6.16)

6.17)

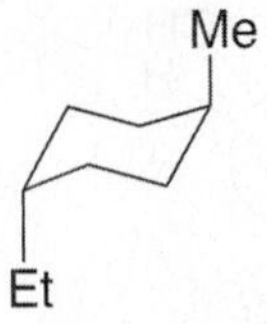

6.18)

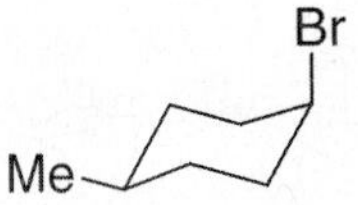

6.19)

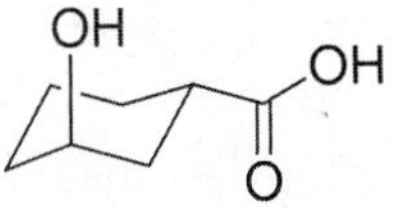

6.20)

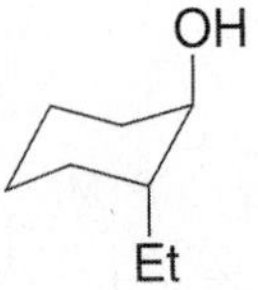

6.21)

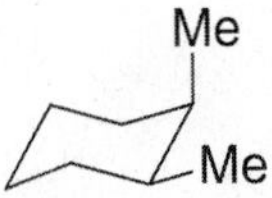

6.24)

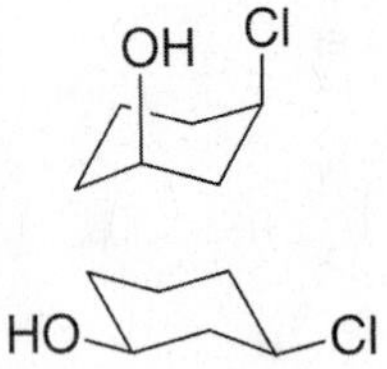

6.25)

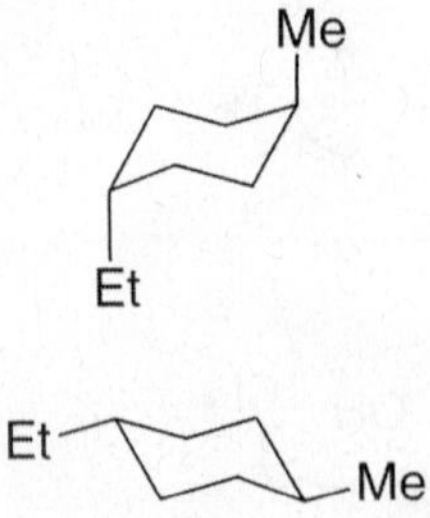

6.26)

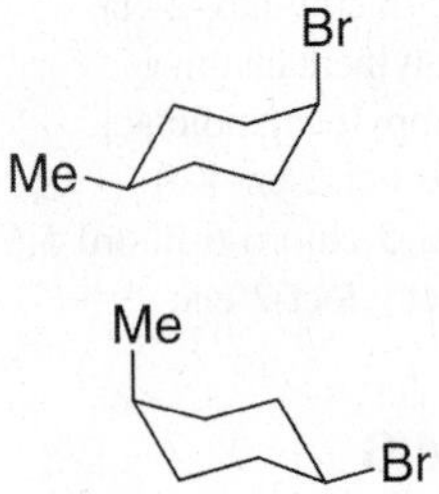

6.27)

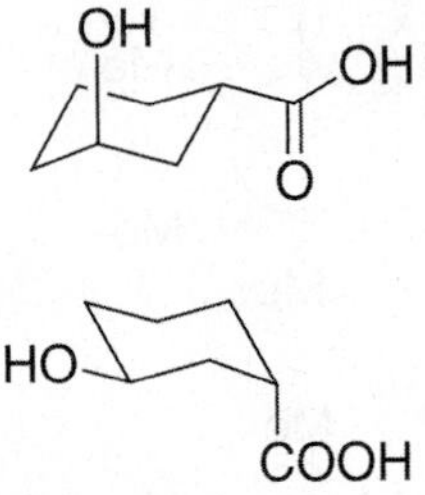

6.28)

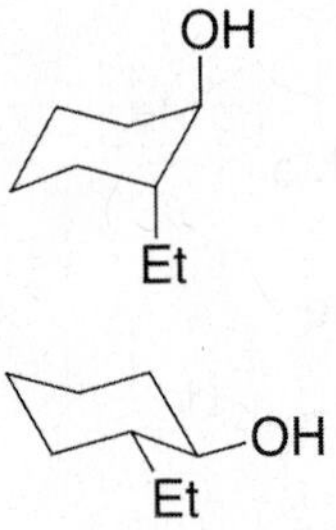

6.29)

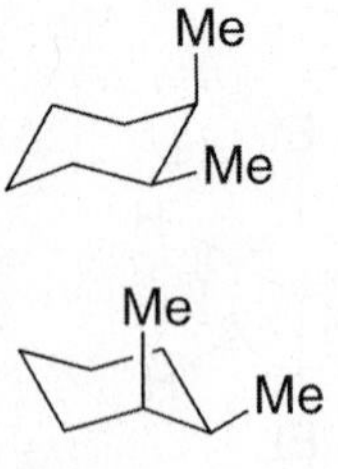

6.31)

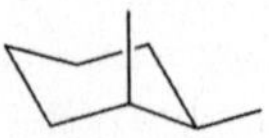

6.32)

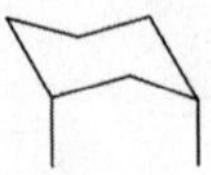

6.33)

6.34)

6.35)

6.36)

6.38)

6.39)

6.40)

6.41)

6.42)

6.43)

6.44)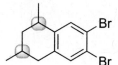

6.45)

Chapter 7

7.2)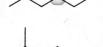

7.3)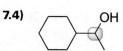

7.4)

7.5)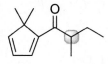

7.6)

7.7)

7.9)

7.10)

7.11)

7.12)

7.13)

7.14)

7.15)

7.17)

7.18)

7.19)

7.21)

7.22)

7.23)

7.24)

7.25)

7.26)

7.27) *S*
7.28) *R*
7.29) *S*
7.30) *S*
7.31) *R*
7.32) *R*
7.33) *R*
7.34) *R*
7.35) *S*

7.37)

7.38)

7.39)

7.40)

7.41)

7.42)

7.44) (*Z*)-2-fluoropent-2-ene
7.45) (1*R*,3*R*)-3-methylcyclohexanol
7.46) (*S*)-3-methylpent-1-ene
7.47) (*E*)-4-ethyl-2,3-dimethylhept-3-ene
7.48) (2*Z*,4*E*)-hepta-2,4-diene
7.49) (2*E*,4*Z*,6*Z*,8*E*)-deca-2,4,6,8-tetraene

7.51)

7.52)

7.53)

7.54)

7.55)

7.56)

7.58)

7.59)

7.60)

7.61)

7.62)

7.63)

7.65) Enantiomers
7.66) Diastereomers
7.67) Enantiomers
7.68) Enantiomers
7.69) Diastereomers
7.70) Diastereomers
7.72) *Meso*
7.73) Not *meso*
7.74) *Meso*

7.76)

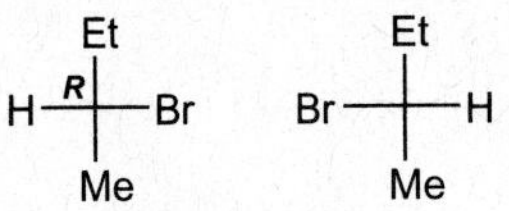

7.77)

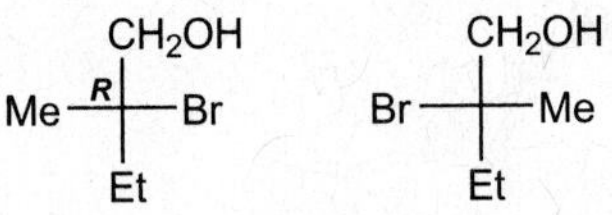

7.78)

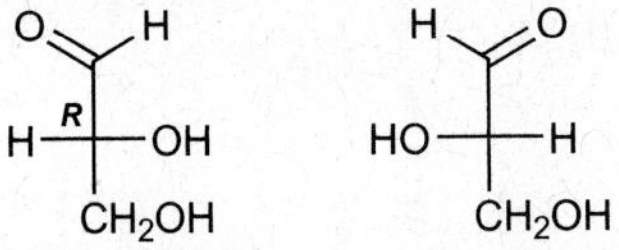

7.79)

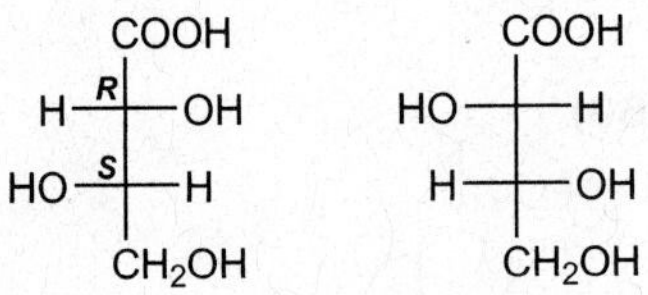

7.80)

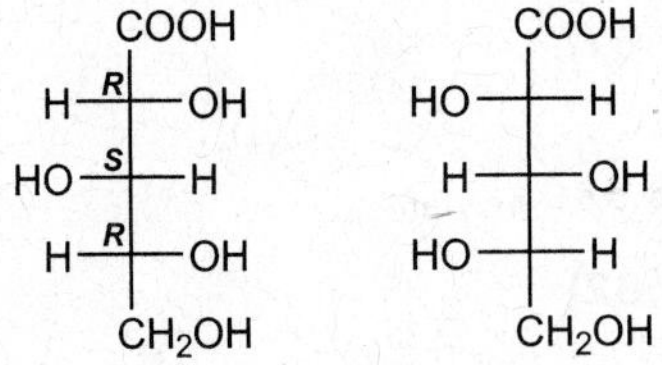

7.81)

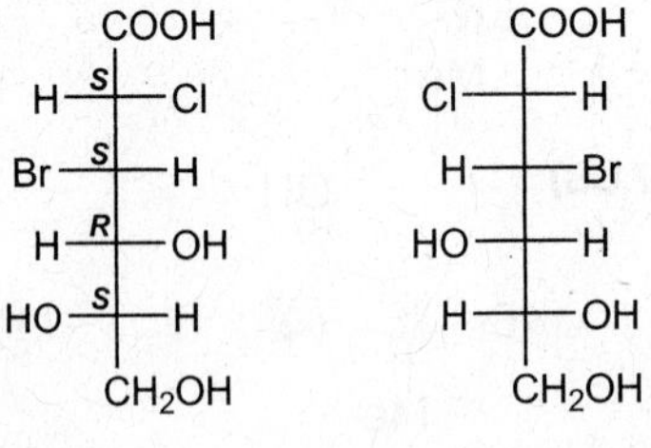

Chapter 8

8.2) Bond → Lone pair

8.3) Lone pair → Bond, then Bond → Lone pair

8.4) Lone pair → Bond, then Bond → Lone pair

8.5) Lone pair → Bond, then Bond → Lone pair

8.6) Lone pair → Bond, Bond → Bond, Bond → Lone pair

8.7) Lone pair → Bond, then Bond → Lone pair

8.9)

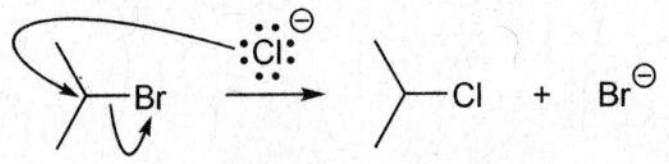

8.10)

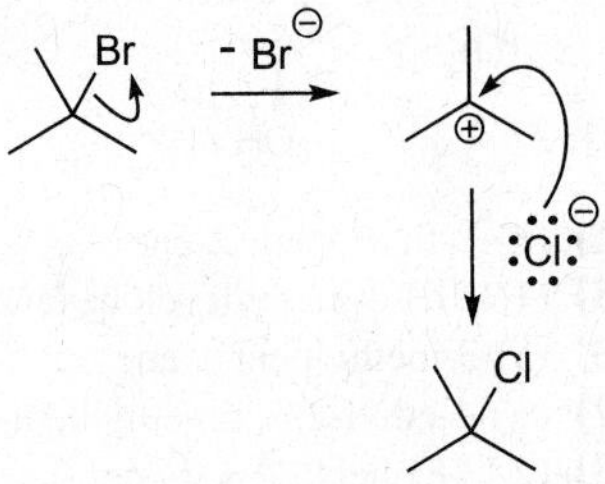

8.11)

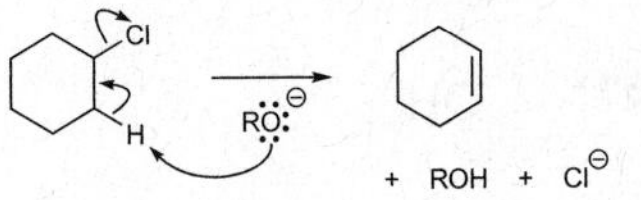

8.12)

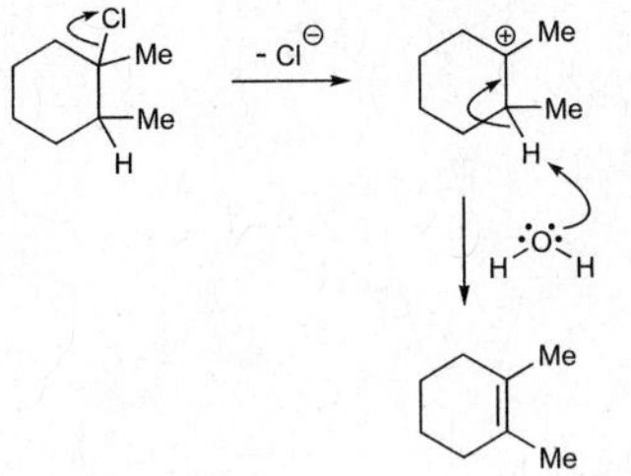

8.14)

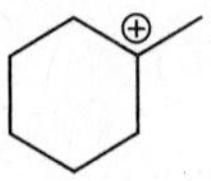

8.15)

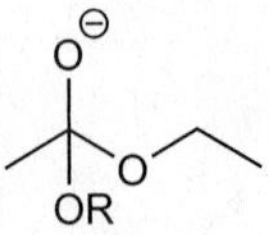

8.16)

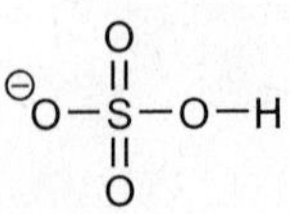

8.17)

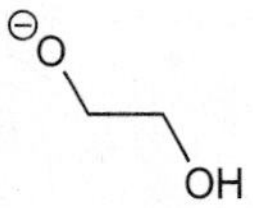

8.18)

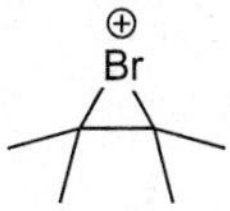

8.19)

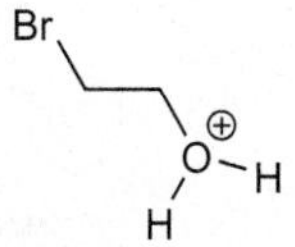

8.21) Hydroxide is the nucleophile
8.22) Water is the nucleophile
8.23) Water is the nucleophile
8.24) MeCl is the electrophile
8.26) Nucleophile
8.27) Base
8.28) Nucleophile
8.29) Base
8.30) Base
8.31) Nucleophile
8.32) Nucleophile
8.33) Base

8.34)

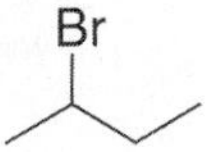

8.35)

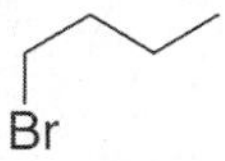

8.36)

trans   *cis*

8.37)

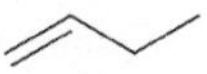

8.39)

8.40)

8.41)

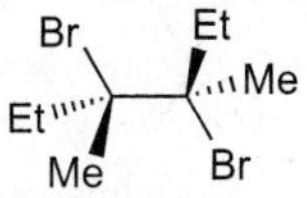

8.42)

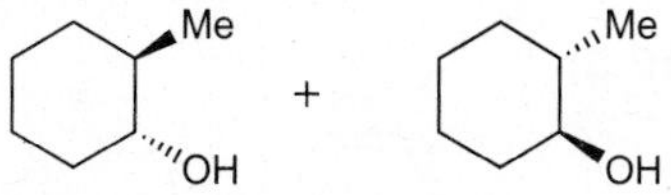

Chapter 9

9.2) Both
9.3) S_N2
9.4) Both
9.5) S_N1
9.7) No
9.8) Yes
9.9) No
9.10) Yes
9.12) S_N2
9.13) S_N1
9.14) S_N1
9.15) S_N2
9.16) S_N2
9.17) S_N2
9.19) mesylate
9.20) iodide
9.21) tosylate
9.22) chloride
9.23) bromide
9.24) bromide
9.25) 3-iodo-3-methylpentane
9.26) Use HCl to protonate OH and turn it into an excellent LG
9.30) S_N2
9.31) S_N1
9.32) S_N1
9.33) Neither
9.34) S_N2
9.35) S_N1

Chapter 10

10.1)

Zaitsev   Hofmann

10.2)

Zaitsev   Hofmann

10.3)

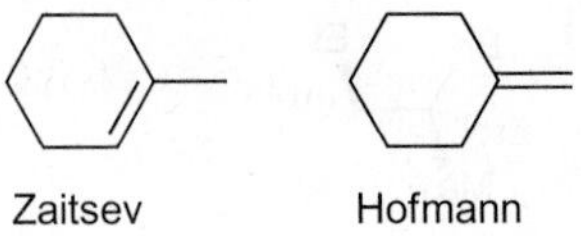

Zaitsev   Hofmann

10.5)

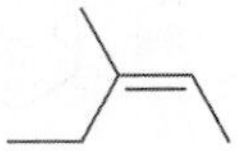

10.6)

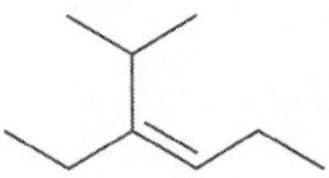

10.7)

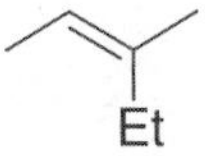

Et

10.8)

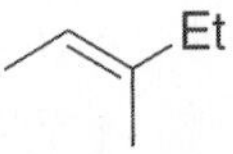

Et

10.9)

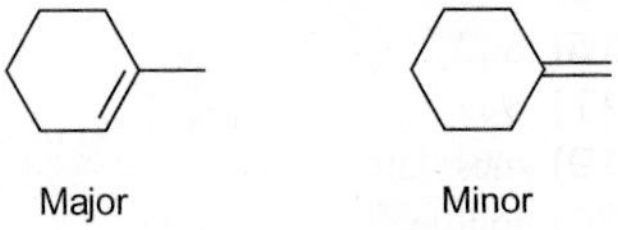

Major   Minor

10.10)

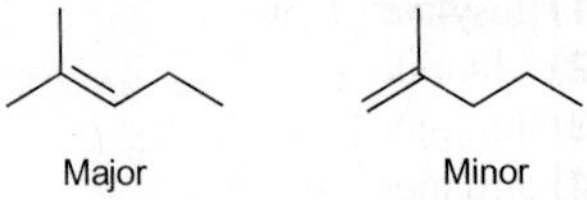

Major   Minor

10.11)

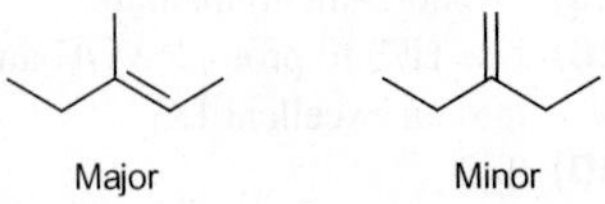

Major   Minor

10.12) weak nucleophile / weak base
10.13) strong nucleophile / weak base
10.14) strong nucleophile / strong base
10.15) strong nucleophile / weak base
10.16) weak nucleophile / weak base
10.17) strong nucleophile / weak base
10.18) weak nucleophile / weak base
10.19) strong nucleophile / weak base
10.21) S_N2 = major, E2 = minor
10.22) E1
10.23) E1+S_N1
10.24) S_N2
10.25) S_N1

10.27)

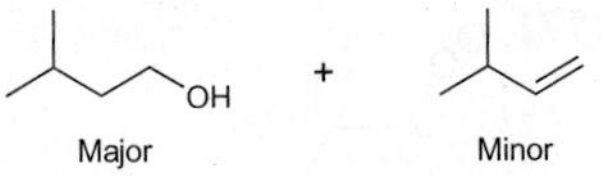

Major   +   Minor

10.28)

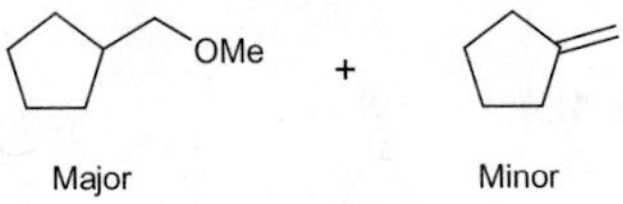

Major   +   Minor

10.29)

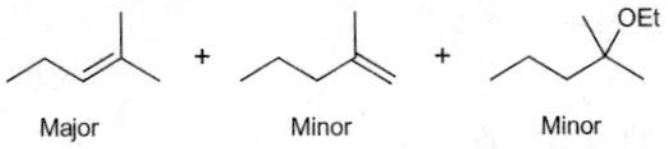

Major   Minor   Minor

10.30)

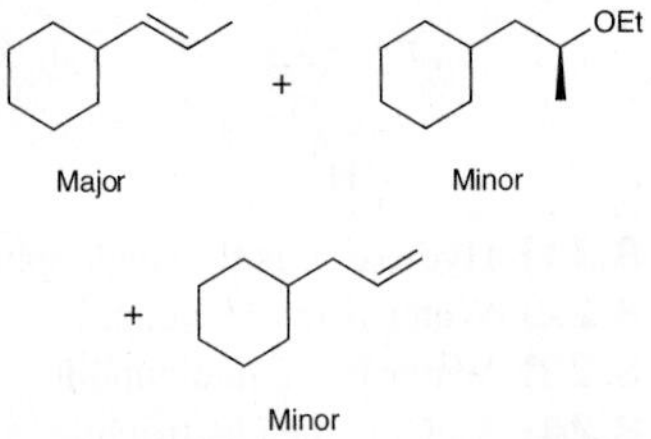

Major   +   Minor

+

Minor

10.31)

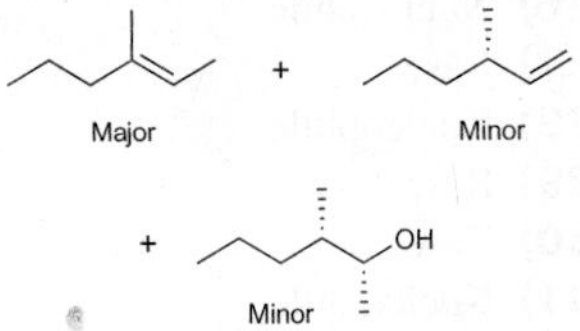

Major   +   Minor

+

Minor

10.32)

Major   +   Minor

10.33)

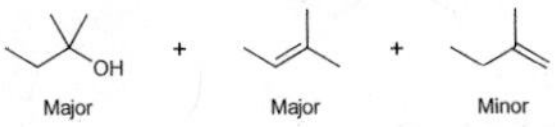

Major   Major   Minor

10.34)

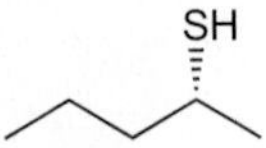

SH

10.35)

Major   +   Minor

10.36)

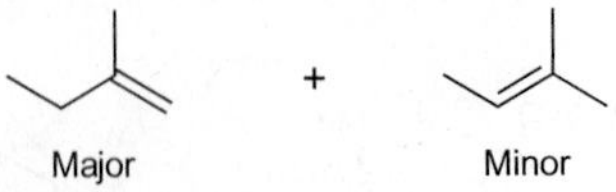

Major   +   Minor

10.37)

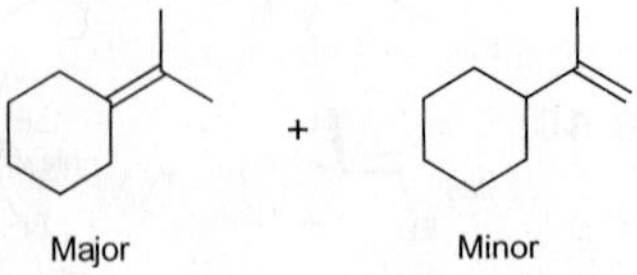

Major   +   Minor

10.38)

10.39)

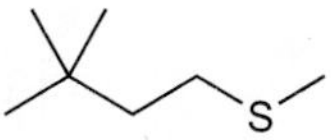

Chapter 11

11.2)

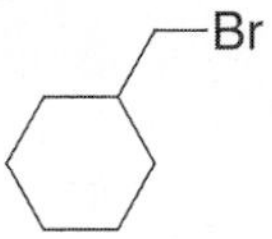

11.3)

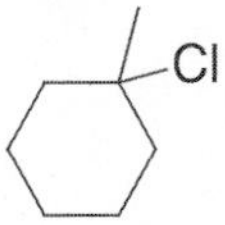

11.4)

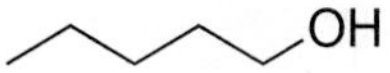

11.5)

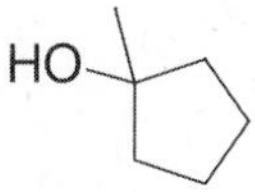

11.7) 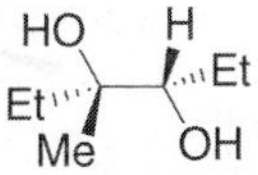 + Enantiomer

11.8) + Enantiomer

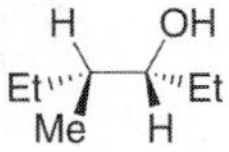

11.9) + Enantiomer

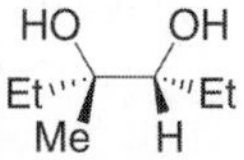

11.10) + Enantiomer

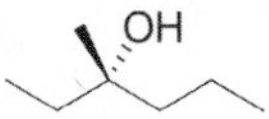

11.12) + Enantiomer

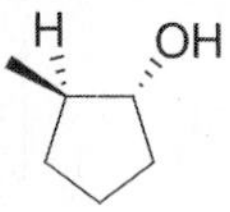

11.13) + Enantiomer

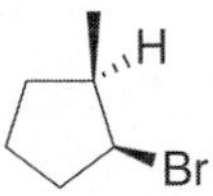

11.14) + Enantiomer

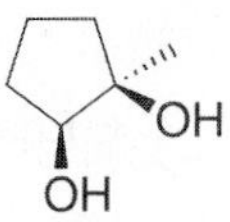

11.15) + Enantiomer

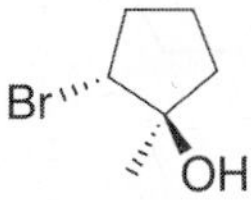

11.17) + Enantiomer

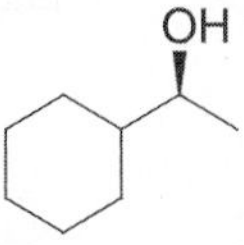

11.18) + Enantiomer

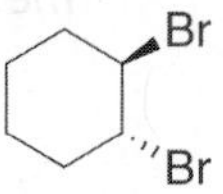

11.19) + Enantiomer

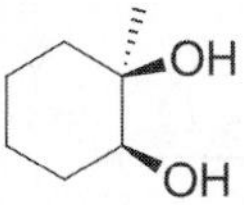

11.20)

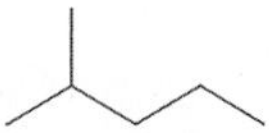

11.22) 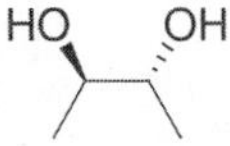 + Enantiomer

11.23) (meso)

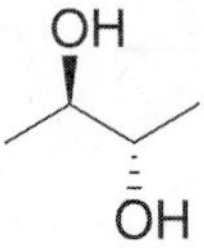

11.24) + Enantiomer

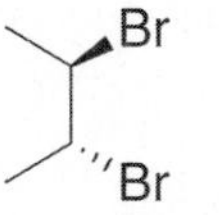

11.25) (meso)

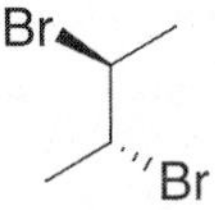

11.27) + Enantiomer

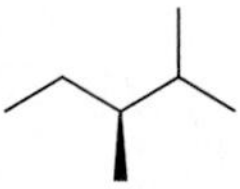

11.28)

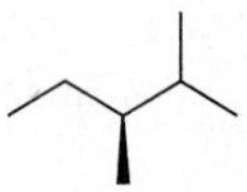

+ Enantiomer

11.29)

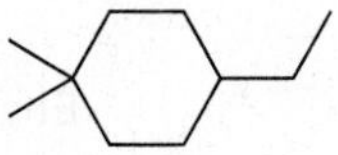

11.30)

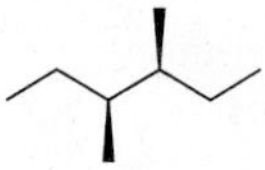

+ Enantiomer

11.31)

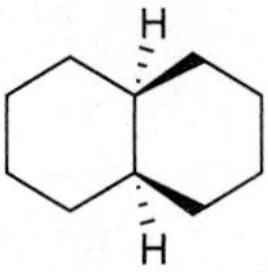

(*meso*)

11.32)

11.34)

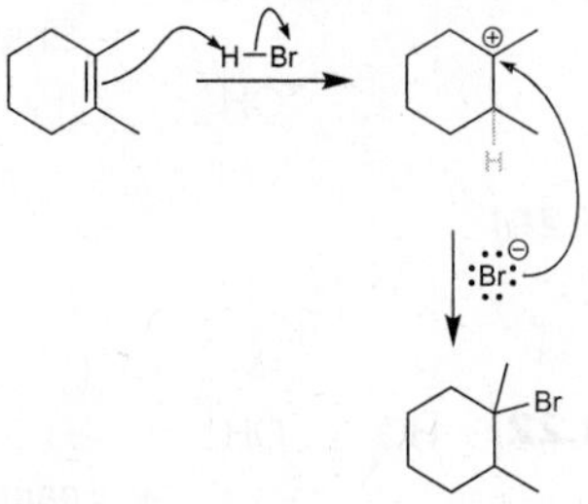

11.35)

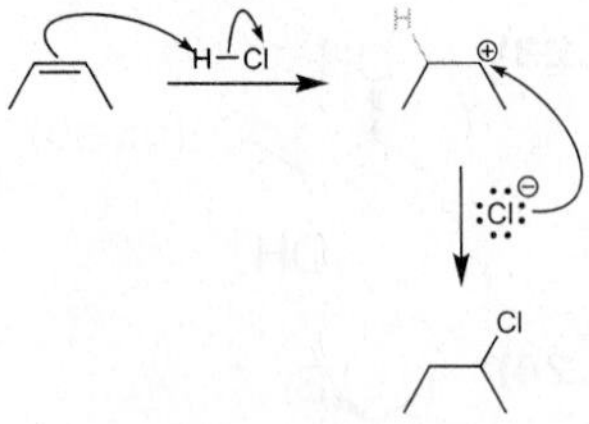

11.36)

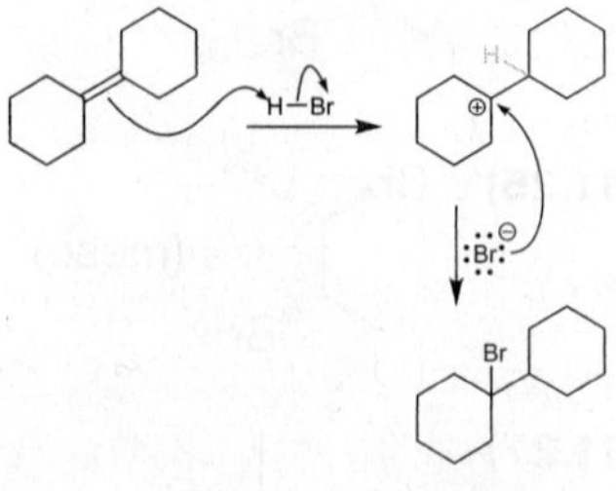

11.37)

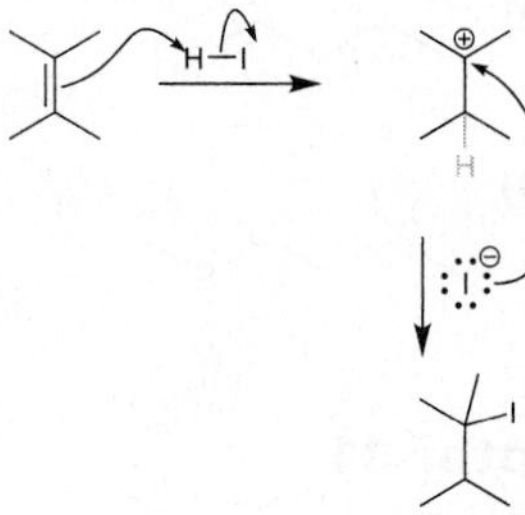

11.39)

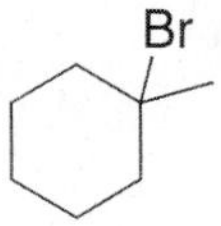

11.40)

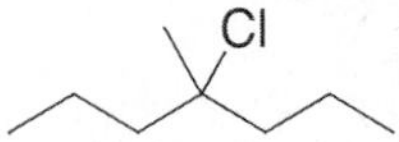

11.41)

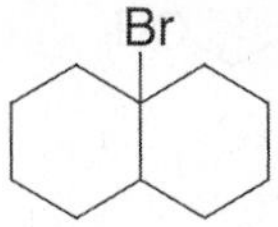

11.42)

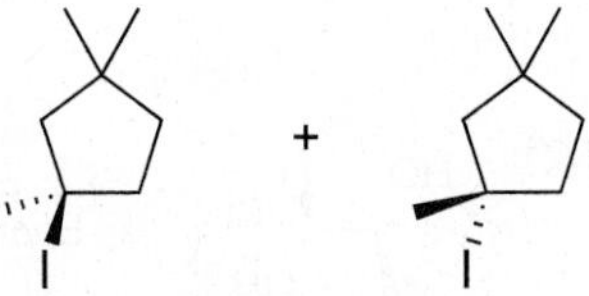

+

11.44)

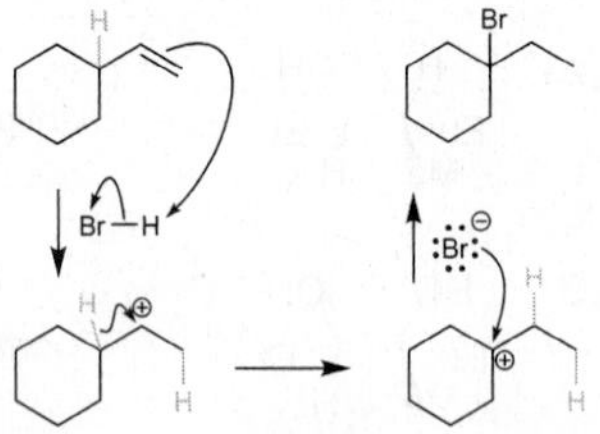

11.45)

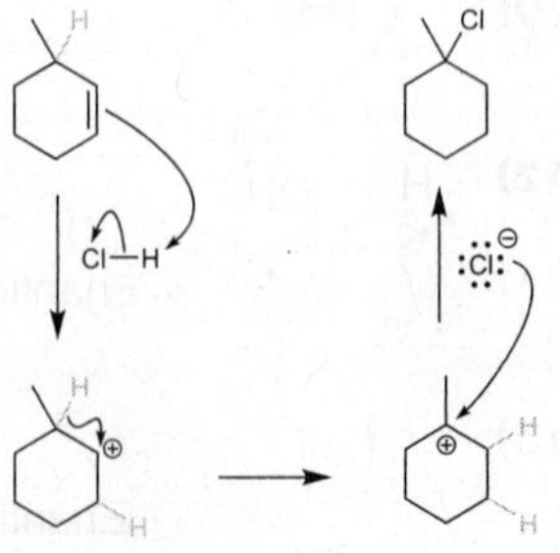

11.46)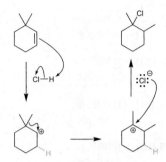

11.47)

11.49)

11.50) + Enantiomer

11.51)

11.52) + Enantiomer

11.54) HBr, ROOR
11.55) HBr, ROOR
11.56) HBr
11.57) HBr, ROOR

11.59)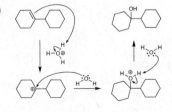

11.60)

11.61)

11.62)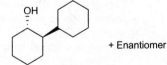

11.64) HO‑ + Enantiomer

11.65) OH

11.66) OH + Enantiomer

11.67) OH + Enantiomer

11.68) HO + Enantiomer

11.69)

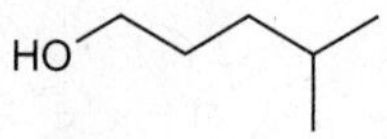

11.71) 1) BH$_3$ • THF
2) H$_2$O$_2$, NaOH

11.72) HBr
11.73) HBr, ROOR
11.74) H$_2$, Pt
11.75) HCl
11.76) 1) BH$_3$ • THF
2) H$_2$O$_2$, NaOH

11.77) NaOEt
11.78) t-BuOK
11.80) 1) t-BuOK or NaH
2) HCl

11.81) 1) NaOEt
2) HBr, ROOR

11.82) 1) t-BuOK
2) BH$_3$ • THF
3) H$_2$O$_2$, NaOH

11.83) 1) conc. H$_2$SO$_4$, heat
2) BH$_3$ • THF
3) H$_2$O$_2$, NaOH

11.84) 1) TsCl, py
2) NaOEt
3) BH$_3$ • THF
4) H$_2$O$_2$, NaOH

In this example, the first two steps of our synthesis (the elimination of water) could alternatively be accomplished in just one step with an E1 process by using concentrated sulfuric acid.

11.85) 1) NaOEt
2) HBr

11.87) 1) HBr, ROOR
2) t-BuOK

11.88) 1) HBr
2) NaOEt

11.89) 1) HBr
2) NaOEt

11.90) 1) HBr, ROOR
2) t-BuOK

11.92) 1) NBS, hv
2) t-BuOK

11.93) 1) NBS, hv
2) NaOEt

11.94) 1) NBS, hv
2) NaOEt

11.95) 1) NBS, hv
2) t-BuOK

11.96) 1) NBS, hv
2) t-BuOK
3) HBr, ROOR

11.97) 1) NBS, hv
2) NaOEt
3) HBr, ROOR
4) t-BuOK

11.99)

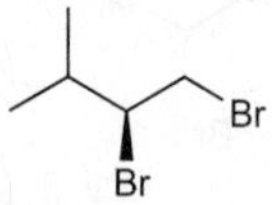

+ Enantiomer

11.100)

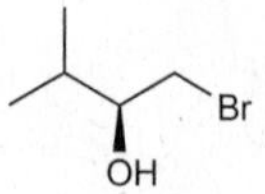

+ Enantiomer

11.101)

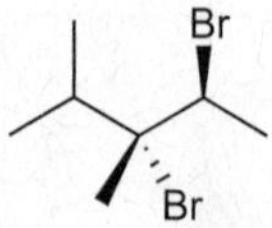

+ Enantiomer

11.102)

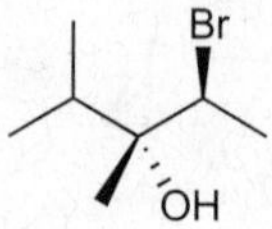

+ Enantiomer

11.103)

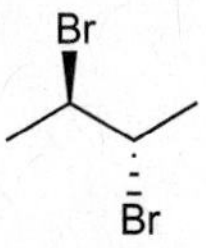

(*meso*)

11.104)

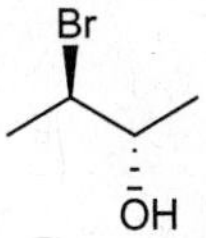

+ Enantiomer

11.106)

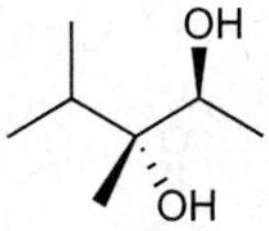

+ Enantiomer

11.107)

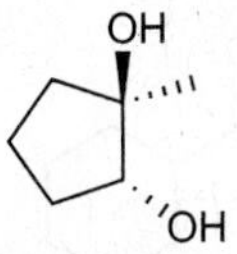

+ Enantiomer

11.108)

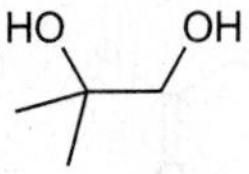

11.109)

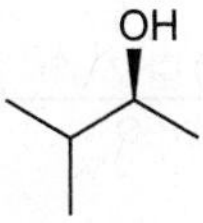

+ Enantiomer

11.111)

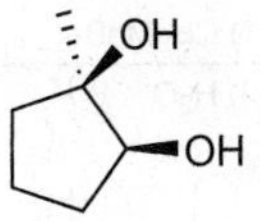

+ Enantiomer

11.112)

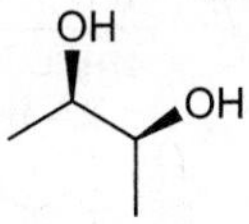

(*meso*)

11.113)

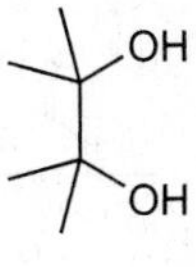

11.114)

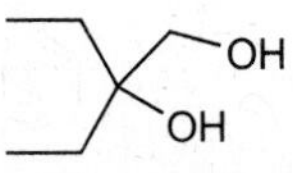

11.115)

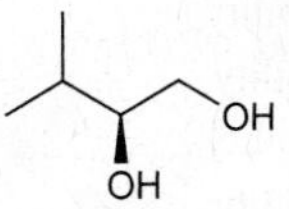

+ Enantiomer

11.116)

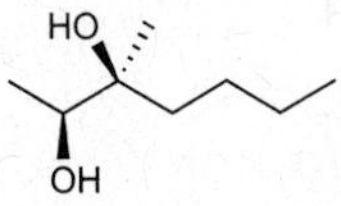

+ Enantiomer

11.118)

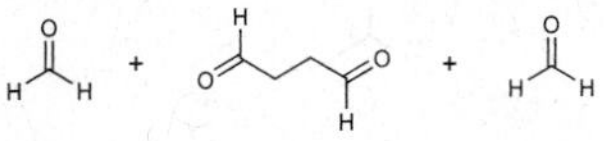

11.119)

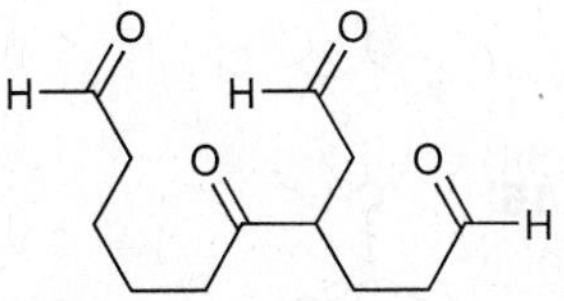

11.120)

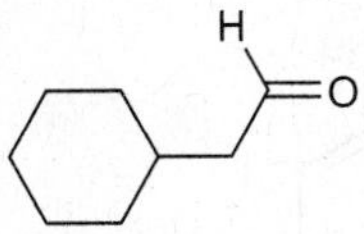

11.121)

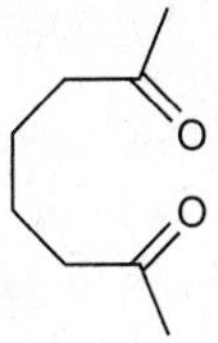

11.122)

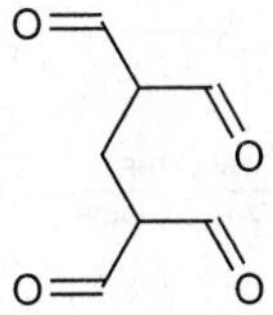

11.123)

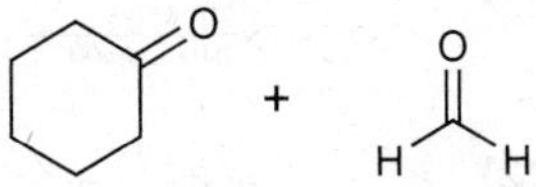

+

Chapter 12

12.2) Tertiary
12.3) Primary
12.4) Primary
12.5) Secondary

12.6) High solubility
12.7) Low solubility
12.8) High solubility
12.9) Low solubility

12.12) (structure: 2-pentanol, OH)

12.13) (structure: 2-pentanol, OH)

12.14) (structure: 1-chloro-2-propanol, OH and Cl)

12.15) (structure: acetic acid, OH and O)

12.16) (structure: phenol, OH)

12.17) (structure: 2-hydroxyacetophenone, OH and O)

12.19) (reaction: alkene $\xrightarrow{H_3O^+}$ tertiary alcohol OH)

12.20) (methylenecyclohexane $\xrightarrow[\text{2) } H_2O_2,\ NaOH]{\text{1) } BH_3 \cdot THF}$ cyclohexylmethanol OH)

12.21) (reaction: $\xrightarrow[\text{3) } H_2O_2,\ NaOH]{\substack{\text{1) } t\text{-BuOK} \\ \text{2) } BH_3 \cdot THF}}$ OH)

12.22) (cyclohexene $\xrightarrow[\text{2) } H_2O_2,\ NaOH]{\text{1) } BH_3 \cdot THF}$ trans-2-methylcyclohexanol OH + En)

12.24) +1
12.25) +3
12.26) +1
12.27) +1
12.28) 0
12.29) +1

12.31) (structure: acetone, O)

12.32) (structure: cyclopentanone, O)

12.33) (structure: cyclobutanecarbaldehyde, O and H)

12.34) (structure: phenylacetaldehyde, O and H)

12.35) (structure: dicyclohexyl ketone, O)

12.36) (structure: pentanal, H and O)

12.38) (acetaldehyde O, H $\xrightarrow[\text{2) } H_2O]{\text{1) } CH_3MgBr}$ 2-propanol OH)

12.39) (cyclopentanone O $\xrightarrow[\text{2) } H_2O]{\text{1) } CH_3MgBr}$ 1-methylcyclopentanol OH)

12.40) (isopentyl MgBr, 1) H—CHO (O, H), 2) H_2O → OH)

12.41) (benzyl MgBr, 1) H—CHO (O, H), 2) H_2O → 2-phenylethanol OH)

12.42)

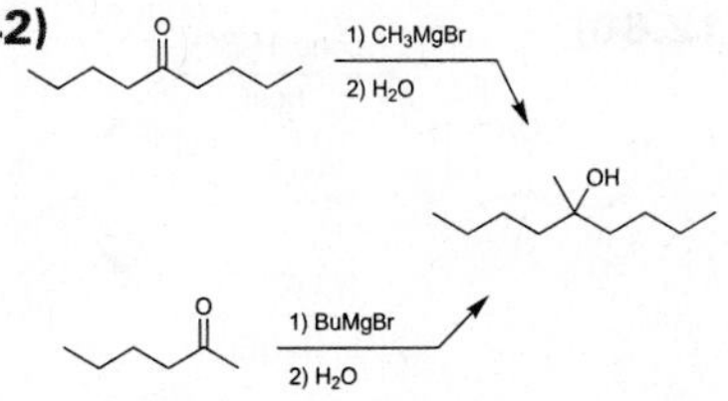

12.43)

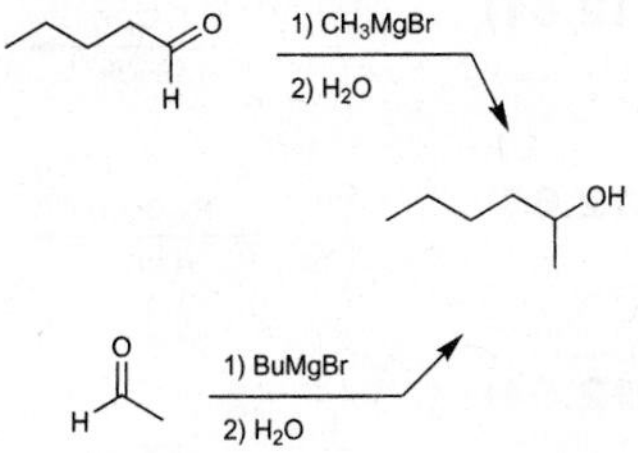

12.44)

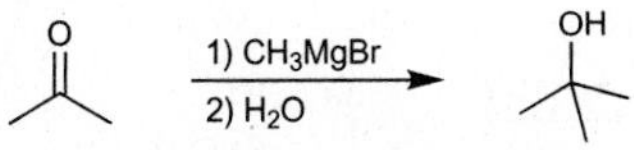

12.45)

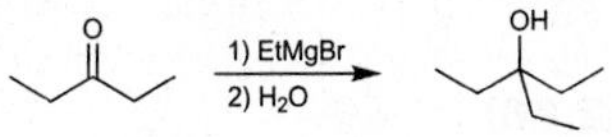

12.46)

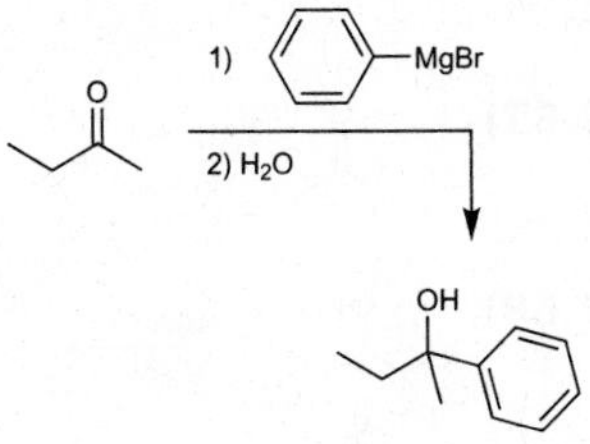

12.47)

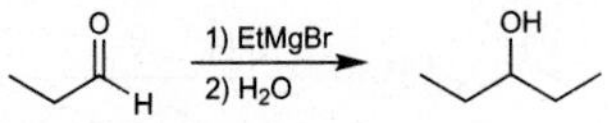

12.48)

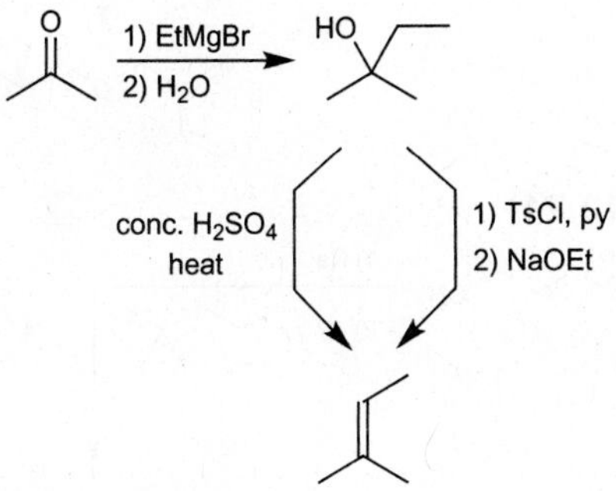

12.49)

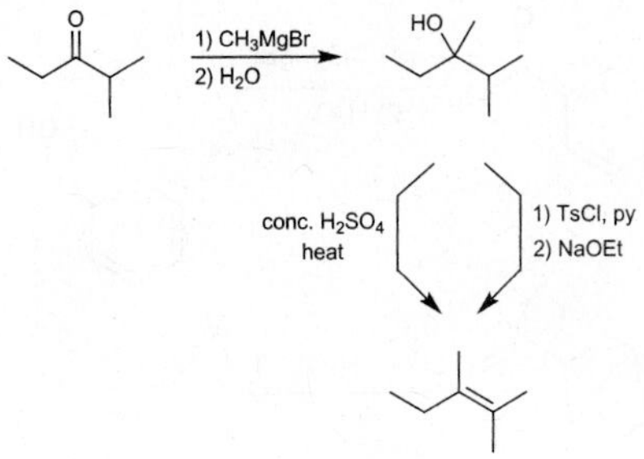

12.51)

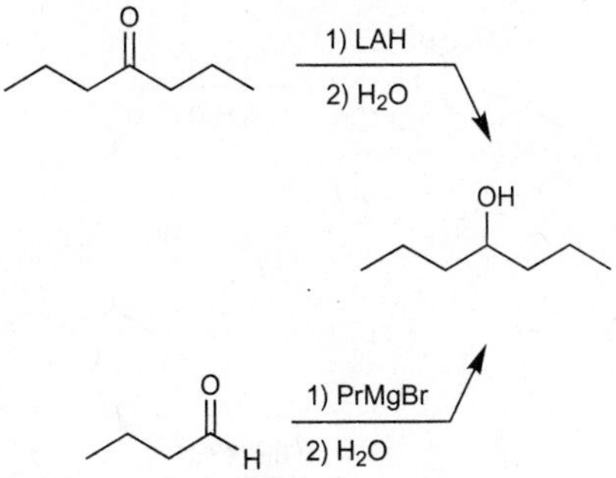

12.52)

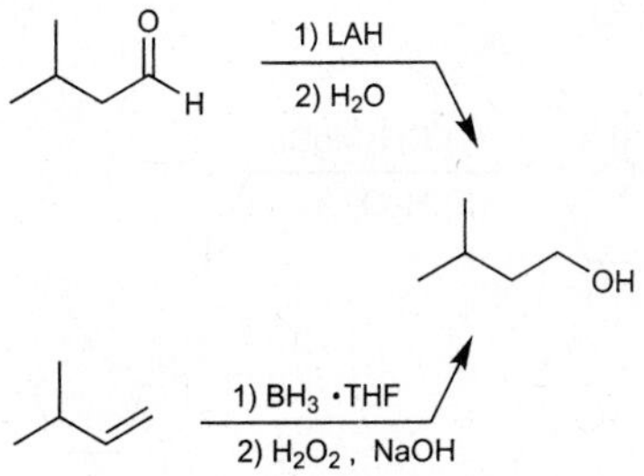

12.53)

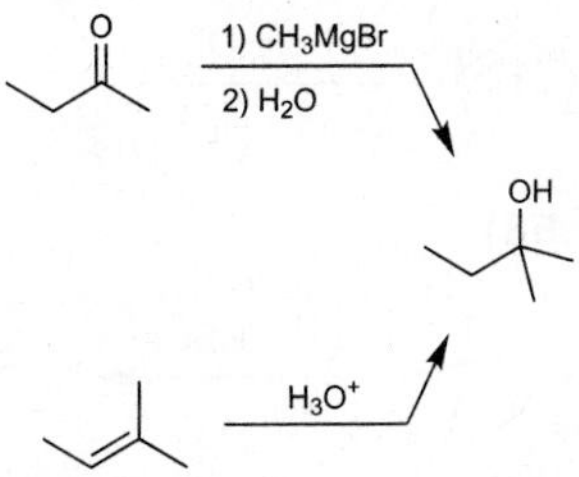

12.54)

12.55)

12.56)

12.58)

12.59)

12.60)

12.61)

12.63)

12.64)

12.65)

12.66)

12.67)

12.68)

12.70)

12.71)

INDEX